COMPUTER-AIDED ELECTRONIC CIRCUIT BOARD DESIGN AND FABRICATION

COMPUTER-AIDED ELECTRONIC CIRCUIT BOARD DESIGN AND FABRICATION

Using OrCAD/SDT and OrCAD/PCB Software Tools

AKRAM HOSSAIN

Purdue University, Calumet

Prentice Hall
Englewood Cliffs, New Jersey   Columbus, Ohio

Library of Congress Cataloging-in-Publication Data

Hossain, Akram.
    Computer-aided electronic circuit board design and fabrication :
using OrCAD/SDT and OrCAD/PCB software tools / by Akram Hossain.
      p.   cm.
    Includes index.
    ISBN 0-13-032095-1 (pbk.)
    1. Printed circuits—Design and construction—Data processing.
2. Computer aided design. 3. CAD/CAM systems. I. Title.
TK7868.P7H66 1996
621.3815'31—dc20   95-7610
                                                                                                               CIP

Cover photo: Campillo/ The Stock Market
Cover Designer: Graphica
Editor: Charles Stewart
Production Editor: Christine Harrington
Production Manager: Laura Messerly
Marketing Manager: Debbie Yarnell

 © 1996 by Prentice-Hall, Inc.
A Simon & Schuster Company
Englewood Cliffs, New Jersey 07632

All rights reserved. No part of this book may be reproduced, in any form or by any means, without permission in writing from the publisher.

This book was set in Times Roman by Carlisle Communications, Ltd. and was printed and bound by Semline, Inc. The cover was printed by Phoenix Color Corp.

Printed in the United States of America

10 9 8 7 6 5 4 3 2 1

ISBN: 0-13-032095-1

Prentice-Hall International (UK) Limited, *London*
Prentice-Hall of Australia Pty. Limited, *Sydney*
Prentice-Hall of Canada, Inc., *Toronto*
Prentice-Hall Hispanoamericana, S. A., *Mexico*
Prentice-Hall of India Private Limited, *New Delhi*
Prentice-Hall of Japan, Inc., *Tokyo*
Simon & Schuster Asia Pte. Ltd., *Singapore*
Editora Prentice-Hall do Brasil, Ltda., *Rio de Janeiro*

Limits of Liability and Disclaimer of Warranty

The author and the publisher of this book have used their best efforts in preparing the content of this book. These efforts include researching and testing of all instructions and procedures presented. The author and the publisher shall not be liable in any event for incidental or consequential damages in connection with or arising out of the furnishing, performance, or use of the material contained in this book.

Notice to the User

The user is expressly advised to adopt all safety precautions indicated in this book regarding chemicals, moving machinery, UV exposure systems, hand tools, electricity, and PCB laboratory waste disposal. The users are especially advised to read OSHA guidelines on laboratory standards and chemical hygiene. The user assumes all risks in connection with such instruction and safety procedures. The user must take all safety measures to safeguard himself/herself from all potential accidents that may result from following the procedures described in this book.

Trademarks

OrCAD® is a registered trademark of OrCAD, Inc.
IBM® is a registered trademark of International Business Machines Corporation.
Microsoft® is a registered trademark of Microsoft Corporation.
MS™ and MS-DOS™ are trademarks of Microsoft Corporation.
Windows™ is a trademark of Microsoft Corporation.
JDR Microdevices® is a registered trademark of JDR Microdevices.
Radio Shack® is a registered trademark of Radio Shack Corporation.
Zenith® is a registered trademark of Zenith Data System.
Kepro™ is a trademark of Kepro Circuit Systems, Inc.
Digi-Key® is a registered trademark of Digi-Key Corporation.
CorelDraw® is a registered trademark of Corel Corporation.

Menu and configuration screens are captured from OrCAD/ESP, OrCAD/SDT, and OrCAD/PCB software tools. Schematics and artworks are designed by using OrCAD/SDT and OrCAD/PCB tools.

This book is dedicated to my beloved wife, Parveen, my two daughters, Shahreen and Farihah, and my parents, who gave so much and asked for so little.

PREFACE

One of the most important steps in the manufacturing of electronic products is the design and fabrication of printed circuit board (PCB). Of all the components that are used to make an electronic product, the PCB requires the greatest number of design and manufacturing operations. The ever-increasing demand for high-quality and low-cost electronic products has made it essential for computer-aided tools to be used for the design and fabrication of electronic circuits. Demand for reliable electronic products has also increased in recent years. Today, most electronic circuit fabrication companies have computerized the fabrication process, and the fabrication technology is much more sophisticated. The computer-aided tools permit detailed evaluations of the fabrication process for a complex electronic circuit at the speed of modern computers. Although electronic fabrication can be accomplished by using hand tools, the process is very repetitive, time consuming, and error prone. The computer-aided tools serve as a unique resource to speed up the process. Also, expensive errors can be corrected early and easily. The enormous storing, retrieval, and editing facility of computers aid the process of fabrication. The use of computer-aided design/computer-aided engineering (CAD/CAE) tools for design is not new. However, the process of designing and fabricating electronic circuit boards in an integrated computer-aided environment is beginning to emerge.

Colleges and universities are beginning to adopt electronic design automation (EDA) environments for providing hands-on experience to their students in this fast-growing area of the technology. The decreasing cost of electronic hardware and software in the past decade served as a unique catalyst for the adaptation of the EDA environment by schools, small businesses, and even hobbyists. However, one factor that is affecting the area is the lack of an appropriate textbook with complete examples and solutions. Most of the books that are currently available do not offer a complete hands-on approach to the design and fabrication of electronic boards in an EDA environment. This textbook provides a complete solution to computer-aided electronic circuit board design and fabrication by using the affordable and widely used EDA software tools designed by OrCAD, Inc. OrCAD EDA tools are of professional quality and are sold at an affordable price.

This textbook is designed for a one-semester course in computer-aided electronic circuit board design and fabrication for undergraduate students in electrical engineering or electrical engineering technology programs. This book can also be used as a self-study guide or reference book by professional engineers or technicians. The prerequisite for understanding this book should be a course in computer programming and a course in basic analog and digital electronics. Students who have advanced knowledge in digital and analog electronics may get added satisfaction by using the book when building projects for their senior design class.

Four projects are described in this book. Each one begins with a raw idea and ends with a complete project that includes printed circuit board design and fabrication, which is the focus of this book. As students are going through each example, they will not only learn the computer-aided skills for the design and fabrication of printed circuit board; they will also develop confidence and technical know-how in completing electronic projects. The safety, skills, and procedures emphasized in this book are necessary to make a complete electronic project. Project planning, printed circuit board design and fabrication, mounting of electronic components, soldering PCBs, and project assembly concerning each project are included. These exercises will provide necessary self-assurance to students who want to head out into the world of electronic circuit board design and fabrication. After going through this book, students will be able to undertake more complex PCB design and fabrication projects.

The following is a brief summary of the book, chapter by chapter.

- Chapter 1 provides an overview of the computer-aided electronic design and fabrication process.
- Chapter 2 leads the students through the sequences of planning, designing, and experimenting with electronic projects.
- Chapter 3 provides information about the hardware platform, the software installation guide, and other related information.
- Chapter 4 provides common functions of the software tools and the steps of PCB layout design.
- Chapter 5 shows how to use OrCAD/Design Management Tools to organize PCB-design-related files.
- Chapter 6 discusses the OrCAD/Schematic Design Tools and their various commands.
- Chapter 7 leads the students step-by-step through various approaches of capturing circuit schematics by using the software tools.
- Chapter 8 takes the students through the OrCAD/Layout Tools and their various commands.
- Chapter 9 leads the students step-by-step through various procedures of PCB layout design and critical tooling for the fabrication process.
- Chapter 10 emphasizes the various safety practices related to PCB fabrication process: safety tests, safety gears, and PCB-fabrication-laboratory waste disposal. Major steps involved in fabricating PCBs in industry and in the laboratory, defects of PCBs and their repair, and quality assessment and assurance of PCBs are also discussed.

- Chapter 11 describes the procedures used for drilling, assembling, and soldering electronic parts on a printed circuit board. The final testing of the project is explained, and some tips about detecting defective parts are offered.
- Appendix A provides a few common DOS commands.
- Appendix B provides information about packages for electronic devices and procedures for hard copy printout.
- Appendix C provides the addresses and telephone numbers of various sources for PCB fabrication needs.
- Appendix D takes the students through OrCAD's most recent SDT 386+ and PCB 386+ tool sets.
- Appendix E is a partial printout of the various packages available in OrCAD's module library.

ACKNOWLEDGMENTS

I would like to thank my students at Purdue University, Calumet, for giving me the impetus for writing this book. Without their constant curiosity and interest, this book would not be possible. I especially would like to thank Suzali Suyut for his generous help in drawing many of the figures in this book. Without his help, it would have been almost impossible on my part to finish this book in time for publication. I would like to thank David McLees for his help in drawing Figures 10–1 through 10–11, and I am grateful to Abdul Mohit for his help in taking photographs of the projects. I would like to thank OrCAD for designing such wonderful software, and I am extremely grateful for their allowing me to use graphic renderings of their software for this book. I want to thank the following reviewers for doing the time-consuming task of reviewing this book and providing me with invaluable suggestions for the development of this book: Paul Ramcy of St. Clair College, Windsor, Ontario; Dan Sea of DeVry Institute of Technology; Harold Hultman, Jr., of Montgomery College; and Mohammed B. Khan of California State University, Long Beach. Finally, I would like to express my deepest appreciation to my wife, Parveen, whose encouragement and generous help kept me going.

CONTENTS

Chapter 1
AN OVERVIEW OF COMPUTER-AIDED ELECTRONIC FABRICATION

1-1.	Why Use Computer-Aided Design, Drafting, and Fabrication?	1
1-2.	Objectives	2
1-3.	A New Operating Environment of OrCAD Software Tool Sets	2
1-4.	Basic Components of the Printed Circuit Board Fabrication Process	3
1-5.	The Computer-Aided Design, Drafting, and Fabrication Process	18

Chapter 2
ELECTRONIC CIRCUITS: PLANNING, DESIGNING, EXPERIMENTING, AND PACKAGING

2-1.	Why Plan?	25
2-2.	Planning a School Laboratory Project	25
2-3.	Practical Electronic Projects	34
2-4.	Project 01-1995: The Deluxe Logic Probe	34
2-5.	Project 02-1995: The Electronic Cricket	40
2-6.	Project 03-1995: The Infrared Object Counter	46
2-7.	Project 04-1995: The Mini Stereo Amplifier	52

Chapter 3
PERSONAL COMPUTER AND OrCAD/SDT AND OrCAD/PCB SOFTWARE TOOLS

3-1.	Hardware Requirements for OrCAD/SDT and OrCAD/PCB Tool Sets	63
3-2.	Installation of the Software Tool Sets	66
3-3.	The Software Tool Operating Environment	70
3-4.	Common DOS Information Needed to Use the Software Tools	72
3-5.	Directory and File Structure	74
3-6.	DOS Environment Variables	76

Chapter 4
THE ELECTRONIC DESIGN AUTOMATION ENVIRONMENT AND OrCAD SOFTWARE TOOL SETS

4-1.	Computer Software Tools in the World of Electronic Design and Fabrication	79
4-2.	OrCAD/Design Management, OrCAD/SDT and OrCAD/PCB Tool Sets	80
4-3.	Printed Circuit Board Master Artwork Design Steps	82
4-4.	Some of the Common Characteristics of OrCAD EDA Tool Sets	84
4-5.	Configuration of the ESP Design Environment	88
4-6.	Configuration of User Buttons	89

Chapter 5
FILE ORGANIZATION AND DESIGN MANAGEMENT TOOLS

5-1.	Design Management Tool Set	95
5-2.	Managing Designs	95
5-3.	Managing Files	106
5-4.	Organization of Design Files for Projects	109

Chapter 6
SCHEMATIC DESIGN TOOL SET

6-1.	What Is an Electronic Circuit Schematic?	115
6-2.	Schematic Graphic Symbols Used by OrCAD	115
6-3.	Structuring Schematic Files	120
6-4.	Description of the Schematic Design Tool Set	122
6-5.	Configuration of the Schematic Design Tool Set	135
6-6.	Common Functions of the Draft Tool	143

Chapter 7
COMPUTER-AIDED SCHEMATIC CAPTURE

7-1.	Drafting and Designing Schematics	173
7-2.	Organizing File and Set Design Environment	173
7-3.	Designing and Drafting Circuit Schematic	175
7-4.	Check Electrical Rules	191
7-5.	Annotate Schematic193	193
7-6.	Select Field View	200
7-7.	Cleanup Schematic	202
7-8.	M2EDIT Text Editor	203
7-9.	Update Field Contents	204

7-10.	Archive Schematic Parts	209
7-11.	Create Netlist for PCB Software Tools	212
7-12.	Create Bill of Materials	217
7-13.	Schematic Hard Copy	221
7-14.	An Example Using the Logic Probe Project	224
7-15.	The Other Projects	236
7-16.	Modify Library Parts	257

Chapter 8
PRINTED CIRCUIT BOARD LAYOUT TOOL SET

8-1.	What Are the Components of a Printed Circuit Board?	269
8-2.	Description of the PC Board Layout Tool Set	270
8-3.	Configuration of PCB Layout Tool Set	273
8-4.	Common Functions of the PCB Tool Set	279

Chapter 9
COMPUTER-AIDED TECHNIQUES FOR PCB LAYOUT DESIGN

9-1.	The PC Board Layout Design Process	309
9-2.	PCB Master Artwork Printing	343
9-3.	Printing Component Legend and Solder Mask	345
9-4.	Create Drill Files for CNC Drill Machine	348
9-5.	Generate Module Report	350
9-6.	The Projects	351

Chapter 10
PRINTED CIRCUIT BOARD FABRICATION PROCESS

10-1.	Printed Circuit Board Fabrication in a School Laboratory	383
10-2.	Common Safety Practices in a School Laboratory	383
10-3.	A Typical Format of a Safety Test	388
10-4.	Printed Circuit Board Materials	389
10-5.	The Manufacturing Process	390
10-6.	Recommended PCB Fabrication Process for a School Laboratory	402
10-7.	Fabricating PCBs by Using the Excel Process	407
10-8.	Quality Assessment of PCBs	411
10-9.	PC Board Fabrication for the Projects	412
10-10.	Defect Repair	412
10-11.	Tools, Machinery, and Chemical Requirements for a Laboratory-Based PC Board Fabrication Facility	415

Chapter 11
ASSEMBLING AND SOLDERING OF ELECTRONIC PARTS ON PRINTED CIRCUIT BOARDS

11-1.	Project Assembly Tools	419
11-2.	PCB Drilling	420
11-3.	Final Shaping of the PC Board	422
11-4.	Drilling Desired Holes and Openings on the Enclosure	425
11-5.	Component Assembly	425
11-6.	Soldering Iron, Solder, and Their Application Techniques	428
11-7.	Assembled Circuit Boards of the Project	435
11-8.	Packaging of the Projects	435
11-9.	Final Testing of the Projects	437
11-10.	A Few Tips about Detecting Defective Parts	440

APPENDIX A	**MICROCOMPUTERS AND DOS COMMANDS**	**443**
APPENDIX B	**OrCAD/PCB MODULES BY THEIR PACKAGE NAMES**	**448**
APPENDIX C	**POTENTIAL SOURCES FOR PC BOARD FABRICATION NEEDS**	**452**
APPENDIX D	**A QUICK USER GUIDE TO OrCAD/ SDT 386+ AND OrCAD/PCB 386+**	**454**
APPENDIX E	**ACTUAL SHAPE AND SIZE OF COMMON PCB MODULES**	**477**
INDEX		**481**

COMPUTER-AIDED ELECTRONIC CIRCUIT BOARD DESIGN AND FABRICATION

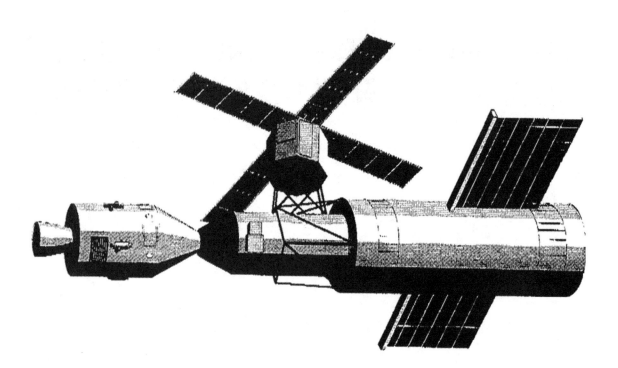

1

AN OVERVIEW OF COMPUTER-AIDED ELECTRONIC FABRICATION

1-1. WHY USE COMPUTER-AIDED DESIGN, DRAFTING, AND FABRICATION?

Electronic circuits, which are extensively used in almost all industries, have become integrated parts of modern technological developments. Modern electronic circuits, which have many advantages—namely, less maintenance, high accuracy, high reliability, cost effectiveness, and ease of interfacing with computers—have replaced many of the traditional electrical, electronic, pneumatic, and mechanical components. In particular, the computer-based system has revolutionized electronic circuits and their fabrication technology. These advancements in the application of electronic circuits depend largely on the technology of electronic circuit fabrication. Since computers have become major tools in all aspects of technology and since modern electronic circuits are becoming extremely complex, computer-aided electronic circuit fabrication has become a necessity.

Today, most of the electronic circuit packaging companies have computerized the development of templates, masks, numeric control files, and other facets of the fabrication process. In addition to this, fabrication technology itself has become much more sophisticated. Computer-aided tools permit detailed evaluations of the fabrication process for a complex electronic circuit at the speed of modern computers. Although electronic fabrication can be accomplished by using hand tools, the process is very repetitive, time consuming, and error prone. The computer-aided tool serves as a unique resource to speed up the process. Also, expensive errors can be easily corrected at an early stage. The enormous storing, quick retrieval, and quick editing facility of a computer aids the process of electronic fabrication. This dramatic advancement, coupled with new innovations in recent years in electronic fabrication technology, requires today's engineers to adopt new methodology and requires technicians to develop new skills.

Objectives

After completing this chapter, you should be able to

1. Describe how computer-aided systems can be used for designing, drafting, and fabricating electronic circuits
2. Identify the steps of fabricating printed circuit boards using OrCAD software tools

1-2. OBJECTIVES

This book will introduce the concept of electronic circuit design and fabrication with computer-aided software and laboratory-based hardware tools. The text starts with the basic concept of electronic projects and ends with full-blown projects. There are many computer-aided tools available for Electronic Design Automation (EDA). Obviously, the one produced by OrCAD Inc. is one of the most powerful, affordable, and complete computer-aided tools for this purpose.

This textbook intends to present materials and concepts to help the student learn OrCAD/Schematic Design Tools (OrCAD/SDT), OrCAD/Printed Circuit Board (OrCAD/PCB) layout design tools, and printed circuit board fabrication by way of a hands-on approach. Students learn to design electronic circuit schematics, design artwork for printed circuit boards by using OrCAD Software Tools, and fabricate printed circuit board (PCB) by using an aqua (water)-based nontoxic chemical process in a school laboratory environment. This book is a complete manual for the use of OrCAD/SDT and OrCAD/PCB software for electronic circuit board design, drafting, and fabrication. A step-by-step procedure starting with schematic design to printed circuit board layout design through complete, worked-out examples will give students hands-on practice using the software tools. Students should use this textbook with their computer turned on and the electronic design environment running. To use this book students do not have to be electronic experts; however, understanding electronic parts and circuits will give added satisfaction to the learning.

1-3. A NEW OPERATING ENVIRONMENT OF OrCAD SOFTWARE TOOL SETS

The operating environment of OrCAD/SDT and OrCAD/PCB tools has changed significantly over the years. These changes have made the software more manageable and user-friendly in an electronic design automation environment. This text is written on the basis of OrCAD/SDT Version 4.22 and OrCAD/PCB Version 2.21. However, since the major changes are mainly in software operating environment, this text can still be effectively used with a few later and earlier versions of the software product. Students who have already learned the earlier versions of the software tools will have no trouble using this textbook. The major change is that the earlier versions were command line-oriented at the DOS prompt, whereas these versions are menu driven. The main advantage of these versions (SDT Ver. 4.22 and PCB Ver. 2.21) is that users do not have to memorize the command lines and switches to work through the software. Thus, there is almost no chance to make a syntax error in writing a command line, so the users can work considerably faster through the software tools. Also, Design Management Tools (provided with the software) help manage various types of design files for a design under this environment. In short, the EDA environment became faster, easier, more user-friendly, and more manageable with these versions of the software. Recently, OrCAD has released another new version of SDT and PCB tools, called OrCAD/SDT 386+ and OrCAD/PCB 386+. These enhanced versions of the tools were not released in time to incorporate them in this edition. However, I have provided the changes that have taken place with these two tools, coupled with an example, in Appendix D.

1-4. BASIC COMPONENTS OF THE PRINTED CIRCUIT BOARD FABRICATION PROCESS

After reading the text and doing the hands-on computer work, students should be able to do the following types of design, drawing, and fabrication proficiently.

Part 1: Circuit Definition and Logical Files

    1-4.1 Block Diagrams and Flowcharts

    1-4.2 Schematics Designed from Hand-Drawn Circuit Diagrams

    1-4.3 Electrical Rules Checks

    1-4.4 Generation of Bill of Materials

    1-4.5 Preparation of Circuit Schematic for PC Board Layout Design

    1-4.6 Printing and Plotting of Schematics

    1-4.7 Creation of a Netlist File for PC Board Layout Tool Sets

Part 2: Logical to Physical Transformation

    1-4.8   Module Parts Placement and Board Routing

    1-4.9   Manual and Automatic Routing

    1-4.10 Printing and Plotting Master Artwork Layers, Silkscreen, Solder Mask, Drill Template, and Other Outputs

Part 3: Board Fabrication

    1-4.11 Printed Circuit Board Fabrication

    1-4.12 Printed Circuit Board Drilling, Assembling, and Testing

Part 4: Advanced Features Available with the Software

    1-4.13 Creation of Custom Symbols and Modules Not Available in the OrCAD Software Library

Part 1: Circuit Definition and Logical Files

1-4.1 Block Diagrams and Flowcharts

A two-dimensional block diagram for electronic projects and a flowchart for computer programs can be drawn by using OrCAD software tools. Alternatively, hand-drawn block diagrams and flowcharts can be used for the project. The conversion of hand-drawn block diagrams and flowcharts to computer-drawn ones is not essential for the electronic fabrication process. However, it is convenient to draw flowcharts and block diagrams for electronic projects by using OrCAD tool sets in the EDA environment. This will also produce industry-accepted documentation for electronic projects. Developing a block diagram and flowchart for a project is part of the planning phase. It is almost impossible to express an electronic project by using only words or diagrams. However, if you combine both, it may be easier and more accurate for your explanation of the idea. Thus, block diagrams and flowcharts play an important role in high-level planning. (Chapter 2 discusses the planning of school laboratory projects.) Block diagrams and flowcharts can be drawn using OrCAD/SDT and OrCAD/PCB tools. Figures 1–1 and 1–2 show a block diagram and a flowchart, respectively, of a regulated power supply electronic project that is drawn by using the software tools.

1-4.2 Schematics Designed from Hand-Drawn Circuit Diagrams

Capturing a computer-drawn schematic is one of the essential steps in the design and fabrication of the printed circuit board. A schematic is a scheme of logical connections among symbols representing electrical parts. OrCAD/SDT tools are used to draw schematics of an electronic circuit. A hand-drawn schematic of an electronic circuit converted into a computer-

Figure 1–1. Functional Block Diagram of a Regulated Power Supply

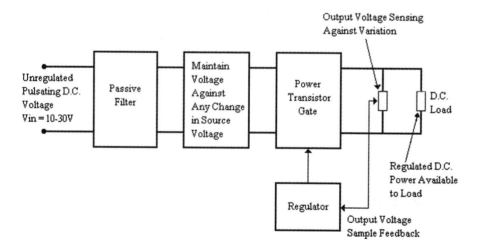

1-4. Basic Components of the Printed Circuit Board Fabrication Processs

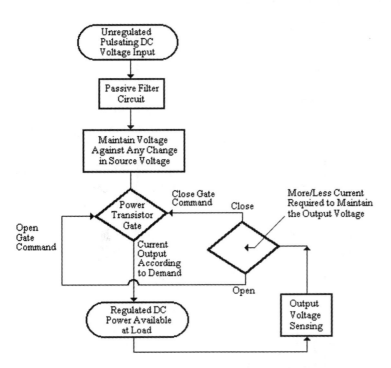

Figure 1-2. Flowchart of a Regulated Power Supply

drawn schematic is shown in Figure 1-3. Depending on the size of the electronic circuit, the schematic can be drawn on various sizes of worksheet. OrCAD offers five standard sheet sizes for capturing a schematic. The sizes are A through E for English and A4 through A0 for metric. A custom page up to 65 inches square is also possible.

1-4.3 Electrical Rules Checks

OrCAD software provides an electrical rule checker that checks basic electrical rules, such as unused inputs, if output pins of two devices are connected together, and if power and ground are connected together. This tool scans an electronic design, checks the electrical rules according to the user-defined processing options and conditions specified in the **Electrical Rule Matrix,** and reports any problems back to a destination file. These rules can be changed or modified by altering the **Electrical Rule Matrix,** located in SDT configuration. Two obvious faults are introduced into the circuit shown in Figure 1-3. Two junction boxes were deleted to introduce the fault. Figure 1-4 shows the circuit schematic of Figure 1-3 before and after **Checking Electrical Rules.** The circuit on the right of Figure 1-4 shows the errors, which are marked by highlighted circles. Errors such as these must be fixed before proceeding further with the design and fabrication process. Removal of two junction boxes in Figure 1-4 produced three error circles. This happened because removal of one junction box may produce more than one error in certain situations.

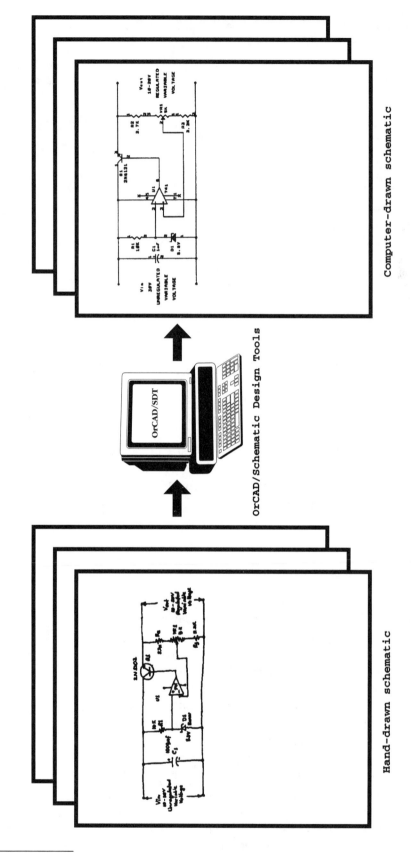

Figure 1-3. Circuit Schematic Capture Using OrCAD/SDT

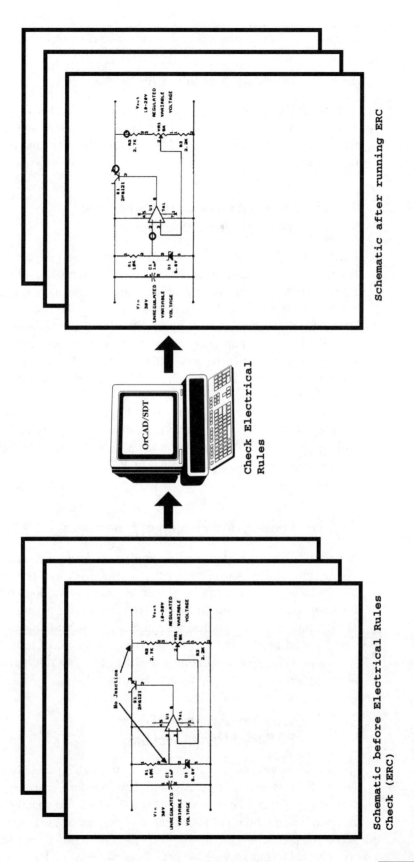

Figure 1-4. Electrical Rule Check of a Circuit Schematic

1-4.4 Generation of Bill of Materials

Create Bill of Materials, available under OrCAD/SDT, will produce a complete list of the materials of the circuit schematic and their associated specifications. Figure 1–5 shows the circuit schematic and Bill of Materials of the circuit shown in Figure 1–3.

1-4.5 Preparation of Circuit Schematic for PC Board Layout Design

Each part (or symbol) in a schematic diagram has ten part-fields. The first two part-fields contain information on part reference and part value. The other eight could contain user-defined information such as types of packaging and other pertinent information of the part. The packaging information is essential and is used later by the PC Board Layout tools. To stuff this information into a part-field of components of a schematic, a tool called **Update Field Content,** available under SDT Tool Set, is used. The schematic shown in Figure 1–3, when appended with packaging information for each component, will look like the schematic shown in Figure 1–6. Each part in the left-hand schematic of Figure 1–6 has only a part reference such as R1, R2, U1, or D1 and an associated part value such as 10K, 2.7K, 741, or 6.8V. After passing the schematic through the **Update Field Content** tool, the package ID, such as RC01, 8DIP300, D07, is affixed to each part respectively.

1-4.6 Printing and Plotting of Schematics

Schematics are usually printed several times during the design process for visual verification. At the end of the schematic design process, schematics usually need to be printed or plotted for hard copy documentation. Final copies of schematics should be printed using a laser printer for quality documentation. If colors are desirable, plotting of schematics can also be done using a color plotter. OrCAD software provides a set of tools for both printing and colorful plotting. Figure 1–7 shows the circuit schematic of the regulated power supply printed by a laser printer.

1-4.7 Creation of a Netlist File for PC Board Layout Tool Sets

In an EDA environment, design tools must be able to exchange information among themselves and with the third-party vendor's software tools. The file that allows this exchange is a netlist file. A netlist file contains all of the connectivity information and packaging information of the electronic parts of a schematic file. Schematic Design Tools can create netlists in a variety of industry-accepted formats. Using the netlist tool available under SDT, PC board layout designers must perform this essential step of creating a netlist file for the

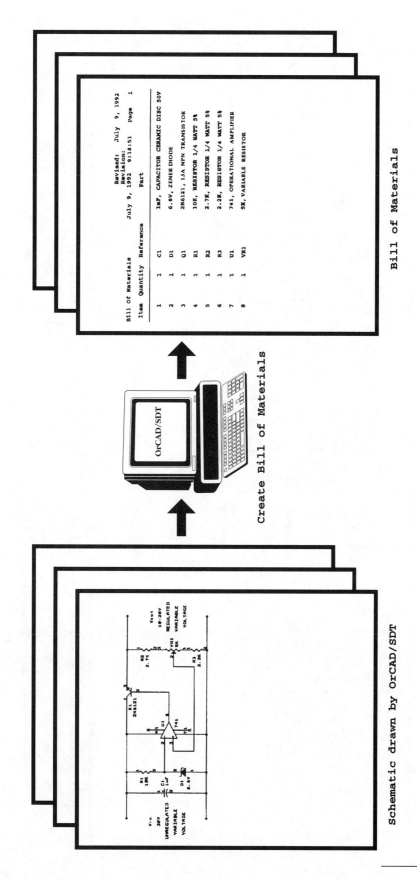

Figure 1-5. Generation of Bill of Materials

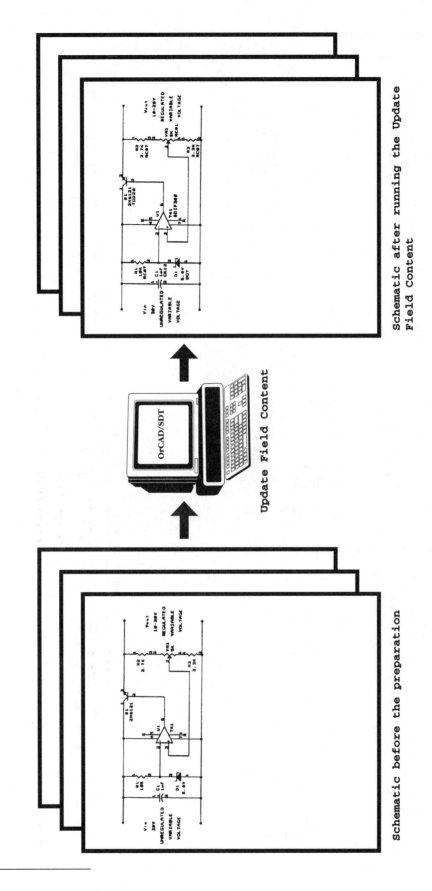

Figure 1-6. Schematic Preparation for PC Board Layout Tools

1-4. Basic Components of the Printed Circuit Boad Fabrication Process

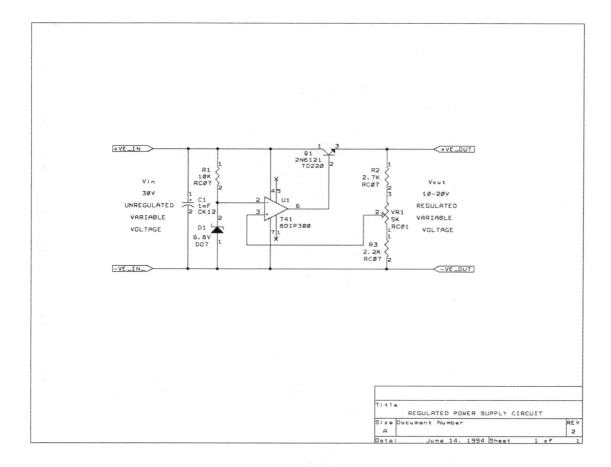

Figure 1-7. The Circuit Schematic as Printed by Laser Printer

schematic before proceeding to use the PC Board Layout tools. Netlist in OrCAD/PCB II format is the input file for the PC Board Layout tool set. Figure 1-8 shows the netlist file of the schematic shown in Figure 1-7. Figure 1-8 also provides information about various parts of the netlist file. More detailed information about netlist files will be provided later in this text.

Part 2: Logical to Physical Transformation

1-4.8 Module Parts Placement and Board Routing

Placing modules on the board is one of the critical steps in PC board fabrication process. Modules are downloaded by using the netlist file, which contains a list of the logical signals and part connections. Modules can also be downloaded manually by the designer from the module library available in OrCAD/PCB. Figure 1-9a shows a PC board on which the modules have just been downloaded from the PCB library by using the netlist file. Once the modules are downloaded onto the board, they need to be rearranged for two reasons: first, to reflect the architectural and functional flow of the

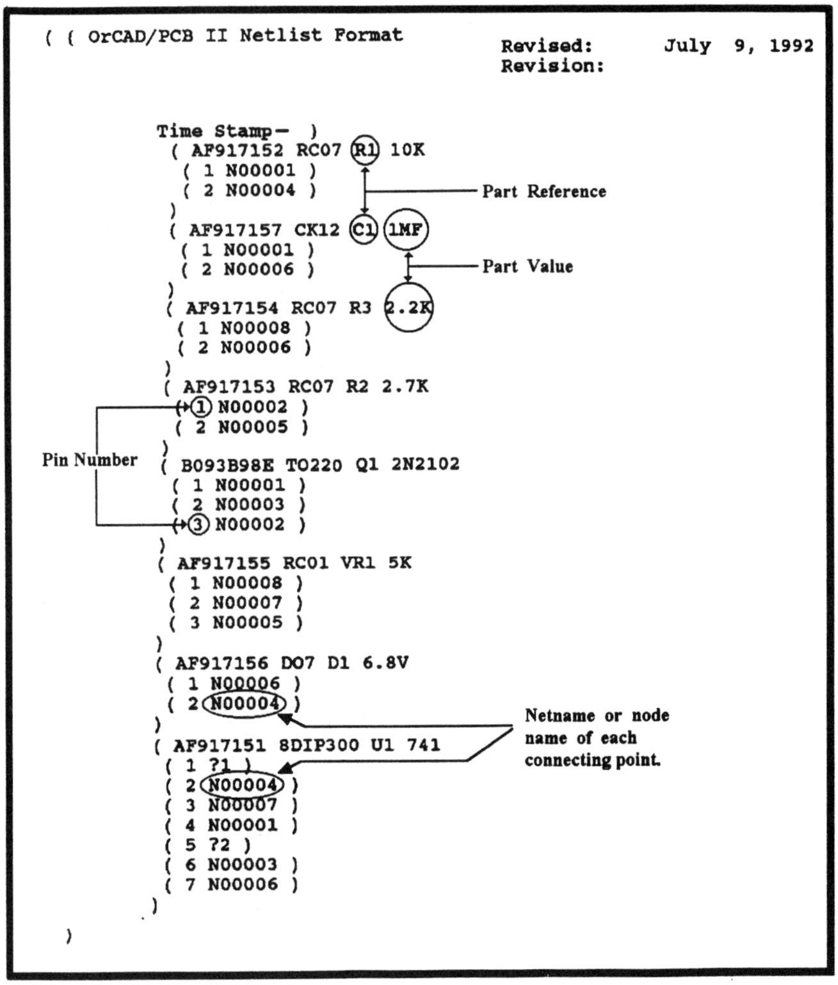

Figure 1-8. Netlist File of the Schematic Shown in Figure 1-7

```
( ( OrCAD/PCB II Netlist Format        Revised:     July 9, 1992
                                       Revision:

    Time Stamp- )
   ( AF917152 RC07 R1 10K          ──── Part Reference
    ( 1 N00001 )
    ( 2 N00004 )
   )
   ( AF917157 CK12 C1 1MF
    ( 1 N00001 )                    ──── Part Value
    ( 2 N00006 )
   )
   ( AF917154 RC07 R3 2.2K
    ( 1 N00008 )
    ( 2 N00006 )
   )
   ( AF917153 RC07 R2 2.7K
    ( 1 N00002 )
    ( 2 N00005 )
   )
   ( B093B98E TO220 Q1 2N2102
    ( 1 N00001 )
    ( 2 N00003 )
    ( 3 N00002 )
   )
   ( AF917155 RC01 VR1 5K
    ( 1 N00008 )
    ( 2 N00007 )
    ( 3 N00005 )
   )
   ( AF917156 DO7 D1 6.8V
    ( 1 N00006 )
    ( 2 N00004 )                    ──── Netname or node
   )                                     name of each
   ( AF917151 8DIP300 U1 741             connecting point.
    ( 1 ?1 )
    ( 2 N00004 )
    ( 3 N00007 )
    ( 4 N00001 )
    ( 5 ?2 )
    ( 6 N00003 )
    ( 7 N00006 )
   )
)
```

Pin Number (label pointing to pin numbers 1 and 3 in Q1 block)

design, and second, to minimize the total length of the tracks. Placement of the modules can be optimized using force and ratsnest vector functions available in the routing steps. These vectors, when called on an incomplete layout, will guide the designer for best position of the module. Once the modules are reasonably positioned on the board, routing of the board can begin. It is difficult to separate the process of module placement from the process of routing. These two processes are closely interconnected with each other and also greatly depend on each other. Figure 1–9b shows a PC board layout on which modules are arranged by using force vectors and ratsnest vectors.

1-4.9 Manual and Automatic Routing

Routing converts the logical connections among components into physical tracks of desired widths on the board. These tracks are the segments of conducting paths from one electronic component to another electronic compo-

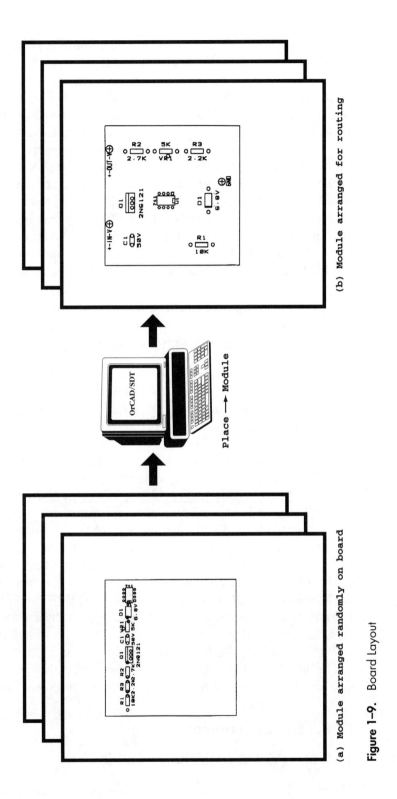

Figure 1-9. Board Layout

nent and cannot overlap each other on a board surface. OrCAD software supports both a manual and an automatic routing process. Rearrangement of the modules and routing the board are successive operations that are done several times to arrive at the final routed condition of the board. Routing can be on two sides of a board. Routes on one side of the board can be connected to the other side of the board by a conducting hole called a **via.** Routes on two sides of the board can also be connected by a component lead. Layers of a multilayer board are connected by special **embedded** or **blinded vias.** The final output of the placement and routing process is a physical drawing of the actual metal interconnects called **artwork.** This artwork is used to set the tooling of the printed circuit board manufacturing process. The artwork for the board representing the circuit of Figure 1–3 is shown in Figure 1–10. Figure 1–10a shows the component side layout of the board before routing, and Figure 1–10b shows the copper side layout of the board after routing.

1-4.10 Printing and Plotting Master Artwork Layers, Silkscreen, Solder Mask, Drill Template, and Other Outputs

Both printing and plotting tools are available under OrCAD/SDT and OrCAD/PCB for printing and plotting of master artwork, silkscreen, solder mask, drill template, and other essential items required for the tooling process of PC board fabrication (Part 3). These must be printed with high-quality laser printers or laser photoplotters. To help this critical tooling process, it is often necessary to print or plot the output in various sizes. OrCAD software provides many of those flexibilities. Master artwork is the physical pattern of metallic conducting paths that connect various components. Silkscreen is the pattern of the physical location of each component on the board identified by its symbol and reference designator. This pattern is transferred onto the individual board by screen printing. The pattern is called silkscreen because screen printing process uses a silkscreen and epoxy ink. Solder mask is the pattern that is used to solder the board selectively while soldering by an automatic machine. Drill template is the pattern that indicates the position and size of each point that needs to be drilled. After the printing or plotting of the master artwork, silkscreen, solder mask, and drill template, the PC board design process virtually ends. Figures 1–11a through 1–11d show the master artwork, silkscreen of the layout, drill template, and solder mask, respectively, of the power supply circuit.

Part 3: Board Fabrication

1-4.11 Printed Circuit Board Fabrication

The printed circuit board fabrication process in a school laboratory environment can be divided into three steps. In the first step, master artwork is transferred onto photosensitized, copper-cladded board through photography, screen printing, or other methods. In a laboratory environment it is more convenient to use the photographic method as compared to screen

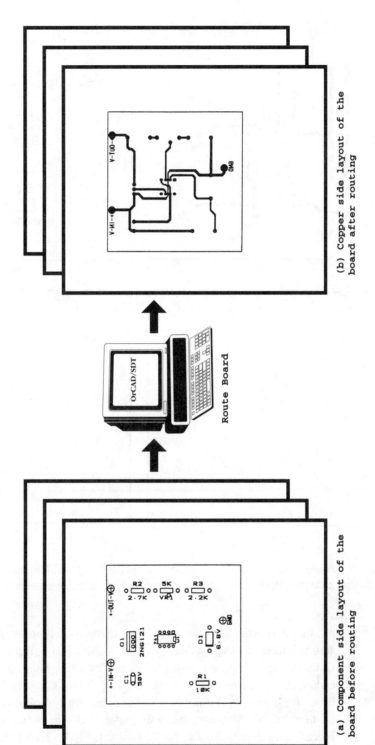

Figure 1-10. Final Board Layout

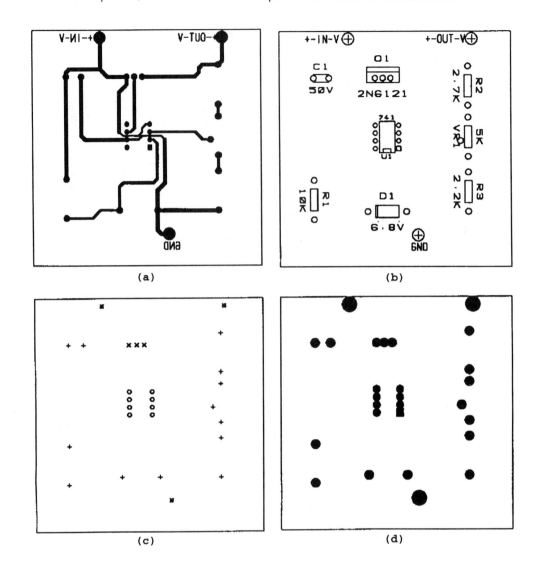

Figure 1-11. (a) Master Artwork, (b) Silkscreen, (c) Drill Template, and (d) Solder Mask

printing and other commercially used methods. Therefore, in the laboratory, an ultraviolet exposure system is used for the photographic transfer process. In the second step, the board is etched by using laboratory-based desktop etcher. A chemical process developed by Excel Circuits is commonly used for etching and photographic transfer of artwork to photosensitized, copper-cladded board. Although there are several dozen industrial and laboratory processes available for PC board fabrication, the Excel chemical process is especially suitable for a laboratory environment because photographic and etching chemicals are water soluble and do not emit toxic fumes. Thus, the processing laboratory does not need any forced ventilation. In addition to this, the process uses a positive-acting photoresist for which it is easy to make the artwork. Complete PC board fabrication process will be described in detail later in this book.

1-4.12 Printed Circuit Board Drilling, Assembling, and Testing

Printed circuit board can be drilled either by a manual high-speed drill press or a high-speed computer-controlled drill press called Computer Numerical Control (CNC). For commercial grade and mass scale production, CNCs are used. OrCAD software generates numeric control data for board drilling and routing in various industry-accepted formats. For example, it can generate drill information in a widely used format called EXCELLON. These data can be sent to computer numerical-control drill and router machines. Hard-copy printout of this drilling information can also be generated. Since CNC machines are expensive and boards are not produced in mass scale in school laboratory environments, boards can be drilled using high-speed drill presses with carbide drill bits. Figure 1–11c shows the drill template, but the board is not of actual size. Figure 1–12 shows the drill template in actual board size. Figure 1–13 shows an ASCII file listing of drill sizes and locations of pads and vias on the board. Figure 1–14 shows the module report of the printed circuit board. This module report is useful because, for a large board, module count and location of certain modules need to be verified. Commercial component assembly and testing are done by automatic machines. However, in a school laboratory, manual process is sufficient.

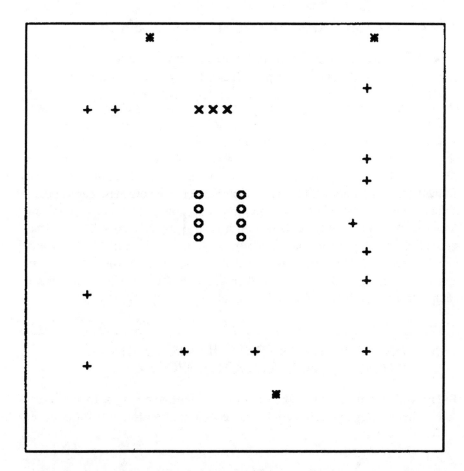

Figure 1–12. Drill Template in Actual Board Size

Figure 1-13. ASCII Drill File

```
ASCII(0.035,1.7,1.6)
ASCII(0.035,1.7,1.7)
ASCII(0.035,1.7,1.8)
ASCII(0.035,1.7,1.9
ASCII(0.035,2.0,1.9)
ASCII(0.035,2.0,1.8)
ASCII(0.035,2.0,1.7)
ASCII(0.035,2.0,1.6)
ASCII(0.045,0.9,1.0)
ASCII(0.045,1.1,1.0)
ASCII(0.045,0.9,2.3)
ASCII(0.045,0.9,2.8)
ASCII(0.045,1.6,2.7)
ASCII(0.045,2.1,2.7)
ASCII(0.045,2.9,2.7)
ASCII(0.045,2.9,2.2)
ASCII(0.045,2.9,2.0)
ASCII(0.045,2.8,1.8)
ASCII(0.045,2.9,1.5)
ASCII(0.045,2.9,1.35)
ASCII(0.045,2.9,0.85)
ASCII(0.054,1.7,1.0)
ASCII(0.054,1.8,1.0)
ASCII(0.054,1.9,1.0)
ASCII(0.13,1.35,0.5)
ASCII(0.13,2.95,0.5)
ASCII(0.13,2.25,3.0)
```

Part 4: Advanced Features Available with the Software

1-4.13 Creation of Custom Symbols and Modules Not Available in the OrCAD Software Library

OrCAD/SDT version 4.20 provides about 20,000 part symbols in its library, and this number increases with every version of the software. Thus, very seldom is it necessary for a designer to create a new schematic symbol of a component or a module for a printed circuit board. The **Librarian** tool can create desired symbols or modules and store them in the library database under user-defined name and library. This means that the user can store these parts under a library called CUSTOM or under any library such as TTL, Moto, or Intel that is provided with the software. However, it is strongly recommended that users store all the symbols they create in a separate library and not in the library provided by OrCAD. Otherwise, user-created symbols may be lost when library updates provided by OrCAD are loaded. Figure 1–15 shows hand-drawn and computer-drawn symbols and modules of electronic components.

1-5. THE COMPUTER-AIDED ELECTRONIC DESIGN, DRAFTING, AND FABRICATION PROCESS

Today, with the decreasing cost of computer hardware and software, computer-aided design and fabrication has become more affordable for colleges

1-5. The Computer-Aided Electronic Design, Drafting, and Fabrication Process

Module Refer.	Module Value	Module File Name	Orient.	Pin	X	Y	Net Name
R1	10	RC07	(90)	1	0.900	2.300	+_IN_V
				2	0.900	2.800	N00002
R3	2.2K	RC07	(90)	1	2.900	2.200	N00005
				2	2.900	2.700	GND
R2	2.7K	RC07	(90)	1	2.900	0.850	+_OUT_V
				2	2.900	1.350	N00003
Q1	2N6121	TO220	(0)	1	1.700	1.000	+_IN_V
				2	1.800	1.000	N00001
				3	1.900	1.000	+_OUT_V
C1	50V	CK05	(0)	1	0.900	1.000	+_IN_V
				2	1.100	1.000	GND
VR1	5K	RC01	(270)	1	2.900	2.000	N00005
				3	2.900	1.500	N00003
				2	2.800	1.800	N00004
D1	6.8V	DO7	(0)	2	1.600	2.700	N00002
				1	2.100	2.700	GND
U1	741	8DIP300	(270)	1	2.000	1.900	
				2	2.000	1.800	N00002
				3	2.000	1.700	N00004
				4	2.000	1.600	+_IN_V
				5	1.700	1.600	
				6	1.700	1.700	N00001
				7	1.700	1.800	GND
				8	1.700	1.900	
M1		MHOLE1	INPUT	(0)	1	1.350	0.500	+_IN_V
M2		MHOLE1	GRD	(0)	1	2.250	3.000	GND
M3		MHOLE1	OUTPUT	(0)	1	2.950	0.500	+_OUT_V

Figure 1-14. Module Report

and universities. The computer-aided electronic design, drafting, and fabrication process starts with a hand-drawn sketch of a system and a few circuit diagrams. This hand-drawn sketch usually consists of several parts of the electronic circuit, such as power supply, input and output devices, and the main electronic circuit, which will actually perform the desired electronic operation. Once the circuit schematic of each part of the system is available in hand-drawn form, it must be converted to a computer-drawn schematic for further processing.

After the schematics are drawn with computer software tools, they are usually printed or plotted to check for any errors and for the purpose of documentation. The design rule checker available with the tool is used to check

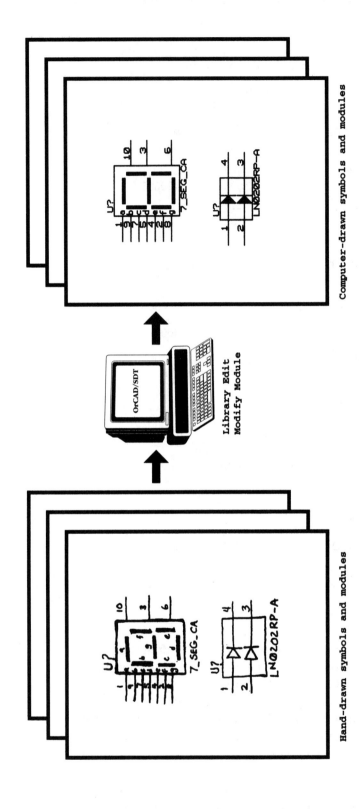

Figure 1-15. Custom Symbols Created by OrCAD/SDT

1–5. The Computer-Aided Electronic Design, Drafting, and Fabrication Process

for errors in reproducing the circuits. This checker looks for unconnected parts, ground to power shorts, and other problems. The circuit may be prototyped by using a solderless breadboard or by wire wrapping to verify its desired function. In many instances, a computer may be used to simulate the whole circuit or part of the circuit to verify its operation before prototyping. At this time, changes in schematics that may be the result of breadboarding can be incorporated to produce the final version of the schematic.

Before proceeding with the printed circuit board design, circuit schematics have to be incorporated with packaging information for each component in the schematic. Packaging information and interconnection among components are extracted from the schematics in the form of a netlist file. Although there are many parts of the circuit (often in multiple sheets), there will be only one netlist file as long as all these parts are linked. This netlist file is the input file for the printed circuit board software tools. To ensure electrical accuracy, you should run the electrical rule checker before generating the netlist file for a schematic.

In PCB Layout Tools, PC board size is defined first. Then the components are unloaded with their interconnections (net information) within the specified boundary of the board. When all the components are available on the board, they are rearranged and placed according to the desired board layout. It may be necessary to retrieve the edge connector, or other single and multi-lead connector, from the PCB module library in order to connect the board circuit to external parts or another circuit. Board layout is primarily governed by the components' interconnection scheme. After the completion of the board layout, the board is routed manually, automatically, or a combination of both to produce master artwork for the printed circuit board. The design conditions for the printed circuit board can be defined for the router to control the route thickness, number of vias, and number of routing layers, etc. When all the components are routed properly, master artwork should be printed or plotted.

The master artwork is the most critical part of the graphical documentation. Since master artwork is used to produce the most precise manufacturing tooling in the production of the PC boards, extreme overall accuracy is essential. To achieve high accuracy, a process called laser photoplotting is utilized. For the laboratory-based process, master artwork is printed directly onto a Mylar (vellum) paper by using a high-resolution laser printer. The silkscreen of the board is also printed on Mylar paper by a laser printer. Once the master artwork and the silkscreen of the circuit are available on a Mylar paper, the rest of the process in a school laboratory environment is manual. However, in a modern electronic manufacturing facility, the process after artwork is semi-automatic.

A complete flowchart of the printed circuit board fabrication process starting with the hand-drawn schematic is shown in Figure 1–16. The flowchart is a simplified version of a commercial fabrication process. The fabrication process described later in this text for school laboratory fabrication of PC boards is somewhat different from the commercial process and will be discussed in depth in Chapter 10 of this text. Whatever the electronic

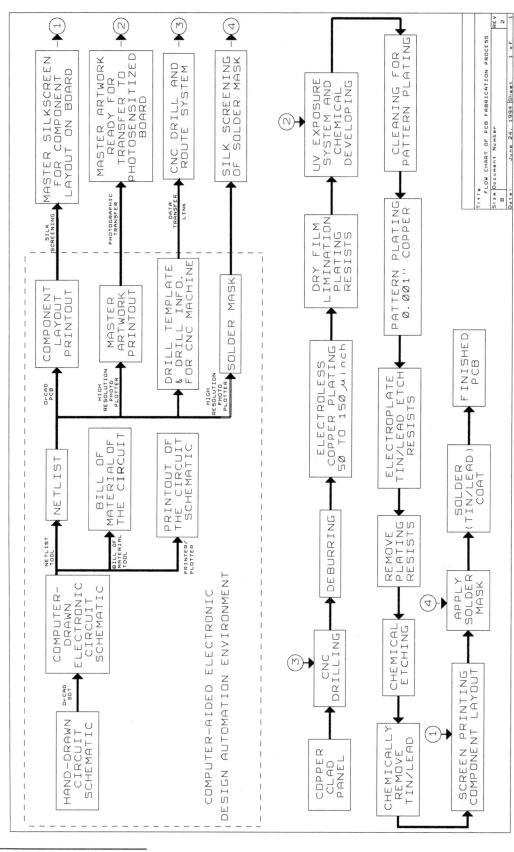

Figure 1-16. Printed Circuit Board Fabrication Process

project and its design and fabrication process, careful planning is essential. Chapter 2 will discuss the planning of four electronic projects.

QUESTIONS

1. Why have computers revolutionized PCB fabrication?
2. Why have computer-aided electronic design and fabrication become necessities?
3. What is the biggest disadvantage of using hand tools and templates for designing and fabricating electronic circuits?
4. Why is the use of computer-aided tools for electronic design and fabrication economical for a manufacturer?
5. Why should engineers and technicians adopt this new methodology of designing and fabricating electronic circuits?
6. What does EDA mean?
7. Why should EDA be a good environment for electronic design and development?
8. What information is contained in an electronic schematic?
9. What is the maximum schematic drawing size that OrCAD software supports?
10. Why should electrical rules checker be a useful tool for an EDA environment?
11. What is the Bill of Materials of a schematic?
12. What should be done to a schematic if it is intended to be used for PC board layout design?
13. What information is the link between OrCAD/SDT and OrCAD/PCB?
14. Can other PCB software use schematics drawn by using OrCAD/SDT?
15. What information is important to a schematic design tool?
16. What is the artwork of an electronic circuit?
17. Why is printing the hard copy of an artwork so critical for fabricating printed circuit boards?
18. What kind of a PC board fabrication process is preferred by factories in the United States?
19. Can you make special schematic symbols by using OrCAD/SDT tools?
20. What is the main difference between an industrial PCB fabrication process and the process described in this book?

2

ELECTRONIC CIRCUITS: PLANNING, DESIGNING, EXPERIMENTING, AND PACKAGING

2-1. WHY PLAN?

Planning for an electronic project in a school laboratory environment is as important as planning for an electronic project in industry. The basic purpose of planning is to ensure accurate project implementation and to avoid any catastrophic failure in the final product. A high-quality and reliable electronic project can result only from thorough, careful planning and development of the project.

2-2. PLANNING A SCHOOL LABORATORY PROJECT

Before proceeding with the actual implementation, analysis of the following areas will result in a fruitful benefit even for a simple project in a school laboratory environment:

    2–2.1 Project Description

    2–2.2 Time and Resource Requirement and Analysis

    2–2.3 Level of Effort Requirement and Distribution

    2–2.4 Designing and Drawing

    2–2.5 Packaging Scheme

    2–2.6 Material Procurement

    2–2.7 Tool and Machinery Requirement Analysis

    2–2.8 Developmental Plan

    2–2.9 Experimenting

At the end of this chapter, definition and analysis of four electronic projects are provided. These will ultimately be developed into full-blown projects in this book.

Objectives

After completing this chapter, you should be able to

1. Describe your chosen electronic project, its rationale, and your objectives

2. Estimate the time, resources, and level of effort required for the project

3. Integrate circuits and modules for the project

4. Develop a packaging scheme

5. Select type, thickness, and size of board substrate material required for the project

6. Prepare a material procurement, tools, and machinery requirement list

7. Write a developmental plan for the project

8. Perform experiments to verify the operation of the circuit

2-2.1 Project Description

A detailed project description is essential before one should proceed with any work concerning the project. Three fundamental items are usually discussed:

Definition: Brief description of the project

Rationale: Why such a project should be undertaken

Objectives: Goals of the project

An electronic project description should also have information about input/output user interface, availability of power source, and types of applications desired. A system functional diagram should be drawn at this time to support the verbal description of the project. The system diagram also will remain as a high-level functional description of the project for effective implementation of the design effort. The blank worksheet provided in Figure 2–1a may be used for drawing a functional diagram for the project of your choice.

2-2.2 Time and Resource Requirement and Analysis

Time and resources are two important factors that need to be analyzed before you start. There is only a limited amount of time and resources available for a project. Therefore, careful analysis of these two factors is necessary for every project before its initiation. If the analysis shows that the project will involve more time than is available, then you should define only a part of the project or choose another project. If the project requires lots of money and resources, you may define a small part of it or choose another project.

2-2.3 Level of Effort Requirement and Distribution

If the time and resources necessary are such that the project requires more than one student, the work must be shared by the student workers. The most difficult task is to partition the work involved such that everyone gets the chance to contribute. Therefore, it is neither expected nor is it possible to divide the work equally among the project workers. Often, one worker may carry more of the workload than the others do.

2-2.4 Designing and Drawing

Designing the electronic circuit is the most critical part of the whole project. Accurate design is the foundation for a successful project. The design process intrinsically involves implementing a concept (project definition) into actual electronic circuits. It is the framework of the project. The design process does not always have to start from the transistor or integrated circuit. In other

PROJECT NO.:	SYSTEM FUNCTIONAL DIAGRAM:	DATE:	STUDENT'S NAME:	COMPUTER FILENAME:

Figure 2–1a. Blank Worksheet for Functional Block Diagram

words, one does not have to reinvent the wheel every time one needs to implement an idea.

Designing may involve an integration of various circuits and modules. High-level designs are often object oriented, where the objects are described in the form of a block. To explain the design only by words or only through drawings could be a nightmare. Thus, initial design report is a combination of both drawings and explanations. Design should clearly define signal input and output devices, power and ground interfaces, hand-drawn circuit diagrams that perform the desired electronic functions, and hand-drawn packaging illustrations of the project. All these information and drawings may be modified and ultimately be a part of the final project report. Theory and operational characteristics of the circuit should also be a part of this portion of the report. A written description of the circuit operation will provide clear understanding about the project to every person involved. The blank worksheet provided in Figure 2–1b may be used to draw the initial schematic diagram for the project of your choice.

PROJECT NO.:	CIRCUIT SCHEMATIC:	DATE:	STUDENT'S NAME:	COMPUTER FILENAME:

Figure 2–1b. Blank Worksheet for the Hand-Drawn Schematic

2-2.5 Packaging Scheme

Packaging is also an important part of a project. High-quality packaging improves the value and beauty of a project. Packaging illustrations are sets of drawings showing the external, internal, and finished appearance of the project. Packaging enclosures should be chosen with various factors in mind, such as functionality, maintenance, durability, space and weight, quality, environment, safety, and aesthetics. A hand-drawn packaging plan must be created to show the basic idea. However, for an electronic project at school, a packaging enclosure is generally chosen from the ones available, keeping some of the above factors in mind.

Three other important aspects of the project need to be determined as a part of the packaging plan: (1) type and thickness of PC board substrate material, (2) thickness of the copper foil, and (3) board size. The blank worksheet provided in Figure 2–1c can be used to draw the packaging scheme for the project of your choice.

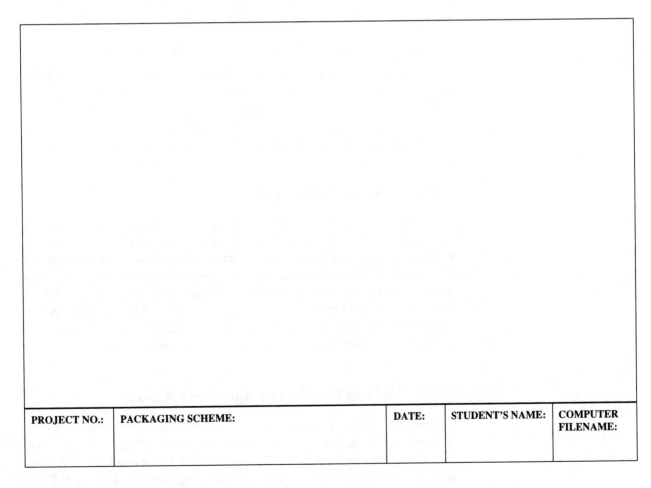

Figure 2–1c. Blank Worksheet for Packaging Scheme

Type and Thickness of PC Board Substrate Material. Commercially, there are various types of substrate material available. However, NEMA grade FR-4 is suitable for most electronic applications. It can be used for applications up to 40MHz of frequency, with medium power and low voltage. For the projects in this book, NEMA grade FR-4 of 0.062 (1/16") and 0.031 (1/32") will be used.

Thickness of Copper Foil. Thickness of copper foil is generally specified in ounces per square foot. Foil of one ounce per square foot is approximately 0.0014 in. thick. In some cases the thickness of the copper foil can be less than 0.001 inches. Thickness of copper foil is determined by the maximum continuous current in the circuit. For the projects in this book, boards with 1-oz. copper foil will be used.

Board Size for the Project. Board size determination for a particular project can be quite time consuming. For a marketable commodity, board size and shape are driven by many other criteria. However, for determining the board size for the projects other than those described in this book, the following rule of thumb may be used:

Actual Board Size = 4 × Total Area of the Board Components

The smaller the board, the more difficult it is to route. A large board with a few components on it does not look nice and compact. It is not cost effective, and it is difficult and expensive to package. Practically speaking, it could be quite expensive to stock boards of various sizes. Therefore, the projects in this book use only two different sizes: 4" × 4" and 3" × 6".

2-2.6 Material Procurement

Material procurement is a time-consuming task because one has to abide by the timing of the material vendors and the timing of the mail carrier. In addition, economy is an important factor in a student laboratory project. One-stop procurement can often be very expensive, yet it is convenient, considering the short time available for student projects. A list of parts should be generated prior to the first trip to the store. A project parts list will minimize the number of trips. The blank worksheet provided in Figure 2–1d can be used for preparing a list for the project of your choice.

2-2.7 Tool and Machinery Requirement Analysis

Different projects have different tool and machinery requirements. However, most electronic projects need common hand tools, such as soldering irons, rosin-cored (60/40 tin/lead) solder, wire cutters, pliers, wire strippers, screwdrivers, "third hand," de-soldering devices, high-speed drill machines, and high-speed carbide drill bits of various sizes. A drill-bit size

Project Title:						
PC Board Dimensions:				No. of Etched Sides:		
Edge Connector:						
Item No.	Reference Designator as in Schematic	Qty.	Description			

Remarks:

PROJECT NO.:	PART LIST AND COMPONENT DIMENSIONS:	DATE:	STUDENT'S NAME:	COMPUTER FILENAME:

Figure 2–1d. Blank Worksheet for Preliminary List of Parts

is generally provided by a number instead of the actual diameter of the drill bit. The size of a hole is important but not critical. Forcing a component lead through a small hole may damage the board while too large of a hole will cause sloppy mounting. Common sizes of bits required for printed circuit board drilling are provided in Table 2–1.

Carbide drill bits for PCB drilling are available with an enlarged shank of approximately 1/8 in. in diameter, regardless of their sizes. These carbide drill bits can withstand heat at speeds of 15,000 to 80,000 rpm, which are typical speeds of a high-speed drill press or a computer numerical control machine. For PC board drilling, specially made bits are generally used for accurate drilling.

Special tools are required for inserting via eyelets and pins for double-sided applications. These tools can be purchased from an eyelet manufacturer.

Other special tools include light tables, drafting lamps, ultraviolet (UV) exposure systems, photo developer systems, etcher systems, computer numerical control machines, and, most of all, computer software tools for schematic and artwork design. The following is a list of specialized tools, machinery, and chemicals required for the PC board projects in this chapter:

- Computer and CAD/CAE software package
- High-resolution laser printer
- UV exposure system
- Desktop etcher
- High-speed drill or CNC machine
- Soldering iron
- "Third hand"
- Digital multimeter
- Logic analyzer and logic probe
- Oscilloscope
- Regulated power supply
- Photosensitized double- and single-sided copper-clad board

Table 2–1. Drill Bit Equivalents

Drill Bit Size in Number	Equivalent Size in Inches
55	0.0520
64	0.0360
66	0.0330
67	0.0320
69	0.0292
70	0.0280
72	0.0250
75	0.0210

- Chemicals for etching, photo developing, and tin plating
- Screwdriver set
- Wire stripper and cutter
- Tin-lead solder (60/40)
- Solder sucker or de-soldering bulb
- High-speed carbide drill bits, sizes 64, 70, 72, 75

2-2.8 Developmental Plan

The developmental plan is a written scheme that must be used as a broad guideline when developing the project. If the project team consists of a single person, he or she can keep the plan in his or her head. But this is very inefficient, because vital issues regarding the project may be forgotten or missed by the developer. Thus, it is always advisable to write out the developmental scheme of the project. If the project involves more than one person, the plan must be distributed so that all those involved can follow through exactly as described. The plan, if written well, will maintain the continuity of the development process in case of change in personnel or even leadership. Even for fairly simple projects, a developmental plan is also important and may play a vital role. Although a developmental plan for a school class project is not as critical as for a project in an industrial setting, it should still be produced.

2-2.9 Experimenting

Very few designs work the first time. Thus, experimenting is critical as well as essential to prove a design. Since many changes may need to be made to the circuit during experimentation, it is recommended that a quick and easy circuit assembly method be used to check the design. For a school laboratory project, two types of methods are generally used. Solderless breadboarding is used for comparatively smaller circuits, and wire wrapping is generally used for more complicated circuits. However, for electronic projects where evaluating the electrical as well as the mechanical performance is essential, a prototype model of the complete project needs to be built.

Prototyping is more involved experimentation and is usually practiced in industrial environments. Prototype models of a product are built so that the product can be fully evaluated and any catastrophic failure in the final product can be avoided. Once this step is done, fabrication of the electronic project begins. The fabrication of the project starts with capturing the schematic by using computer-aided tools. Output from this step should be considered as a fundamental building block of the whole design process. High-level planning and experimenting are crucial to the overall success of the project.

2-3. PRACTICAL ELECTRONIC PROJECTS

The following four electronic projects will be developed as full-blown projects in this book. Circuit diagrams of these projects have been obtained from various books and magazines. Sources of these projects are provided in their respective figures.

Project No.: 01-1995
Project Title: The Deluxe Logic Probe
Project by: (Student's Name)
Computer Filename: LOG_PRO

Project No.: 02-1995
Project Title: The Electronic Cricket
Project by: (Student's Name)
Computer Filename: CRICKET

Project No.: 03-1995
Project Title: The Infrared Object Counter
Project by: (Student's Name)
Computer Filename: IR_COUNT

Project No.: 04-1995
Project Title: The Mini Stereo Amplifier
Project by: (Student's Name)
Computer Filename: AMPLIFIE

2-4. PROJECT 01-1995: THE DELUXE LOGIC PROBE

2-4.1 Project Description

Definition. An inexpensive and portable digital logic probe is required to debug both TTL and CMOS logic circuits. A system functional diagram of the proposed logic probe is shown in Figure 2–2.

Rationale. Digital circuits are not always perfect and often require some troubleshooting. A logic probe is a test instrument designed specifically for testing a digital circuit. The logic tester will indicate present logic level (voltage level) at a desired point in a digital circuit. To test a digital circuit thoroughly, other test instruments, such as a logic analyzer or digital multimeter, may be used. These instruments are often too expensive, bulky, and time-consuming to use for determining voltage level. Thus, for most of the problems in a comparatively simple digital circuit, a logic probe could be used for debugging.

Objectives. The main objective of this effort is to build an inexpensive, audible, and visually indicative logic probe. The prospective logic probe must be portable, convenient, have different color LED's to indicate different logic levels, and be able to test both TTL and CMOS logic. An audible

2-4. Project 01-1995: The Deluxe Logic Probe

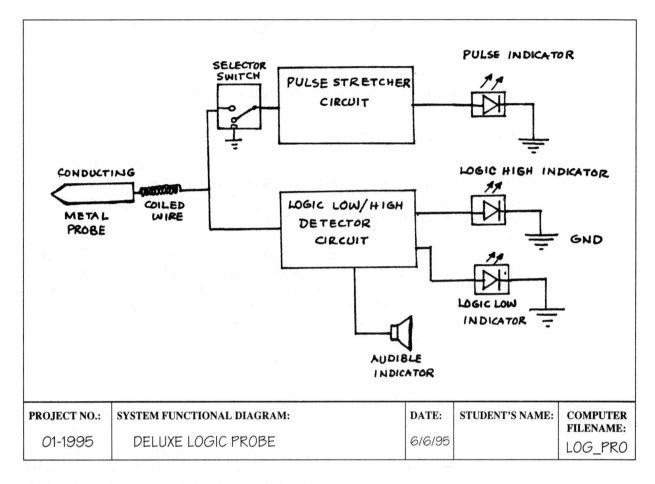

Figure 2-2. System Functional Block Diagram of the Deluxe Logic Probe

signal in addition to the LED indicator will be an added feature of the tester. The logic probe should also be able to indicate short pulses that are not indicated by high or low LED.

2-4.2 Time and Resource Requirement and Analysis

There are many standard logic probe circuits available in electronic project-building books, so designing the circuit is the same as reinventing the wheel. Therefore, the logic probe circuit can be chosen from the many already available. Considering that the probe circuit does not have to be designed from the fundamental electronic components, the project should take a total estimated time of 3 to 4 weeks (approximately 8 hours/week) to build. Including packaging, estimated material cost would be approximately $15.00. Estimated time spent on tools and machinery such as CAD tools for artwork design, etching facility, high-speed drilling machine (or computer numerical controlled machine), and soldering would be approximately 2 to 3 weeks out of the total time allowed for the project.

2–4.3 Level of Effort Requirement and Distribution

The project is expected to be a simple one. Therefore, it is a one-person project. A simple project like this is generally best performed by a single person doing the project from the beginning to the end.

2–4.4 Designing and Drawing

Since there are many standard logic probe circuits available, it is convenient to select one that could be tailored to fit the need. Figure 2–3 shows a hand-drawn logic probe schematic, obtained from a book of electronic projects, that satisfies the basic project requirements.

Theory and operation. The chosen logic probe circuit has four outputs, three inputs, and one selector switch; three LED's to indicate high, low, and pulse; and a piezoelectric buzzer for audible output. The three inputs are power, ground, and the test probe. The selector is used to switch in and out of the pulse stretcher circuit. Due to the use of CMOS gates, the input power may be as low as 3V and as high as 18V. Therefore, the prospective logic probe may be used for both CMOS and TTL logic.

2–4.5 Packaging Scheme

Since the probe is not intended to be marketed as a product, high-quality commercial packaging is not necessary. The 6" × 3" and 0.062 in. thick Excel board will be used. (Excel boards are photosensitized, copper cladded, and made by Excel Circuits.) Considering the available board size, it has been tentatively decided to package the probe in a 6.5" × 3.5" × 1.5" plastic box with detachable probe connections. Figure 2–4 shows the packaging scheme of the probe.

2–4.6 Material Procurement

Since the dimensions of the PC board are a big factor in determining the size of the enclosure, they need to be chosen before proceeding with the design process. Electronic parts and other hardware materials should be arranged to test the operation of the circuit. Electronic surplus stores in your area may be an inexpensive source of parts for the project. Good parts out of a defective, unrepairable electronic product may be another quite inexpensive source of parts. Producing a list of parts will reduce the number of trips to the store. Although a bill of materials for the circuit will be produced later by computer software, it is a good idea to make an initial list. Figure 2–5 is the preliminary list of parts for the project.

2–4.7 Tool and Machinery Requirement Analysis

Initial analysis indicated that the project does not require any special tools. Common hand tools such as soldering iron, 60/40 tin/lead solder, soldering

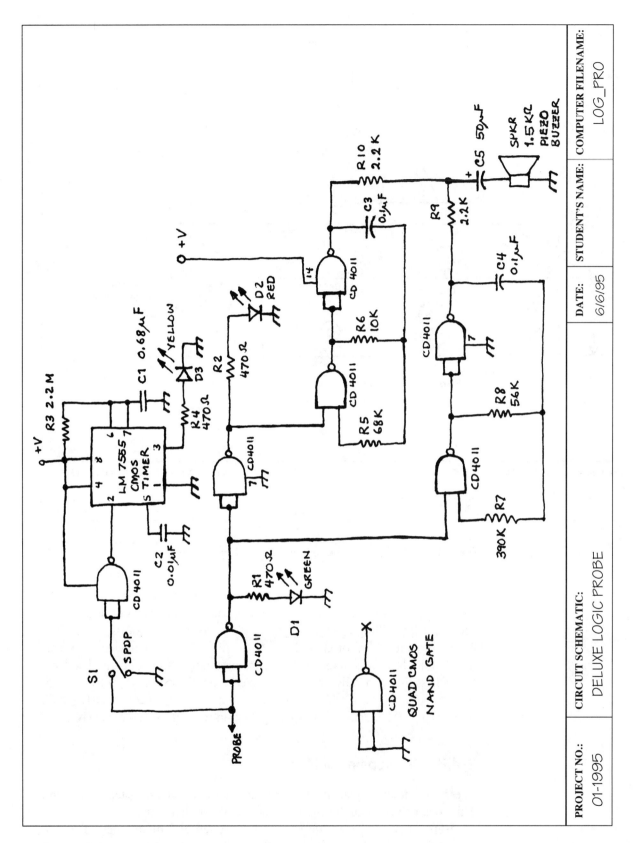

Figure 2-3. Hand-Drawn Schematic of the Deluxe Logic Probe
Source: Delton T. Horn, *50 CMOS IC Projects*, Blue Ridge Summit, Penn.: TAB Books, 1988, p. 64. Used with permission from TAB Books Inc.

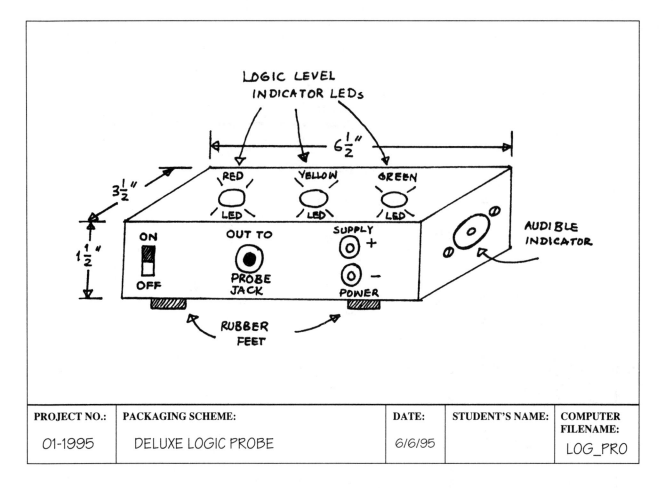

Figure 2-4. Packaging Scheme of the Deluxe Logic Probe

flux, wire cutter, pliers, wire stripper, screwdriver, "third hand," and high-speed carbide drill bits will suffice. Three or four different sizes of drill bits are necessary for drilling holes through the PC board for component leads. Drill bits required for this project are No. 64, 70, 72, and 75. You may need a sheet-metal bit to drill holes in the enclosure for mounting off-board parts. The sizes (in inches) that generally are required are 0.062, 0.078, 0.094, 0.109, 0.125, 0.140, 0.156, 0.172, 0.187, 0.203, 0.219, 0.234, 0.250, 0.265, 0.281, 0.297, 0.312, 0.359, 0.375, and 0.390. Buying them in sets will be less expensive.

2-4.8 Developmental Plan

The developmental plan for the project involves a step-by-step procedure. This scheme is crucial during the project developmental period.

Item 1: A computer file should be opened and maintained. The file should contain all the details about the project, including references, circuit diagrams, developmental plan, initial parts list, and anything that will serve as a source of potential information. Correspondence, if any, about the project and sources of parts should be part of this file.

Project Title:	DELUXE LOGIC PROBE		
PC Board Dimensions:	L= 6", W =3", Th = 1/16"	No. of Etched Sides:	1 (ONE)
Edge Connector:	NONE		

Item No.	Reference Designator as in Schematic	Qty.	Description
1	C1	1	0.68 μF Ceramic Capacitor
2	C2	1	0.01 μF Ceramic Capacitor
3	C3, C4	2	0.1 μF Ceramic Capacitor
4	C5	1	50 μF Electrolytic Capacitor (35V)
5	D1, D2, D3	3	LED, Silicon, RED, YELLOW, GREEN
6	U1, U2	2	CD4011 QUAD CMOS NAND GATE
7	U3	1	LM7555 CMOS TIMER
8	R1, R2, R4	3	470 Ω 1/4 W 5% RESISTOR
9	R3	1	2.2 MΩ 1/4 W 5% RESISTOR
10	R5	1	68 KΩ 1/4 W 5% RESISTOR
11	R6	1	10 KΩ 1/4 W 5% RESISTOR
12	R7	1	390 KΩ 1/4 W 5% RESISTOR
13	R8	1	56 KΩ 1/4 W 5% RESISTOR
14	R9, R10	2	2.2 KΩ 1/4 W 5% RESISTOR
15	S1	1	SPDP SWITCH
16	SPKR	1	1.5 KΩ PIEZO BUZZER
17	—	4	RUBBER FEET WITH NUTS, BOLTS, AND WASHERS
18	—	2	TERMINAL POSTS, RED, AND BLACK
19	—	8	1/4" NONCONDUCTING SPACER
20	—	8	ALLIGATOR-TO-ALLIGATOR CLIP LEADS
21	—	1	NO.16 GAUGE ALUMINUM OR PLASTIC BOX L = 6.5", W = 3.5", H = 1.5"
22	—	1 SET	MALE AND FEMALE SMALL AUDIO JACK.
23	—	1	ON/OFF SWITCH
24	—	1	ASSORTED COLOR WIRE (INSULATED)

Remarks:

PROJECT NO.: 01-1995	PART LIST AND COMPONENT DIMENSIONS:	DATE: 6/6/95	STUDENT'S NAME:	COMPUTER FILENAME: LOG_PRO

Figure 2–5. Preliminary List of Parts for the Deluxe Logic Probe
Source: Delton T. Horn, *50 CMOS IC Projects.* Blue Ridge Summit, Penn.: TAB Books, 1988, p. 63. Used with permission from TAB Books Inc.

Item 2: Books and application notes must be consulted to understand the theory and operation of the logic probe circuit. Reference material must be recorded completely. Copyright issues must be resolved before proceeding further with the design and fabrication process. If necessary, permission must be solicited from the original circuit designer for building the project.

Item 3: Estimated time to complete the project is 3 to 4 weeks, with approximately 8 hours per week of actual work.

Item 4: Estimated cost is $15.00.

Item 5: The project can be completed by one person within a reasonable time frame. Thus, this is a single-person project.

Item 6: A list of parts for the project, including any special tools, drill bits, and packaging materials, must be generated.

Item 7: Estimated time spent on tools and machinery would be approximately 2 to 3 weeks out of the total time for the project.

2-4.9 Experimenting

Experimentation with the circuit using solderless breadboard or wire wrapping will help improve the ultimate success of the project. Since the circuit is a simple one, solderless breadboarding would be the most convenient way to do the experimentation with the circuit. After procuring all the parts, arrange them in a plastic box especially used for CMOS IC chips. CMOS chips are very susceptible to static electricity. While working with CMOS chips, grounding straps around the wrist are appropriate. If grounding straps are not available, extreme care should be taken against static electricity buildup in the experimenter's body. Desired operation of the logic probe can be verified thoroughly by using the breadboard version of the probe for various voltage levels and by confirming the voltage levels using a digital multimeter and an oscilloscope. Figure 2–6 shows the final circuit diagram of the logic probe after experimenting and testing on a breadboard.

2-5. PROJECT 02-1995: THE ELECTRONIC CRICKET

2-5.1 Project Description

Definition. This project will create an inexpensive compact electronic circuit that will behave and sound like a cricket. A system functional diagram of the proposed electronic cricket circuit is shown in Figure 2–7.

Rationale. The male cricket produces the distinctive chirping sound by rubbing together especially modified parts of its forewings. By closely imitating the behavior of a cricket, we can have some fun.

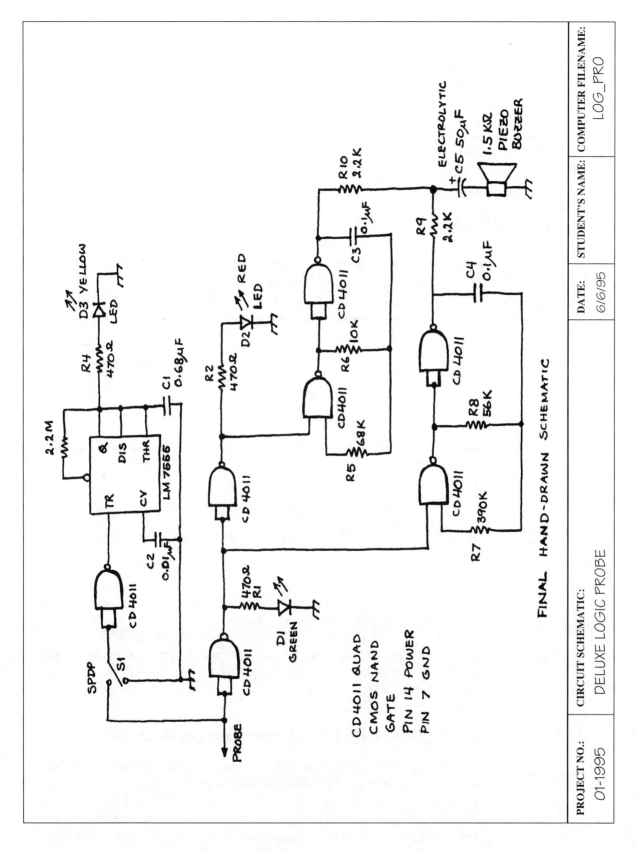

Figure 2-6. Final Hand-Drawn Schematic of the Deluxe Logic Probe
Source: Delton T. Horn, *50 CMOS IC Projects*, Blue Ridge Summit, Penn.: TAB Books, 1988, p. 64. Used with permission from TAB Books Inc.

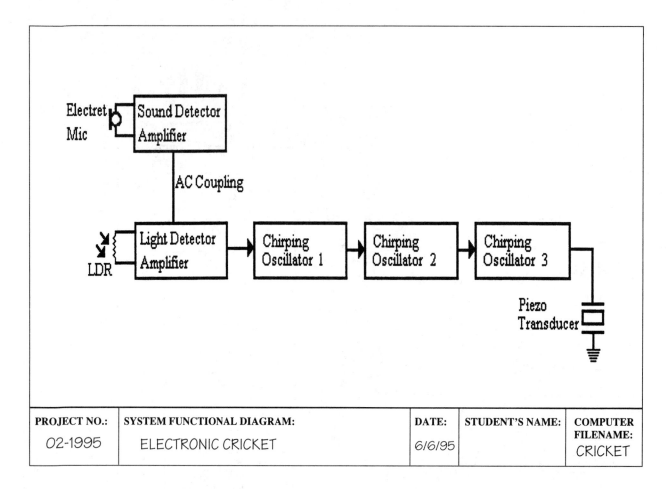

Figure 2-7. System Functional Block Diagram of the Electronic Cricket

Objectives. Our main objective is to imitate the chirruping sound and the unusual characteristics of a male cricket. The light- and sound-sensitive electronic creature must be built in a compact form so that it can be hidden in a small shoe box or something similar. It should create an entertaining reaction in children. The sensitivity to light and the time it takes to resume its operation must be adjustable. Frequency of the chirping sound should also be controllable.

2-5.2 Time and Resource Requirement and Analysis

There are electronic circuits that are sensitive to both light and sound. You may find them in electronic project building books. Searching may take several hours in a library. However, if you have the circuit, it should not take more than 4 to 5 weeks (approximately 8 hours/week) to build this project. Including packaging, the entire project should not cost more than $20. Estimated time to design artwork, fabricate PC board, and put the circuit together should be approximately 3 to 4 weeks out of the total time available for the project.

2-5.3 Level of Effort Requirement and Distribution

The project is a simple one, and there is no major electronic circuit design involved. Therefore, it is a one-person project.

2-5.4 Designing and Drawing

Before starting, you must know the behavior of a cricket. The clearer the outcome of the circuit to you at this stage, the better your design will be. A cricket responds to both sound and light. This means that if it sees a light or hears any sound, it stops chirping. It resumes its activity when no sound and light are present for about 30 seconds.

2-5.5 Packaging Scheme

Since the cricket is intended to be hidden inside a small box, attractive packaging is not required. However, light and sound sensors and the chirper must be placed conveniently so that they can do their work. Light sensor, sound sensor, chirper, and battery must be off-board parts. The enclosure for the project has been tentatively decided to be a 5.5" $\times$ 4.5" $\times$ 2" plastic or sheet-metal box. Figure 2–8 shows the packaging scheme of the electronic cricket. The extra 1.5 in. in length is for a 9V battery.

2-5.6 Material Procurement

Since the circuit is from an electronic project-building book, a parts list comes with it. Electronic surplus stores may be inexpensive sources for parts for the project. You will need a 4" $\times$ 4" and 0.062" thick Excel board. Figure 2–9 shows the preliminary list of parts for the project.

2-5.7 Tool and Machinery Requirement Analysis

Initial analysis indicates that the project does not require any special tools. Common hand tools such as soldering iron, rosin-cored 60/40 tin/lead solder, wire cutter, pliers, wire stripper, screwdriver, "third hand," and high-speed carbide drill bits will suffice. Three or four different sizes of drill bits are necessary for drilling holes through the PC board for the component leads. Drill bits required for this project are No. 64, 70, 72, and 75.

2-5.8 Developmental Plan

The developmental plan for the project involves a step-by-step procedure. This scheme is crucial during the project developmental period, outlining the basic steps of the project.

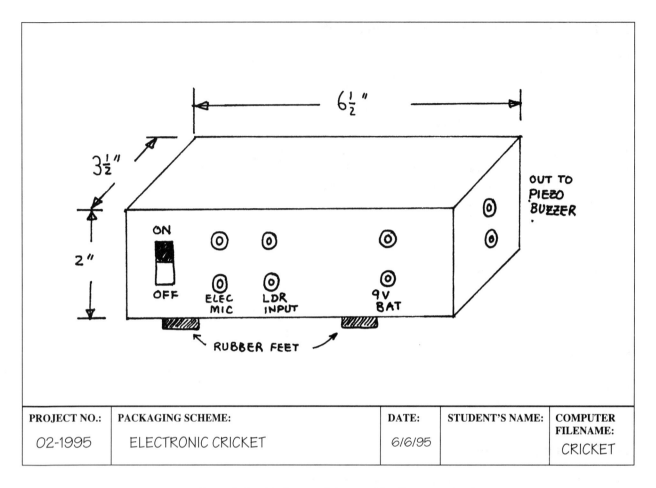

Figure 2-8. Packaging Scheme of the Electronic Cricket

Item 1: A computer file should be opened and maintained. The file should contain all the details about the project, including references, circuit diagrams, developmental plan, initial parts list, and anything that will serve as a source of potential information. Correspondence, if any, about the project and source of parts should be part of this file.

Item 2: Books and application notes must be consulted to understand the theory and operation of the cricket's circuit. Reference material must be recorded completely. Copyright issues must be resolved before proceeding further with the design and fabrication process. If necessary, permission must be solicited from the original circuit designer for building the project.

Item 3: Estimated time to complete the project is 4 to 5 weeks, with approximately 8 hours per week of actual work.

Item 4: Estimated cost is $20.00.

Item 5: The project can be completed by one person within a reasonable time frame. Thus, this is a single-person project.

Item 6: A list of parts for the project, including special tools, drill bits, and packaging materials, must be generated.

Project Title:	ELECTRONIC CRICKET		
PC Board Dimensions: L = 4", W = 4", Th = 1/32"			No. of Etched Sides: 2 (TWO)
Edge Connector: NONE			

Item No.	Reference Designator as in Schematic	Qty.	Description
1	R5	1	1 KΩ 1/4 W 5% RESISTOR
2	R17	1	4.7 KΩ 1/4 W 5% RESISTOR
3	R3	1	8.2 KΩ 1/4 W 5% RESISTOR
4	R1, R2, R13	3	10 KΩ 1/4 W 5% RESISTOR
5	R7, R9	3	22 KΩ 1/4 W 5% RESISTOR
6	R4, R6, R10, R16	4	47 KΩ 1/4 W 5% RESISTOR
7	R15	1	100 KΩ 1/4 W 5% RESISTOR
8	R18	1	220 KΩ 1/4 W 5% RESISTOR
9	R8, R11, R12, R18	4	1 MΩ 1/4 W 5% RESISTOR
10	TRIMPOT	1	1 KΩ
11	C8	1	0.0047 μF Ceramic Capacitor
12	C1	1	0.01 μF Ceramic Capacitor
13	C3, C4	2	0.1 μF Ceramic Capacitor
14	C6	1	1 μF Tantalum Capacitor
15	C2, C7	2	2.2 μF ELECTROLYTIC CAPACITOR
16	C5	1	33 μF ELECTROLYTIC CAPACITOR
17	U1, U2	2	741 Op-Amp
18	U3	1	4069 CMOS INVERTER (SIX DEVICES PER PACKAGE)
19	D1....D5	5	SILICON DIODE 1N4148
20	PIEZO	1	PIEZO BUZZER (TYPE L7022)
21	MIC	1	ELECTRET MICROPHONE (C-1160)
22	LDR	1	LIGHT DETECTING RESISTOR (TYPE Z-4801)
23	BATTERY	1	9V BATTERY
24	—	4	RUBBER FEET WITH NUTS AND BOLTS
25	—	8	TERMINAL POSTS: RED, BLACK, GREEN, AND YELLOW
26	—	4	1/4" NONCONDUCTING SPACERS
27	—	1	PLASTIC BOX L = 4.5", W = 5.5", H = 2"
28	—	8	ALLIGATOR-TO-ALLIGATOR CLIP LEADS

Remarks:

| PROJECT NO.: 02-1995 | PART LIST AND COMPONENT DIMENSIONS: | DATE: 6/6/95 | STUDENT'S NAME: | COMPUTER FILENAME: CRICKET |

Figure 2–9. Preliminary List of Parts for the Electronic Cricket
Source: Adapted from Dick Smith, SAMS Fun Way into Electronics. Indianapolis: Howard W. Sams & Co, 1986, p. 46.

Item 7: Estimated time spent on tools and machinery would be approximately 3 to 4 weeks out of the total time allowed for the project.

2-5.9 Experimenting

The success of the project greatly depends on the desired operation of the circuit. If any of the light or sound sensing parts are not working, the effect of the project will be greatly reduced. To be sure that the circuit is working, you must build it on a solderless breadboard first. You must investigate and experimentally determine the values of some of the critical resistors, capacitors, and variable resistors for the circuit to give you the optimum performance. The final version of the circuit is shown in Figure 2–10.

2-6. PROJECT 03-1995: THE INFRARED OBJECT COUNTER

2-6.1 Project Description

Definition. An infrared (IR) counter will count the number of objects passing on a conveyer system through an IR beam. Every time an object interrupts the beam, the counter will increment by one. At any instant of time, the two-digit LED will display the total number of objects passed through the beam from the time of reset. A system functional diagram of the circuit of the proposed project is shown in Figure 2–11.

Rationale. Counting objects passing on a conveyer belt is a common feature required in the product manufacturing industry. There are many counting systems that have been designed using infrared sensors. Consider a beverage plant; it needs to count the number of bottles passing through for final packaging. IR counting is touchless. Therefore, it does not disturb the object passing on the conveyer. It will be great fun to see this counter working as the object breaks the beam.

Objectives. Our main objective is to interface the two-digit BCD counter and the IR detection system. The audio alarm for each object as it passes the beam and overflow alarm sounds when the counter reaches 99. By using the overflow alarm, the operator can be notified about this occurrence.

2-6.2 Time and Resource Requirement and Analysis

There are many BCD counter circuits available in electronics books and magazines. You may even find a dedicated IC chip that will do the job for you. A circuit for an IR emitter and detector system is available in many places. A parts store may provide you the circuit diagram with the matching IR emitter

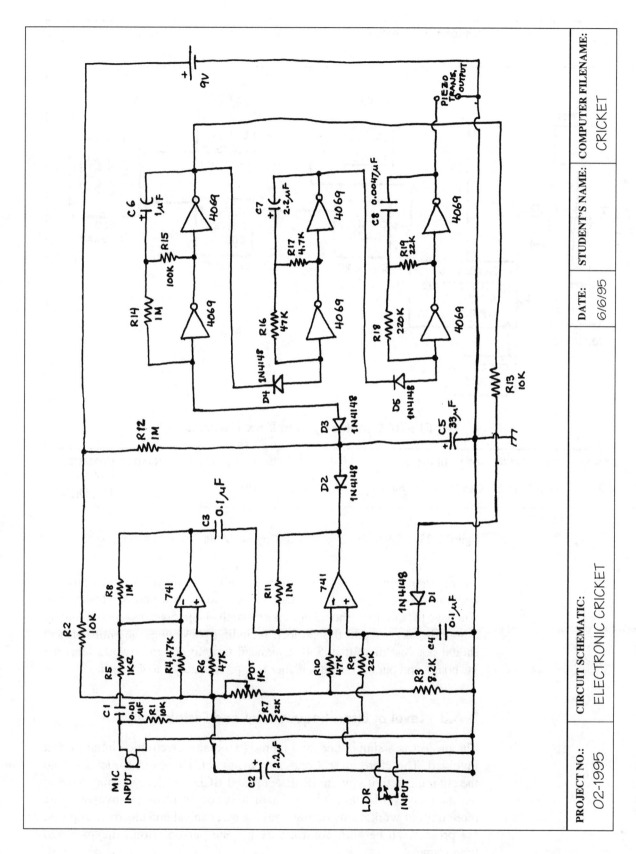

Figure 2-10. Final Hand-Drawn Schematic of the Electronic Cricket
Source: Adapted from Dick Smith, SAMS *Fun Way into Electronics*. Indianapolis: Howard W. Sams & Co, 1986, p. 47.

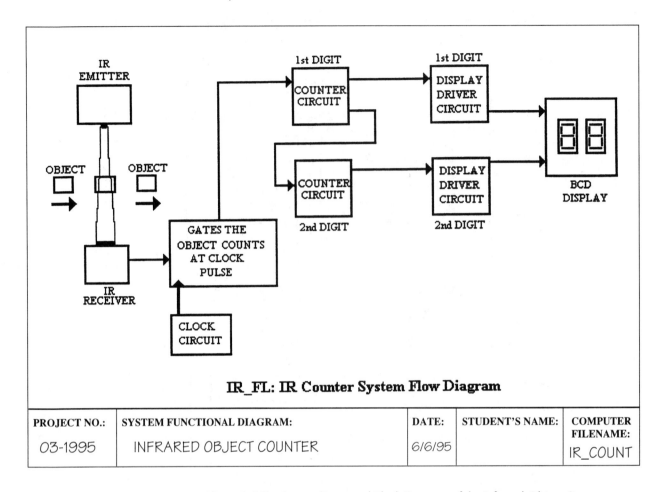

Figure 2-11. System Functional Block Diagram of the Infrared Object Counter

and detector device. You may also find circuit diagrams in electronic project-building books. Searching may take several hours in a library. However, if you have the circuit, it should not take more than 4 to 5 weeks (approximately 8 hours/week) to build this project. Including packaging, the entire project should not cost more than $30. Estimated time to design artwork, fabricate PC board, and put the circuit together is approximately 3 to 4 weeks.

2-6.3 Level of Effort Requirement and Distribution

The project is a simple one, and there is no major electronic circuit design involved. Therefore, it is a one-person project. However, to demonstrate the counting activity by the IR detector and BCD counter, you may have to build a conveyer system. The mechanical aspect of this may involve a few more days of work. Considering both the mechanical and electrical aspects, the project can be built comfortably by one person within the specified time frame.

2-6.4 Designing and Drawing

Before you start designing, you must visualize the operation of the system. To interface the counter and the detector circuit, you must know the type of output from the IR detector circuit and input required for the counter to work. The BCD counter circuit consists of two seven-segment LED displays, two BCD-to-seven-segment decoder drivers, and two decade counters. There are two types of LED display available, common cathode and common anode. Either type will do the job; however, common anode is preferable for this project. The IR detector system has one IR emitter and one detector. A D-type positive edge triggered flip-flop gates the count at the positive edge of the clock pulse.

Theory and operation. The IR emitter is a very simple device; it is basically a forward-biased diode that emits an IR signal. The detector is an IR-sensitive device. The emitter is always emitting an IR signal, and the receiver is receiving it. This means D-input will not create any low-to-high transition at the output Q of the flip-flop. This in turn will keep the counter inactive. Any interruption in the beam will store a logic-0 at the input of the D-type flip-flop. At the positive edge of the clock, this logic-0 will create a low-to-high transition at the input of the counter. This action will trigger the counter to count up by one.

2-6.5 Packaging Scheme

Although the basic electronic circuit needs to be interfaced with many off-board parts, such as audible buzzers, IR emitter and detector, and seven-segment LED display devices, the basic electronics circuit can be easily put on a 4" × 4" PC board. The IR detector and emitter need to be placed face to face so that the beam will be interrupted by the objects passing on the conveyer belt. The whole project can be built on a wooden board. The conveyer belt can be driven by a small DC motor. The buzzers can be screwed down on the board. The PC board can be placed on the wooden board by using insulated or non-insulated standsofts. If you wish to package the PC board in an enclosure, use a 4.5" × 4.5" × 2.5" plastic or sheet-metal box. Figure 2–12 is the packaging scheme for the infrared object counter.

2-6.6 Material Procurement

Since the circuit is extracted from a student's electronic project, a parts list comes with it. Electronic surplus stores in your area may be inexpensive sources for parts for the project. You will need a 4" × 4" and 0.062" thick Excel board. Figure 2–13 shows the preliminary list of parts for the project.

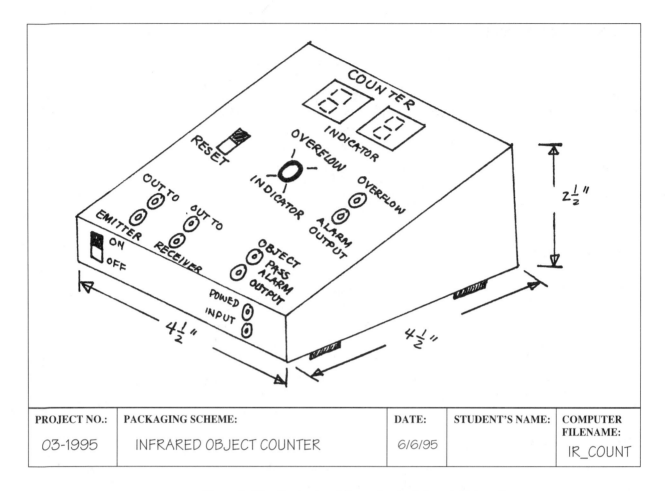

Figure 2–12. Packaging Scheme for the Infrared Object Counter

2–6.7 Tool and Machinery Requirement Analysis

Initial analysis indicates that the project does not require any special tools. Common hand tools such as soldering iron, soldering flux, wire cutter, pliers, wire stripper, screwdriver, "third hand," and high-speed carbide drill bits will suffice. Three or four different sizes of drill bits are necessary for drilling holes through the PC board for the component pigtail. Drill bits required for this project are No. 64, 70, 72, and 75.

2–6.8 Developmental Plan

The developmental plan for the project involves a step-by-step procedure. This scheme is crucial during the project developmental period, outlining the basic steps of the project.

Item 1: A computer file should be opened and maintained. The file should contain all the details about the project, including references, circuit diagrams, developmental plan, initial parts list, and anything that will serve

Project Title:	INFRARED OBJECT COUNTER		
PC Board Dimensions:	L= 4", W= 4", Th = 1/16"		No. of Etched Sides: 2 (TWO)
Edge Connector:	NONE		

Item No.	Reference Designator as in Schematic	Qty.	Description
1	C1, C2	1	1 μF Electrolytic Capacitor
2	C3	1	0.1 μF Ceramic Capacitor
3	D1	1	SILICON DIODE: 1N4001
4	IS1	1	OPTO COUPLER 3086
5	Q1	1	NPN TRANSISTOR 2N2222
6	JP1, JP2	2	8 PIN HEADER TO CONN. LED, 7 SEG.
7	R1, R2	2	4.5 KΩ 1/4 W 5% RESISTOR
8	R3, R7	2	470 Ω 1/4 W 5% RESISTOR
9	R4	1	270 Ω 1/4 W 5% RESISTOR
10	R5	1	10 KΩ 1/4 W 5% RESISTOR
11	R6	1	220 KΩ 1/4 W 5% RESISTOR
12	R8, R9	2	100 Ω 1/4 W 5% RESISTOR
13	R10	1	220 Ω 1/4 W 5% RESISTOR
14	U1	1	555 TIMER
15	U2	1	7474 DUAL D-TYPE FLIP-FLOP
16	U3, U4	2	7490 DECADE COUNTERS
17	U5, U6	2	7447 BCD 7-SEGMENT DECODERS/DRIVERS
18	U7	1	7420 NAND, 4 INPUT.
19	BLU, RED, BLK	1	PIEZO BUZZER
20	RED, BLK	1	BUZZER
21	S1, S2, S3	1	SPDP SWITCH
22	7-SEG. LED	2	7-SEGMENT LED DISPLAY
23	IR EMITTER	1	INFRARED EMITTER
24	IR RECEIVER	1	INFRARED RECEIVER
25	—	4	RUBBER FEET WITH NUTS, BOLTS, AND WASHERS
26	—	10	TERMINAL POSTS, RED AND BLACK
27	—	1	NO. 16 GAUGE ALUMINUM OR PLASTIC BOX
28	—	4	1/4" NON CONDUCTING SPACER

Remarks: IR EMITTER AND RECEIVER MUST BE A MATCHED PAIR — MEANS FREQUENCY MUST BE SAME

PROJECT NO.: 03-1995	PART LIST AND COMPONENT DIMENSIONS:	DATE: 6/6/95	STUDENT'S NAME:	COMPUTER FILENAME: IR_COUNT

Figure 2–13. Preliminary List of Parts for the Infrared Object Counter
Source: Adapted from Edward Cherbak, EET250, Student Project, Fall 1991.

as a source of potential information. Correspondence, if any, about the project and source of parts should be part of this file.

Item 2: Books and application notes must be consulted to understand the theory and operation of the infrared emitter and receiver, pulse gating, and BCD counter circuit. Reference material must be recorded completely. Copyright issues must be resolved before proceeding further with the design and fabrication process. If necessary, permission must be solicited from the original circuit designer for building the project.

Item 3: Estimated time to complete the project is 4 to 5 weeks, with approximately 8 hours per week of actual work.

Item 4: Estimated cost is $30.00.

Item 5: The project can be completed by one person within a reasonable time frame. Thus, this is a single-person project.

Item 6: A list of parts for the project, including any special tools, drill bits, and packaging materials, must be generated.

Item 7: Estimated time spent on tools and machinery would be approximately 3 to 4 weeks out of the total time allowed for the project.

2-6.9 Experimenting

The success of the project greatly depends on the appropriate operation of the circuit. If the BCD counter system does not work properly, the effect of the project will be greatly reduced. You must also provide additional hardware that will interrupt the IR beam, just like objects passing on a conveyer belt. To be sure that the circuit is working, you must build it on a solderless breadboard first. You must investigate and experimentally determine the alignment of the IR emitter and detector. To be able to function properly, they must see each other appropriately. You should verify the operation of the BCD counter system separately. Figure 2–14 shows the final circuit schematic of the infrared object counter.

2-7. PROJECT 04-1995: THE MINI STEREO AMPLIFIER

2-7.1 Project Description

Definition. An inexpensive and portable stereo amplifier is desired for interfacing with AM/FM radio, cassette player, or compact disk player. The proposed stereo amplifier will drive two 8Ω, or ohm, speakers and will receive input from a miniature stereo radio, cassette player, or CD player using a jack. A system functional diagram of the proposed amplifier is shown in Figure 2–15.

Rationale. These miniature radio, cassette players, or CD players have a stereo output but do not have enough power to drive speakers. The proposed amplifier could be useful for amplifying the sound of the miniature

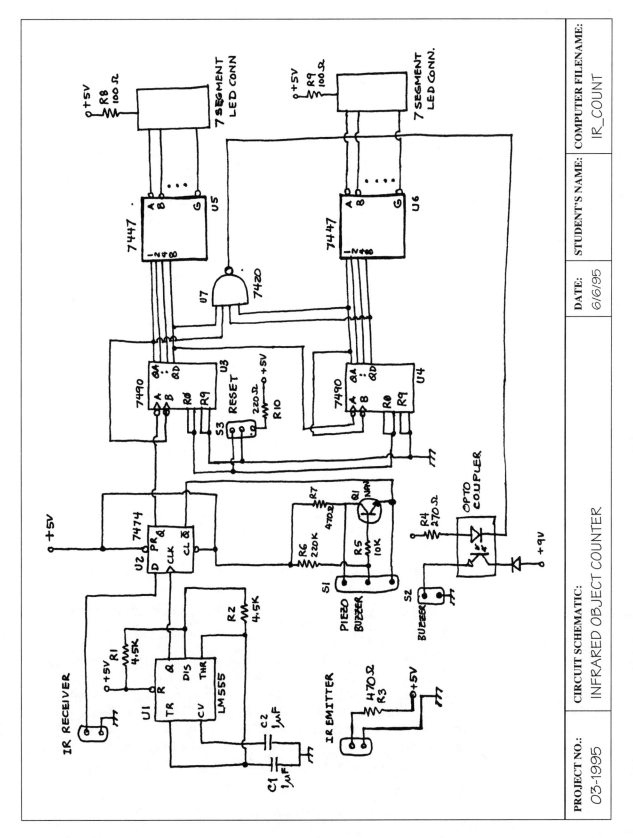

Figure 2-14. Final Hand-Drawn Schematic of the Infrared Object Counter
Source: Adapted from Edward Cherbak, *EET250*, Student Project, Fall 1991.

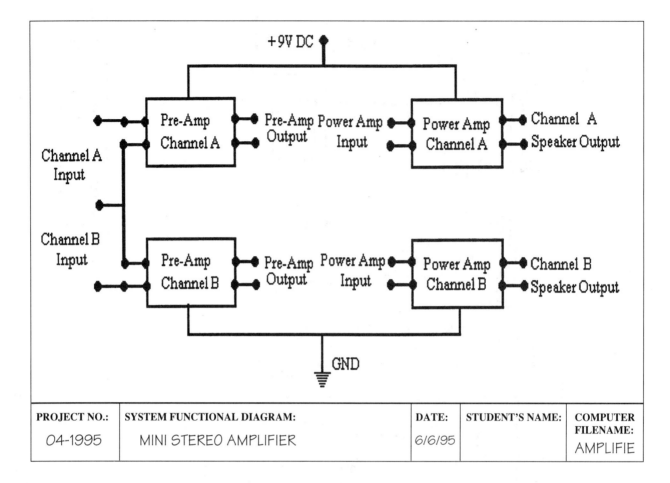

Figure 2-15. System Functional Block Diagram of the Mini Stereo Amplifier

stereo devices so that they can be heard through a pair of speakers rather than through a headphone. In a home setting, it is desirable to use speakers instead of headphones. Just unplugging the headphones and plugging in the proposed amplifier would be of great convenience.

Objectives. Our main objective is to build a portable, high-quality amplifier that can be easily interfaced with a miniature radio, cassette player, or CD player for amplifying its sound. Input to the amplifier should be a stereo-type jack input for most miniature cassette and CD players. The amplifier should be able to drive at least two 8Ω, or ohm, speakers. Power input to the stereo amplifier should be a coaxial DC power plug. Speaker output should be a shielded phone jack for minimized interference.

2-7.2 Time and Resource Requirement and Analysis

There are many stereo amplifier circuits available in electronics books and magazines. However, you may need to fine-tune one that will do the job for you. If you find a circuit in an electronic magazine, it might include the name and address for the source of parts. You should be able to buy the parts from an electronic parts store. Searching for the circuit may take several hours in a library or bookstore. However, if you have a circuit that closely matches the requirement, it should not take more than 4 to 5 weeks (approximately 8 hours/week) to build this project. Including packaging, the entire project should not cost more than $40. Estimated time to design artwork, fabricate PC board, and put the circuit together is approximately 3 to 4 weeks.

2-7.3 Level of Effort Requirement and Distribution

The project is a simple one, and there is no major electronic circuit design involved. Therefore, it is a one-person project. However, to demonstrate the practical performance of the circuit, you may need to interface it with a miniature stereo radio, cassette player, or CD player at the input and speakers at the output. The packaging of the project may involve a few more days of work. Considering both the mechanical and electrical aspects, the project can be built comfortably by one person within the specified time frame.

2-7.4 Designing and Drawing

The circuit must first be built on a solderless breadboard. Generally, the amplifier circuit needs tuning before it functions properly. Input and output impedance matching is a critical factor.

Theory and operation.   The amplifier is an active electronic circuit by which an input signal is amplified by furnishing the DC power from the battery. Amplifiers in a broader sense have two stages, pre-amplifier and power amplifier. A small input signal from a microphone or other source is first amplified by a pre-amplifier circuit. The pre-amplifier output is amplified to a big signal by the power amplifier circuit. Depending on the required output power, the amplifier may have several stages. Output of the pre-amplifier is fed into the power amplifier through a potentiometer for controlling the overall output level of the amplifier. For a stereo amplifier circuit, there has to be two pre- and two power amplifier circuits for driving the two speakers.

2-7.5 Packaging Scheme

The basic electronic circuit needs to be interfaced with a few off-board parts: volume control potentiometer, speaker output, and stereo jack input. The basic electronic circuit must be packaged in an enclosure. Power to the circuit can be provided very easily from a compact, plug-in type power supply unit with a coaxial plug. The circuit can be easily put on a 6" $\times$ 3" PC board. A 6.5" $\times$ 3.5" $\times$ 2" plastic or sheet-metal box can be used for packaging. The PC board must be placed inside the box using insulated standsofts. The packaging scheme is shown in Figure 2–16.

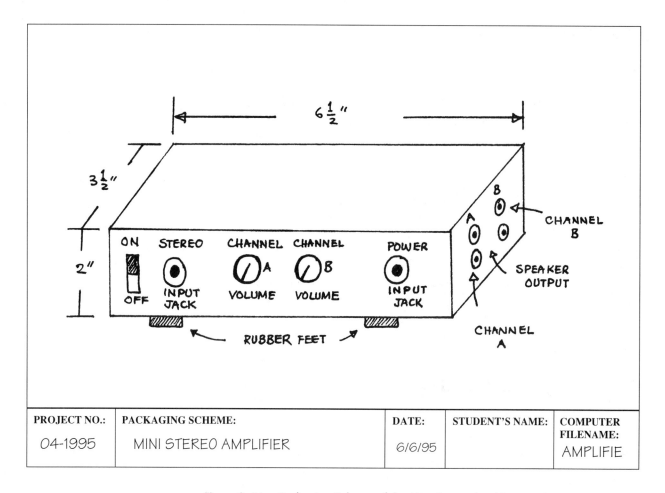

Figure 2–16. Packaging Scheme of the Mini Stereo Amplifier

2–7.6 Material Procurement

Since the circuit is from an electronic project-building book, a parts list comes with it. Electronic surplus stores in your area may be inexpensive sources for parts for the project. You will need a 6" × 3" and 0.062" thick Excel board. Figure 2–17 is the preliminary list of parts for the project.

2–7.7 Tool and Machinery Requirement Analysis

Initial analysis indicates that the project does not require any special tools. Common hand tools such as soldering iron, soldering flux, wire cutter, pliers, wire stripper, screwdriver, "third hand," and high-speed carbide drill bits will suffice. Three or four different sizes of drill bits are necessary for drilling holes through the PC board for the component pigtail. Drill bits required for this project are No. 64, 70, 72, and 75.

Project Title:	MINI STEREO AMPLIFIER		
PC Board Dimensions: L = 6", W = 3", Th = 1/16"			No. of Etched Sides: 2 (TWO)
Edge Connector: NONE			

Item No.	Reference Designator as in Schematic	Qty.	Description
1	C11, C12	2	0.1 µF CERAMIC CAPACITOR
2	C1, C2	2	0.1 µF TANTALUM CAPACITOR
3	C3, C4, C5, C6, C7, C8	6	2.2 µF ELECTROLYTIC CAPACITOR
4	C15	1	100 µF ELECTROLYTIC CAPACITOR
5	C9, C10, C13, C14	4	470 µF ELECTROLYTIC CAPACITOR
6	R13, R14	2	2.7 Ω 5% 1/4 W RESISTOR
7	R1, R2	2	2.2 KΩ 5% 1/4 W RESISTOR
8	R3, R4, R5, R6, R7, R8, R11, R12	8	47 KΩ 5% 1/4 W RESISTOR
9	R9, R10	2	1 MΩ 5% 1/4 W RESISTOR
10	RV1, RV2	2	50 KΩ POTENTIOMETER
11	U1, U2	2	741 Op-Amp.
12	U3, U4	2	LM380 AUDIO-AMP.
13	—	1	MALE AND FEMALE SMALL SIZE AUDIO JACK (3.5mm)
14	—	2	8 Ω SPEAKER
15	—	8	1/4" NONCONDUCTING SPACER
16	—	4	RUBBER FEET WITH NUTS AND BOLTS
17	—	1	PLASTIC BOX L=6.5", W=3.5", H=2"
18	—	2	KNOB
19	—	1	ON/OFF SWITCH
20	—	4	TERMINAL POST (FOR AUDIO APPLICATIONS)
21	—	1	ASSORTED COLOR INSULATED WIRE
22	—	8	ALLIGATOR-TO-ALLIGATOR CLIP LEADS

Remarks:

PROJECT NO.: 04-1995	PART LIST AND COMPONENT DIMENSIONS:	DATE: 6/6/95	STUDENT'S NAME:	COMPUTER FILENAME: AMPLIFIE

Figure 2–17. Preliminary List of Parts for the Mini Stereo Amplifier
Source: Adapted from Dick Smith, *SAMS Fun Way into Electronics*. Indianapolis: Howard Sams & Co., 1986, p. 50.

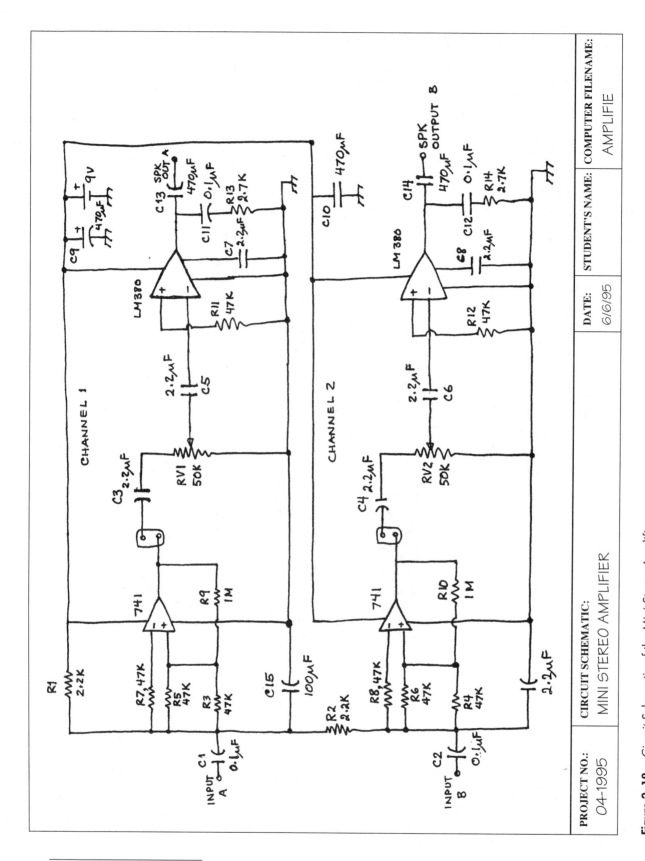

Figure 2-18. Circuit Schematic of the Mini Stereo Amplifier
Source: Adapted from Dick Smith, *SAMS Fun Way into Electronics.* Indianapolis: Howard Sams & Co. 1986, p. 51.

2-7.8 Developmental Plan

The developmental plan for the project involves a step-by-step procedure. This scheme, outlining the basic steps of the project, is crucial during the project developmental period.

Item 1: A computer file should be opened and maintained. The file should contain all the details about the project, including references, circuit diagrams, developmental plan, initial parts list, and anything that will serve as a source of potential information. Correspondence, if any, about the project and source of parts should be part of this file.

Item 2: Books and application notes must be consulted to understand the theory and operation of the stereo amplifier circuit. Reference material must be recorded completely. Copyright issues must be resolved before proceeding further with the design and fabrication process. If necessary, permission must be solicited from the original circuit designer for building the project.

Item 3: Estimated time to complete the project is 4 to 5 weeks, with approximately 8 hours per week of actual work.

Item 4: Estimated cost is $40.00.

Item 5: The project can be completed by one person within a reasonable time frame. Thus, this is a single-person project.

Item 6: A list of parts for the project, including any special tools, drill bits, and packaging materials, must be generated.

Item 7: Estimated time spent on tools and machinery would be approximately 3 to 4 weeks out of the total time allowed for the project.

2-7.9 Experimenting

The success of the project greatly depends on the operation of the circuit. If performance of the circuit is not satisfactory, the project will have little value. Generally, all amplifier circuits need tuning; therefore, the circuit must be built on a breadboard first. Figure 2–18 is the final circuit schematic of the mini stereo amplifier.

QUESTIONS

1. Why is planning important for a school laboratory project?

2. Write a project description of a power supply unit for a stereo cassette player that you and your classmate want to build. Use the blank worksheet provided in Figure 2–1a.

3. Concerning the above power supply project, analyze the requirements for time, resources, and level of effort.

4. Provide your design, operational characteristics, packaging scheme, materials requirement, and tools and machinery requirement for the power supply project.

5. Write your developmental plan for the power supply project.

6. If you need to do experiments to meet the desired specifications of your power supply, write about the modifications from the original design.

7. Define a project of your choice and write a project description and other necessary analyses.

3

PERSONAL COMPUTER AND ORCAD/SDT AND ORCAD/PCB SOFTWARE TOOLS

3-1. HARDWARE REQUIREMENTS FOR ORCAD/SDT AND ORCAD/PCB TOOL SETS

The following is a typical personal computer system configuration on which OrCAD/SDT and OrCAD/PCB software tool sets can be used conveniently. Figure 3–1 shows the typical personal computer setup.

- 3–1.1 Computer
- 3–1.2 Floppy Disk Drive
- 3–1.3 Hard Disk Drive
- 3–1.4 Display Monitor and Driver Card
- 3–1.5 Keyboard
- 3–1.6 Mouse
- 3–1.7 Disk Operating System (DOS)
- 3–1.8 Printer
- 3–1.9 Plotter

3–1.1 Computer

The computer should be of reasonable capacity such that OrCAD software can run easily. To install and run the software comfortably, users need an IBM PC/AT, PS/2, or PC/AT-compatible with at least 80286 processor and 640 KB of RAM. A personal computer with 80386DX 33MHz or 80486DX 50MHz processor with 2 MB of RAM will speed up the command execution time. A math co-processor is not required, but it is desirable if circuit simulation is to be performed prior to completion of the netlist finalization.

Objectives

After completing this chapter, you should be able to

1. Describe the minimum hardware requirement for OrCAD/SDT and OrCAD/PCB tool sets
2. Install OrCAD/SDT and OrCAD/PCB software in your microcomputer
3. Get a little more insight about the OrCAD EDA tools and their menu structure
4. Access the OrCAD software tool sets
5. Describe the directory structure of OrCAD/SDT and OrCAD/PCB tool sets
6. Gather information about DOS environment variables for OrCAD/SDT and OrCAD/PCB

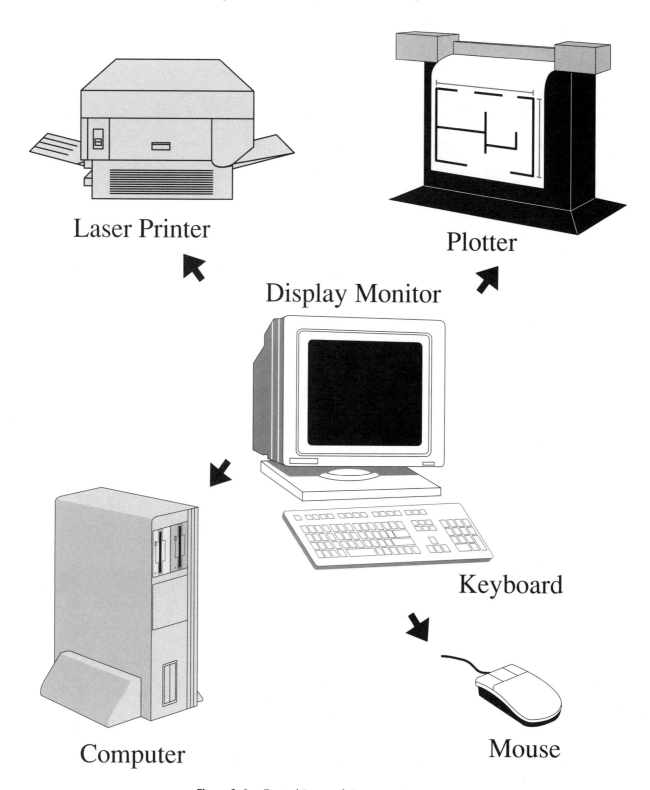

Figure 3–1. Typical Personal Computer Setup
Source: Drawn by using CorelDraw Clipart Image.

3-1.2 Floppy Disk Drive

At least one floppy drive is needed to move data in and out of the computer. A high-density 5.25 in. or 3.5 in. floppy drive is required. The former provides storage capacity of approximately 1.2 MB and the latter a capacity of approximately 1.44 MB.

3-1.3 Hard Disk Drive

A hard disk of at least 40 MB storage capacity is recommended for smoother operation. The version of OrCAD software with ESP environment will require about 10 MB of disk space. Software can also run from floppy drives, but it is not recommended. Larger hard disks, such as 60 MB or 80 MB, will certainly add flexibility to the environment.

3-1.4 Display Monitor and Driver Card

Although OrCAD supports monochrome monitors, a color display monitor is essential for the software to utilize its full power. At least an EGA color monitor with a driver card of 640X350 pixel resolution is required. A VGA color monitor with a driver card of 640X480 pixel resolution would be a more convenient environment. However, a Super VGA color monitor with a driver card of 1024X768 pixel resolution creates a little inconvenience when reading the ESP screen menus. OrCAD supports numerous display drivers. OrCAD Inc. may be contacted to make sure that it has a display driver for the display driver card being used.

3-1.5 Keyboard

A keyboard with alphanumeric and function keys is sufficient for OrCAD software. Keyboards with 101 keys are the most common type used with most personal computers.

3-1.6 Mouse

A mouse is not required, but a mouse driver must be installed. Although arrow keys can be used instead of a mouse, they are significantly slower than a mouse and thus very inconvenient. Therefore, it is strongly recommended to use a mouse while using the software.

3-1.7 Disk Operating System (DOS)

DOS version 3.3 or above is necessary to exploit the full power of OrCAD software. For the software release IV, a high memory command can be provided in the autoexec.bat file to enhance the RAM use.

3-1.8 Printer

Although OrCAD supports many types of dot-matrix printers, output from a lower resolution printer will be difficult to read and certainly will be of marginal quality. Also, PCB artwork printed by a dot-matrix printer will not provide enough resolution to be used for photographic transfer of the artwork to photosensitized board. Usually, a photoplotter or at least a high-resolution laser printer is required as a printing device for the master artwork.

3-1.9 Plotter

A simple plotter can be used for colorful plotting of circuit schematics. However, photoplotters are generally used for industrial quality high-resolution plotting of master artwork and silkscreen. OrCAD software tool sets can generate outputs in Hewlett-Packard HP-GL language and in many other industry-standard photoplotter language formats.

3-2. INSTALLATION OF THE SOFTWARE TOOL SETS

Before installing the software, you need to know the type of display driver for your computer system. If you select one that is incompatible with your display driver card, the software will not be able to display the ESP screens and other menu screens. Nowadays, most software manufacturers provide installation disks with their software, and OrCAD is no exception to that. To install the OrCAD/SDT and OrCAD/PCB software, use the steps provided below. The following installation procedure is for destination drive D, source drive A, and under MYORCAD subdirectory in D drive. However, you may easily change the source and destination drive by replacing A and D with your desired drives. If you are installing PCB II, use only ESP version 4.10 and not higher. If you use a higher version of ESP, you may have problems with local configuration of the PCB tools. Also remember to read the explanation provided by OrCAD for installation. The following keystrokes are used by OrCAD for installation purposes:

← → ↑ ↓/Page Up/Page Down: Move the selection bar
Space Bar: Toggle selection
Y/N: Set all selections

[Enter ↵] : Accept Settings

[ESC] : Abort installation

If you encounter any problems during installation, you need to consult the directions provided in the installation guide that came with your software. If you encounter problems running some part of the software, you need to contact the OrCAD software technical support department.

Installation procedure

Step 1: Place the installation disk marked **Install** in drive A.
Step 2: At the DOS prompt (C:>), type A: [Enter ↵]

3–2. Installation of the Software Tool Sets

Step 3: At the DOS prompt (A:>), type Install [Enter ↵]

Step 4: The software will prompt the following options:
- Change target hard drive C: or....YES/NO ➪ Toggle to YES
- Change executable program directory from \ORCADEXE .... YES/NO ➪ Toggle YES
- Change main driver/library directory from \ORCADESP .... YES/NO ➪ Toggle to YES
- To accept the above selection, press the [Enter ↵] key.

Step 5: Select the drive D: and press the [Enter ↵] key.

Step 6: Using the ← → ↑ ↓ and alphanumeric keys, enter \MYORCAD \ORCADEXE and press the [Enter ↵] key.

Step 7: Using the ← → ↑ ↓ and alphanumeric keys, enter \MYORCAD \ORCADESP and press the [Enter ↵] key.

Step 8: Select the video driver you want to load and toggle the YES/NO to YES. You may want to load a couple of them that you think would be compatible with your monitor and its display driver card. Some of the common ones are VGA640.DRV, EGA16E.DRV, VESA800.DRV, ATI1K.DRV, 8514A.DRV, and VESA.DRV. Accept the selection by pressing the [Enter ↵] key.

Step 9: Select plotter drivers by toggling the YES/NO to YES for the desired drivers. If your computer system has enough hard drive space, load all the plotter drivers by toggling all, using the Y key. This will enable you to produce a plotter file of various formats. Select DXF for AutoCAD, GERBER, HP, etc. Accept the selection by pressing the [Enter ↵] key.

Step 10: Select printer drivers by toggling the YES/NO to YES for the desired drivers. If your computer system has enough hard drive space, load all the printer drivers by toggling all, using the Y key. Accept the selection by pressing the [Enter ↵] key. At this time the software will load all the selected drivers for the floppy drive at A drive.

Step 11: The software will prompt for the OrCAD tool set you want to install at this time. Toggle only the following to YES.
- OrCAD ESP Design Environment YES/NO ➪ Toggle to YES
- OrCAD Schematic Design Tools YES/NO ➪ Toggle to YES
- OrCAD PC Board Layout Tools YES/NO ➪ Toggle to YES

Accept the above selections by pressing the [Enter ↵] key.

Step 12: The software will prompt the following changes:
- Change Source Floppy drive from: A:.... YES/NO ➪ Keep at NO
- Change Target Floppy drive from C:.... YES/NO ➪ Toggle to YES

- Change main Driver/Library directory from: \ORCADESP .... YES/NO ⇨ Toggle to YES
- Change Project directory from: \ORCAD .... YES/NO ⇨ Toggle to YES
- Change Executable Program directory from: \ORCADEXE .... YES/NO ⇨ Toggle to YES
- Check and/or modify AUTOEXEC.BAT:.... YES/NO ⇨ Toggle to YES

Accept the above selections by pressing the [Enter ⏎] key.

Step 13: Select the drive D: by using the ↑ ↓ keys and pressing the [Enter ⏎] key.

Step 14: Using the ← → ↑ ↓ and alphanumeric keys, enter \MYORCAD\ORCADESP and press the [Enter ⏎] key.

Step 15: Using the ← → ↑ ↓ and alphanumeric keys, enter \MYORCAD\ORCAD and press the [Enter ⏎] key.

Step 16: Using the ← → ↑ ↓ and alphanumeric keys, enter \MYORCAD\ORCADEXE and press the [Enter ⏎] key.

Step 17: OrCAD will prompt for ESP Disk. Put it in drive A: and press the [Enter ⏎] key.

Step 18: Software will prompt: Is this a new installation of the ESP Design Environment?

YES/NO ⇨ Type Y and press the [Enter ⏎] key.

Step 19: Software will display all the current installation parameters you just selected as follows:

- Target Hard Drive D: YES/NO ⇨ NO
- Driver and Libraries \MYORCAD\ORCADESP YES/NO ⇨ NO
- Programs \MYORCAD\ORCADEXE YES/NO ⇨ NO
- Projects \MYORCAD\ORCAD YES/NO ⇨ NO
- Text Editor D: \MYORCAD\ORCADEXE\M2EDIT.EXE YES/NO ⇨ NO
- Video Driver EGA16E.DRV YES/NO ⇨ NO

If you want to make any changes, you may do so now or else press the [Enter ⏎] key. The software will install the ESP Design Environment and will display: ESP *Installation Complete*.... Press the [Enter ⏎] key, and the software will prompt you for the SDT Disk 1.

Step 20: Insert the SDT disk 1 of 3, as the prompt for it comes on the screen in drive A: and press the [Enter ⏎] key. Select: *Standard Installation*.... by the ↑ ↓ keys and press the [Enter ⏎] key.

Step 21: Select the VGA or EGA that is compatible to your computer system by using the ↑ ↓ keys, and accept the selection by pressing the [Enter ⏎] key.

3-2. Installation of the Software Tool Sets

Step 22: Software will list all the libraries for schematic parts you want to load: Libraries for Schematic Parts...YES/NO ⇨ Toggle all to YES by typing Y. Accept the selection by pressing the [Enter ↵] key.

Step 23: Software will list all the netlist formats you want to use: Netlist Formats. . . . YES/NO ⇨ Toggle all to YES by typing Y. Accept the selection by pressing the [Enter ↵] key. Software will display. . .*OrCAD Schematic Design Tools V 4.20n installation is complete.* . . . Press the [Enter ↵] key to proceed for PCB software tool set installation.

Step 24: Software will prompt for the PCB II Disk 1. Put the PCB II Disks in drive A: and press the [Enter ↵] key. Software will prompt the following: *Do you want to install PCB II for use under ESP?*. . . . . . . Select YES using the ↑ ↓ keys and accept the selection by pressing the [Enter ↵] key.

Step 25: The software will display the following: *Do you want to install PCB II for use under ESP?* Select YES using the ↑ ↓ keys, and accept the selection by pressing the [Enter ↵] key.

Step 26: The software will prompt the current installation parameters as follows:

- Change Source Floppy Drive from: A: YES/NO ⇨ NO
- Change Target Hard Drive from: D: YES/NO ⇨ NO
- Change Main Driver/Library directory from: \MYORCAD \ORCADESP. . . . YES\NO ⇨ NO
- Change Project directory from: \MYORCAD\ORCAD. . . . YES/NO ⇨ NO
- Change Executable Program directory from: \MYORCAD\ORCADEXE. . . . YES/NO ⇨ NO
- Change Installation Design directory from: \TEMPLATE YES/NO ⇨ NO

Accept the selection by pressing the [Enter ↵] key.

Step 27: Software will prompt as follows: *Do you want to configure PCB II as if it had EMS for the ESP video drivers?* Select YES by using the ↑ ↓ keys, and accept the selection by pressing the [Enter ↵] key.

Step 28: Select *Standard Installation of PCB II Product Files* by using the ↑ ↓ keys, and accept the selection by pressing the [Enter ↵] key.

Step 29: Select the plotter drivers by toggling the desired drivers. Select DXF for AutoCAD, GERBER, HP, etc. Accept the selection by pressing the [Enter ↵] key.

Step 30: Select the printer drivers by toggling the desired drivers. Select HPLASER4, EPSON, etc. Accept the selection by pressing the [Enter ↵] key.

Step 31: Select the Module Group. Type Y and accept the selection by pressing the `Enter ↵` key.

Step 32: Software will load your selections from the floppy disk and prompt for the PCB II Disk 2. Put it in drive A and press the `Enter ↵` key.

Step 33: Software will prompt for manual update of ORCADESP.DAT files and ask you to refer to the PCB manual. Print this page of information by using the **print screen** command from your keyboard. Press the `Enter ↵` key twice. The installation is now complete. You need to reboot your computer so that environment variables written by the software in the AUTO- EXEC.BAT file can be read by the software.

3-3. THE SOFTWARE TOOL OPERATING ENVIRONMENT

With the introduction of release IV of the software, OrCAD Inc. has developed a new operating environment for its computer-aided tools. OrCAD calls this the ESP design environment. With the support of this environment, designers can focus more on design aspects rather than struggling with the tool itself. ESP environment integrates all the electronic design tool sets OrCAD currently provides for the electronic design automation (EDA) environment.

A tool set is a collection of tools designed to perform a set of electronic design automation tasks. OrCAD has five tool sets under the ESP environment. Only the Schematic Design tool set and PC Board Layout tool set that are required for the PC board fabrication process will be discussed in this book.

In this EDA environment, as opposed to other software tools, electronic designers are not required to shuttle back and forth between DOS and the software tool set. In addition to this, you can access, with a click of the mouse, other software tools that are not part of the OrCAD tool set. The OrCAD ESP provides a well-integrated environment for electronic design automation. In this, all the electronic design and testing are integrated under one environment to function in a coordinated manner. One of the major advantages of this is that it eliminates the need to remember strings of command lines, switches, and their arguments. In this environment, all the files that are required for an electronic design process are collected into a single work area, thereby providing an easier and faster file management system.

Almost every tool and its subfunctions have local configurations. These local configurations are particular to a design, so one can change from one design to another without changing the configuration. ESP design environment has several layers of screen display. The first layer is displayed while invoking the OrCAD EDA tool system. The first display consists of names of tool sets, such as Schematic Design Tools, PC Board Layout Tools, Programmable Logic Design Tools, Digital Simulation Tools, and Design Management Tools. Each of these tool sets has a separate screen for second-layer display. The second layer displays subtools organized by their functions. Tools in each of these layers are arranged in such a manner that an appropriate design tool and its configurations can be selected by an electronic circuit designer while working within a tool set. Figure 3–2 shows the first layer of the ESP display screen, which will be referred to as ESP_MENU throughout this text. Figure 3–3 shows the display screen of the Schematic Design tool set menu, which is referred to as SDT_MENU. Figure 3–4 shows the PC Board Layout tool set menu, which is referred to as PCB_MENU. Figure 3–5 shows the Design Management tool set menu, which is referred to as DM_MENU. OrCAD calls this software a world standard for high-performance EDA environments for modern electronic designers.

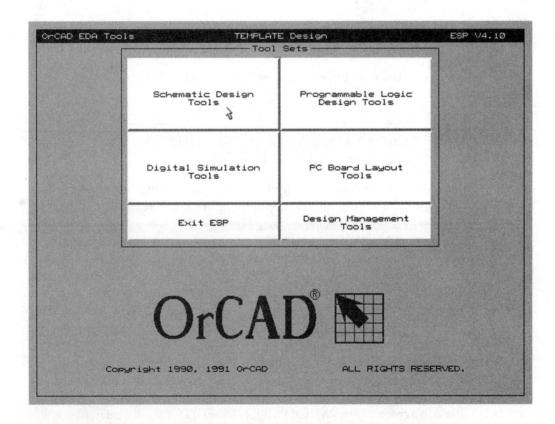

Figure 3–2. First Screen Display of the ESP Display Screen (ESP_MENU) (Courtesy of OrCAD Inc.)

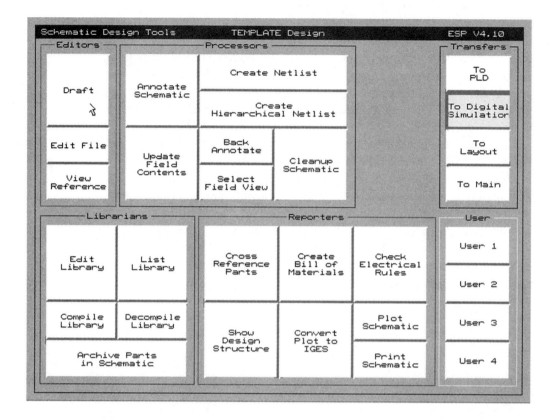

Figure 3-3. Schematic Design Tool Set Menu (SDT_MENU) (Courtesy of OrCAD Inc.)

3-4. COMMON DOS INFORMATION NEEDED TO USE THE SOFTWARE TOOLS

Most microcomputers are controlled by software known as the disk operating system, or simply abbreviated as DOS. This section discusses the most commonly used commands and functions of DOS. More details about DOS can be found in DOS manuals. Major microcomputer manufacturers tailor DOS to fit their needs. Manufacturers may even change some of the instructions. Some of the software developers market DOS in their own names, such as MS-DOS (Microsoft DOS), IBM-DOS, and Zenith-DOS. Although most of the common DOS commands are not different from one developer to another, MS-DOS (Microsoft-DOS) and Zenith-DOS are the two that are used and discussed in this book.

Under the new ESP environment, once the software is installed, very few DOS commands are really necessary to work through the tool sets. The DOS commands are used during software installation and are entered from the edit screen of the M2EDIT editor. M2EDIT is a full-screen text editor provided by the EDA environment. To enter in DOS command mode from the M2EDIT text editor, place the mouse prompt on DOS (across the top border) and click the left button of the mouse. The DOS prompt will appear

3–4. Common DOS Information Needed to Use the Software Tools

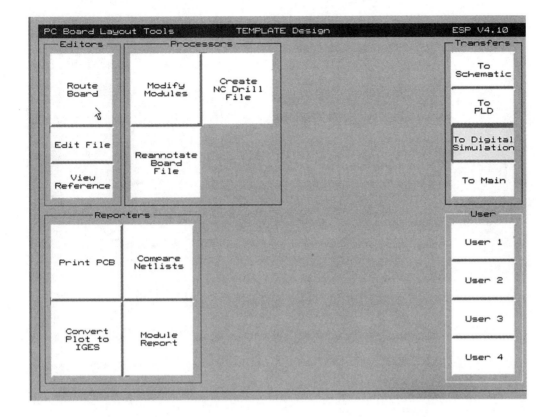

Figure 3–4. PC Board Tool Set Menu (PCB_MENU) (Courtesy of OrCAD Inc.)

in the middle of the screen. Many of the DOS commands, such as COPY, DEL, PRINT, and DIR, can be entered at this prompt. Control is returned to the editor after execution of the DOS command or by pressing the ESC key on the keyboard.

DOS commands can also be used whenever the design environment is suspended in the background. This temporary suspension transfers control to the DOS system. To remind the user that the design environment is suspended in the background, the environment adds another arrowhead to the existing DOS arrowhead. The prompt will look as follows:

C:\MYORCAD\ORCAD\PW_SUPP>>

Any DOS command can be entered at this prompt. Since **Suspend to System** can be accessed from many different places of the environment, the subdirectory \PW_SUPP will indicate the current directory position in the environment. A return to the design environment is accomplished by typing EXIT and pressing ENTER as follows:

C:\MYORCAD\ORCAD\PW_SUPP>>exit [Enter ↵]

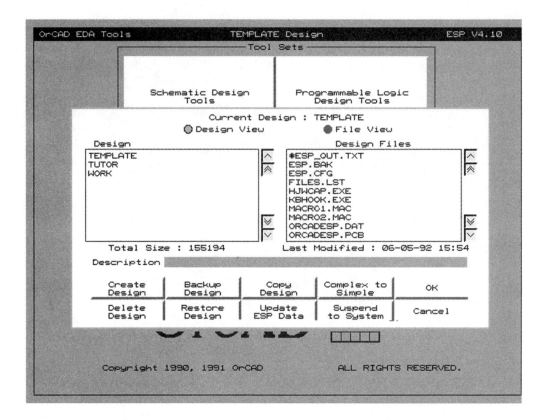

Figure 3-5. Design Management Tool Set Menu (DM_MENU) (Courtesy of OrCAD Inc.)

Exit from the system will return control to the design environment, from which the Suspend to System command was initiated.

Another situation where DOS commands and the OrCAD directory structure are useful is when students save their OrCAD work on a floppy instead of a hard drive. At the end or beginning of a schematic or PC board layout session, students can use DOS commands to retrieve from or store data onto a floppy.

3-5. DIRECTORY AND FILE STRUCTURE

Occasionally, it may be necessary to locate certain files, debug certain problems, store or retrieve data from a floppy, or create a special environment for the software. For this purpose, it is helpful to understand the directory and file structure system of the software tool sets. Figure 3–6 is a brief layout of the directory and file structure of the software. Reviewing the directory structure will show that the files created for a design are stored in DESIGN SUBDIRS within a root directory called OrCAD. If you wish to save your work on a floppy, you may do it in two ways: (1) Use the **Backup**

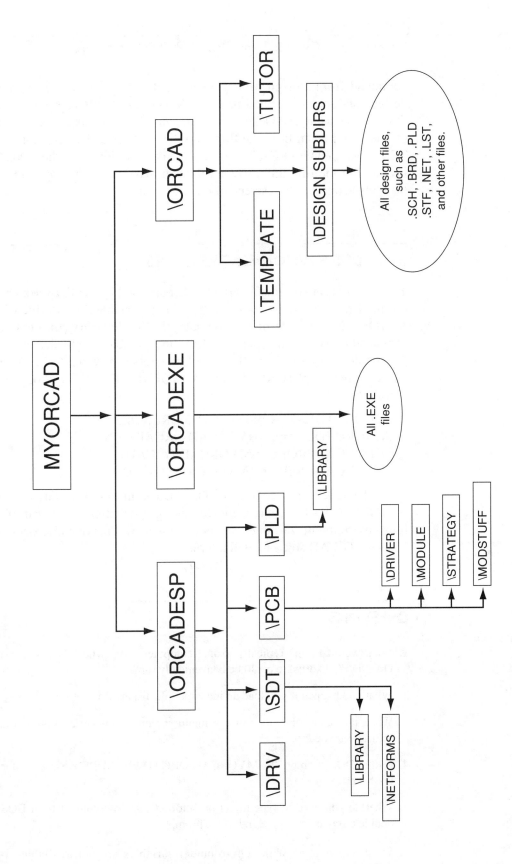

Figure 3-6. Directory Structure of the Software Tool Sets

command from Design Management Tools or (2) go to the appropriate design subdirectory located under C:\MYORCAD\ORCAD\ and copy the desired files using the DOS Copy command. Use the **Restore** command to restore your design from the floppy to the hard drive. The directory named TEMPLATE under OrCAD is always the default directory when OrCAD is invoked for the first time. If you like, you may change this by configuring the global configuration screen of the SDT tool.

3-6. DOS ENVIRONMENT VARIABLES

There are four environment variables that are used by the EDA design environment in conjunction with DOS. These variables are defined in the AUTOEXEC.BAT file of the computer. DOS's **Set Environment (SET)** command is used for this purpose. Thus, the variables are loaded every time the computer is turned on. If the root directory where the OrCAD will be located is named MYORCAD, your AUTOEXEC.BAT file will be modified as follows:

```
SET ORCADEXE=C:\MYORCAD\ORCADEXE\
SET ORCADESP=C:\MYORCAD\ORCADESP\
SET ORCADPROJ=C:\MYORCAD\ORCAD\
SET ORCADUSER=C:\MYORCAD\ORCADESP\
```

The INSTALL utility provided by OrCAD can install these variables in the AUTOEXEC.BAT file automatically during installation, if desired. If these four environment variables are not invoked when your microcomputer boots up, OrCAD software will not run.

QUESTIONS

1. What would be the typical personal computer setup where OrCAD/SDT and OrCAD/PCB software could be conveniently run?

2. What will happen if you load an incompatible display driver for ESP?

3. Write the name or names of the menu screens from which you can access M2EDIT text editor.

4. In this DOS prompt, 'C:\MYORCAD\ORCAD\PW_SUPP>>', what does the double arrow mean?

5. What do you need to type to get back to OrCAD environment from DOS, if a double arrow exists at the end of the prompt?

6. What will happen if the environment variables are not available in your AUTOEXEC.BAT file?

4

THE ELECTRONIC DESIGN AUTOMATION ENVIRONMENT AND ORCAD SOFTWARE TOOL SETS

4-1. COMPUTER SOFTWARE TOOLS IN THE WORLD OF ELECTRONIC DESIGN AND FABRICATION

With the ever-increasing complexity of high density electronic circuits coupled with increased demand, engineers and scientists are forced to replace the conventional method of electronic design and fabrication with computer-aided tools. It is not the design methodology that changed through this computerization; rather, it is the implementation of the design methods. Computer-aided tools can produce quick, accurate, easily modifiable, and automatically documentable outputs for design and produce critical tooling for the electronic circuit fabrication process. However, adequate knowledge of the traditional engineering disciplines is still essential for accurate design work.

In the past, the cost of CAD/CAM/CAE hardware and software systems were so high and the hardware was so large that only big companies and government organizations could afford to use them. However, in the recent years, this situation has dramatically changed as a result of advancement of semiconductor technology. With a significant drop in price coupled with a reduction in size, these remarkable CAD/CAM/CAE tools are within reach of almost any university or college. Because of the immense power of modern desktop computer hardware and software tools, repetitive, laborious, and error-prone manual and semi-manual methods have been replaced by off-the-shelf fully integrated CAD/CAM/CAE systems. OrCAD Electronic Design Automation software is one of the most powerful and affordable tool sets for colleges and universities.

Objectives

After completing this chapter, you should be able to

1. Understand the three software tools of OrCAD required for PC board design
2. Invoke any of the three tool set menus
3. Describe the organization of tools based on their function
4. Describe the PCB artwork design steps
5. Know common characteristics of OrCAD tools such as Transit Menu, Local Configuration, and Dialogue Box
6. Configure the ESP design environment

4-2. ORCAD/DESIGN MANAGEMENT, ORCAD/SDT, AND ORCAD/PCB TOOL SETS

Since the Design Management tool set, the SDT tool set (version 4.20), and the PCB tool set (version 2.21) all run from the ESP environment, it is appropriate to start discussing these tool sets from the ESP menu. OrCAD/ESP's icon-like menu provides a very intuitive and user-friendly interface among various OrCAD software tool sets. The environment enables file management and transfer of information among tool sets and allows access to a common database for organizing, editing, and retrieving required information for an electronic design. A new name is emerging for these kinds of environments: Electronic Design and Automation Bridge (EDA-Bridge).

The OrCAD/ESP tool set menu, called ESP_MENU in this text, can be invoked by typing ORCAD at the DOS prompt within the MYORCAD subdirectory as follows:

C:\MYORCAD.orcad [Enter ↵]

When ESP_MENU is invoked, the tool set menu will be displayed on the screen. The top line of the menu will display the default design name called TEMPLATE Design. If the mouse driver is installed, the mouse arrow prompt will also be active. The tool set screen displays a button-like icon for each tool set under the ESP environment. In this text, only the following tool sets are discussed:

- Schematic Design Tools
- PC Board Layout Tools
- Design Management Tools

To invoke any of the above tools, place the mouse prompt within the boundary of the button-like icon and left-click the mouse, or press the [Enter ↵] key. The appropriate menu will be displayed on the screen.

While in ESP_MENU, the first thing a designer should do is to define the design name of his or her project by using the Design Management Tools. The default design name is always TEMPLATE Design. To keep the TEMPLATE design directory uncorrupted, you should create a design name for each new design activity. It is also recommended that you use a different name for each electronic design project so that it will be easy to manage and document several design projects without any discord. The procedure to use the Design Management Tools will be described in more detail later in this text.

Once the design name is set, the design name will appear at the top of the menu, replacing the name TEMPLATE. Now the environment is ready to invoke the Schematic Design tool set. This is performed by placing the mouse prompt on the Schematic Design tool icon and left-clicking the mouse. This action will invoke the Schematic Design tool set and display an icon-like menu on the screen. This menu is called SDT_MENU in this text. The SDT_MENU is organized into six categories:

- Editors
- Processors
- Librarians
- Reporters
- Transfers
- Users

Each of these subdivisions has several tools, which will be discussed at the appropriate places in the text. Among these subdivisions, Transfers and Users are interesting.

Transfers icons allow the designer to go back and forth among tools under the EDA environment. The tool can be configured to pass variables to the invoked tool when executed. For example, if you transfer from SDT to the PCB tool set, the tool will create a netlist file and thereby transfer the netlist information to the PCB tool.

Users icons can be programmed to switch back and forth among tools outside the EDA environment. Although there are only four of these icons, they can be programmed very easily at the beginning of each session. The procedure of programming one of these icons is provided at the end of this chapter.

In order to go one step further into the SDT_MENU, left-click the mouse on the desired tool function. For example, the tool that will draw the schematic is called **Draft,** located under **Editors** in SDT_MENU. A left click of the mouse on Draft will invoke the schematic design and drafting screen, with the tool menu at the upper left corner of the screen. Figure 4–1 shows the drafting worksheet for schematic design, which is called SDT_MAIN in this text.

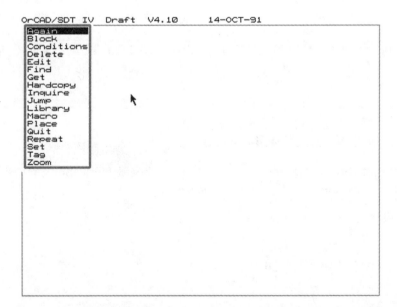

Figure 4–1. Schematic Design Drafting Sheet (SDT_MAIN) (Courtesy of OrCAD Inc.)

Figure 4–2. PC Board Design Work Sheet (PCB_MAIN) (Courtesy of OrCAD Inc.)

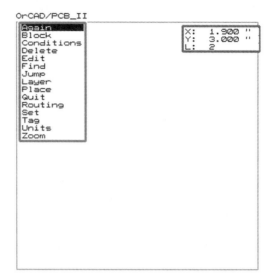

To invoke PC Board Layout Tools, left-click on the PC Board Layout Tools icon under the ESP_MENU. PC Board Layout Tools is organized into five categories:

- Editors
- Processors
- Reporters
- Transfers
- Users.

Designing PC board layout, downloading modules from netlist files, and defining board parameters all are done from the tool called ROUTE BOARD located under Editors. In order to invoke ROUTE BOARD, left-click the mouse on the icon. The board window will appear on the screen, with the menu at the upper left corner. Figure 4–2 shows the board design screen, called PCB_MAIN. More details about Draft, Route Board, and other tools under SDT and PCB Layout will be discussed later.

4–3. PRINTED CIRCUIT BOARD MASTER ARTWORK DESIGN STEPS

Master artwork design for the printed circuit board fabrication process is done by using the three following basic steps:

A. File Organization Using Design Management Tools

　1. Create design

　2. Load existing design

4-3. Printed Circuit Board Master Artwork Design Steps

    3. Set design environment

    4. Delete design

    5. Suspend to system

B. Designing and Drafting Circuit Schematic
    1. Configure schematic design tool
        a.) Library
        b.) Worksheet option
        c.) Key fields
    2. Draft local configuration invoke worksheet
    Load the desired file, since a design may have more than one file.
    3. Capture circuit schematic
        a.) Get schematic symbol
        b.) Connect symbols with wires, busses, etc.
    4. Complex to simple for complex hierarchical file structure only
    5. Annotate schematic
    6. Check electrical rules
    7. Update field content
        a.) Select field view
        b.) Edit field content
        c.) Stuff field content
    8. Archive schematic parts
    9. Create netlist

C. Designing PC Board Layout
    1. Configure PC board layout tools
        a.) Memory allocation option
        b.) Design conditions
        c.) Net conditions
    2. Place board edge
    3. Read netlist to unload modules and their logical connections
    4. Arrange components on board
        a.) Ratsnest
        b.) Force vectors
    5. Route board
    6. Print master artwork, silkscreen, solder mask, etc.

4-4. SOME OF THE COMMON CHARACTERISTICS OF ORCAD EDA TOOL SETS

4-4.1 Transit Menu

Each tool set and its subordinate tools have a Transit Menu. Options provided in this menu may vary slightly from one tool to another. However, its sole purpose is to provide the user with the ability to either configure or execute the tool command. The Transit Menu also allows the user to enter into other configuration screens whose configuration may affect the desired operation of the tool. From this menu the user can decide whether to execute or to configure the tool.

The Transit Menu provides two major options: Execute and Local Configuration. The Local Configuration screen of each tool allows the user to configure the tool so that it operates in a desired manner. By changing the local configuration, the tool's action can be altered either temporarily or permanently. This unique feature makes this tool considerably adaptable for electronic design environment. Figure 4–3 shows available transit menus.

4-4.2 Local Configuration

Most of the tools have some kind of local configuration. If a tool does not have any local configurable option, the tool will always behave in the same way. In practice, the local configurable option makes the same tool behave differently under different configurations. Thus, one icon button can be programmed to do several different options. The existence of local configura-

Figure 4–3. Transit Menus

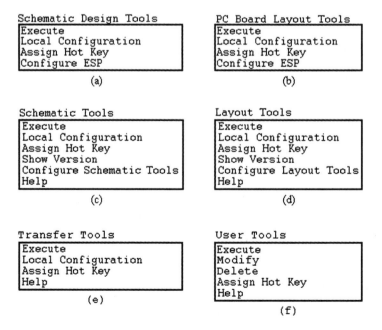

4-4. Some of the Common Characteristics of OrCAD EDA Tool Sets

tion of a tool can be determined by selecting a command option. If the message *Nothing to Configure* appears on the screen, the tool does not have a local configuration. However, for tools that do have a local configurable option, a Local Configuration screen, such as the one shown in Figure 4–4, will be displayed when the command is selected. In Local Configuration the screen file's list box will normally display names of files belonging to the design. Any of these files could be selected by left-clicking the mouse on the filename in the box.

The action of these tools can be altered permanently or temporarily—for example, which file of the design the tool should act upon, which file the result of the action should be stored in, and options of processing a design. The role of Local Configuration will be explained in detail later. The local configuration of a tool is specific to a design, so you can go from one design to another without changing the configuration. The local configuration also eliminates the need of remembering the command line, switches, and arguments. Figure 4–4 shows the Local Configuration window of the Draft tools.

4-4.3 OK and Cancel Icons

A left click of the mouse on the **OK** icon means the acceptance of the configuration and changes made. A left click of the mouse on the **Cancel** icon means the nonacceptance of the changes in configuration and options. These two types of icons often need to be clicked in order to proceed further with the processing. Figure 4–4 also shows the **OK** and **Cancel** icon buttons associated with the Local Configuration window of Draft Tools.

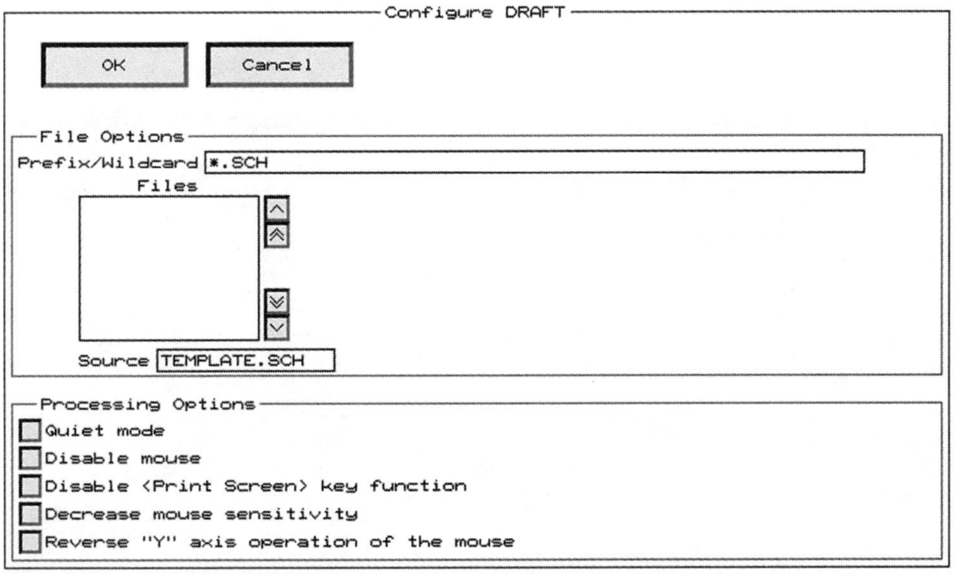

Figure 4-4. Local Configuration Window of Draft Tools (Courtesy of OrCAD Inc.)

4-4.4 Mouse Prompt and Dialogue Box

All of the entry boxes or dialogue boxes in the OrCAD EDA environment work the same way. A left click of the mouse prompt inside the dialogue box will put the prompt in the insert mode. The ⌊Enter ↵⌋ key or left click on the mouse for the second time will bring back the mouse arrow prompt. The insert mode is an underscore bar that will appear inside the box. Pressing the **Insert** key of the keyboard will change the prompt to over-type mode. The over-type mode is a square cursor. Using the **Insert** key, you can toggle back and forth between insert and over-type modes. Any character typed at the prompt is inserted in the existing text.

Whenever a character string for file name, driver name, processing option, or wildcard option is required to be typed inside a dialogue box, the following rules should be remembered:

- The length of the character string is limited by the size of the box.
- The ⌊Enter ↵⌋ key or left click on the mouse will allow the transfer between the insert and the mouse arrow prompt. The key or the mouse operation must be preceded by the prompt pointer being anywhere inside the dialogue box.
- The **Home** key will move the cursor to the beginning of the dialogue box.
- The **End** key will move the cursor to the end of the dialogue box.
- The **Arrow** key or mouse will move the cursor inside the box without erasing the characters.
- The **Backspace** or **Delete** key will delete a character as usual. Figure 4–5 shows the dialogue boxes for worksheet options within the SDT configuration menu.

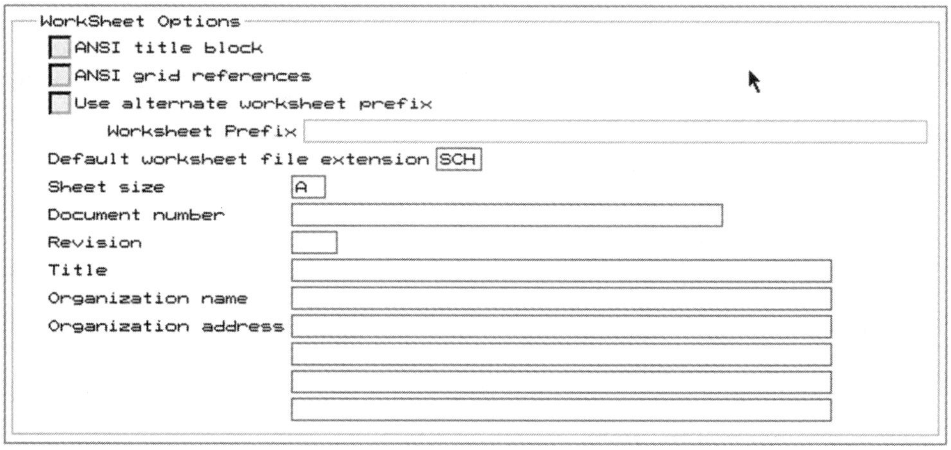

Figure 4–5. Dialogue Boxes for Worksheet Options (Courtesy of OrCAD Inc.)

4–4.5 Buttons for Options and Switches

There are two types of option buttons that are usually used in this software: square and round. These buttons are usually used for various options and switches within configuration menus. They are also used with the various menu windows for going from one group of tools to another group. They are activated by left-clicking the mouse. The green button means the option or switch is active. Note that green is the default color setting. Figure 4–6 shows the buttons for options and switches in the Library Options tool within the SDT configuration. The Library Options tool basically has two functions:

- Insert a Library.
- Remove a Library.

Symbol libraries that are in the Configured Libraries list box are available for use from the SDT_MAIN menu, whereas the libraries in the Available Libraries list box indicate all the schematic symbol libraries available within OrCAD SDT tools. This arrangement is made to manage the system RAM memory efficiently. Configured Libraries are located in the system RAM, and Available Libraries are on hard disk. A library file usually occupies lots of RAM memory. Therefore, if all the available libraries are placed in the Configured Library list box, the system will surely run out of RAM memory space for other activities. Activating the **Insert a Library** button by left-clicking the mouse on the button will allow the Library Options tool to configure libraries. Activating the **Remove a Library** button will allow the tool to remove libraries from the configuration. Removed libraries are always placed in the Available Library box. However, if a schematic requires many different libraries

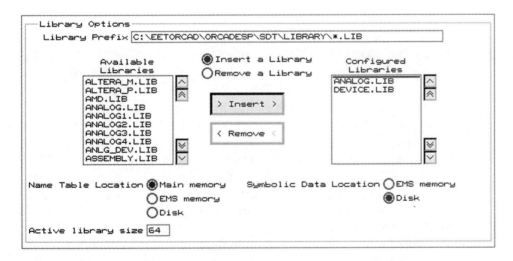

Figure 4–6. Buttons for Options and Switches of the Library Options Tool (Courtesy of OrCAD Inc.)

and memory constraints inhibit loading all the required libraries at the same time, the problem can be solved by using the **Archive Parts in Schematic** function.

4-4.6 Single Arrow and Double Arrow

A left click of the mouse on the single arrow will scroll the List Box by one line in the direction the arrow is pointing. A left click on the double arrow will scroll the List Box by page in the direction the arrow is pointing. Figure 4–6 shows the double arrow and single arrow on the right side of the Available Libraries box and the Configured Libraries box. Left-clicking the mouse on the single arrow or double arrow will move the library list by a single line or by a page, respectively.

4-5. CONFIGURATION OF THE ESP DESIGN ENVIRONMENT

The configuration menu of the ESP environment can be accessed from the Transit Menu of all tool sets. A left click of the mouse on the Schematic Design tool set icon, located in ESP_MENU, will display the Transit Menu on the left corner of the screen. The fourth item in the menu is Configure ESP. A left click of the mouse on this item will display the ESP configuration menu on the monitor screen. The whole menu cannot be seen at one time on a 14-inch monitor screen. The mouse or **PageUp** and **PageDown** keys are used to travel through the configuration items. To explain the configuration, the menu is divided into several figures. Configuration of the ESP menu is not essential because it is configured by the install program. However, configuring this menu will provide a customized design environment. Environmental variables such as color of screen icons and text, print screen and mouse operations, and startup design options can be set through this configuration screen.

The items that can be configured through this screen are display driver, path of text editor, print screen, mouse, name of startup design file, name of redirection file, and screen colors. Figure 4–7 shows the following items of the configuration: Display Driver Prefix, Editor Options, Print Screen Options, Mouse Options, Design Options, Prefix Options, and Redirection Options. Figure 4–8 and Figure 4–9 show the Color Option Tables.

Display Driver Prefix defines the directory path for all the display drivers. Available drivers in the list screen can be scrolled by left-clicking the mouse on the up and down arrow keys. The desired display driver can be configured by left-clicking the mouse on the desired choice.

Editor Options defines the directory path to invoke the text editor. Print Screen Options can enable or disable print screen ability. Mouse Options allows the user to set the direction of mouse operation. Design Options allows the designer to set the design name, which will be invoked at the

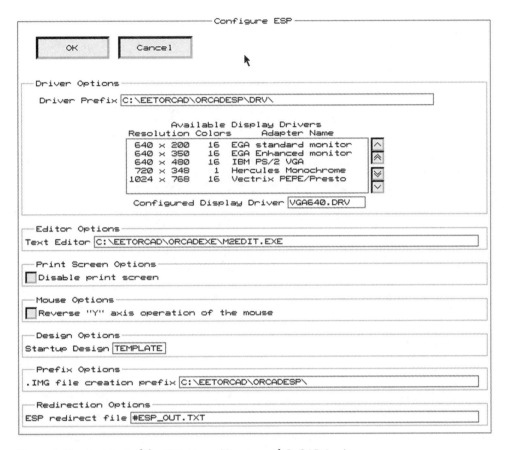

Figure 4-7. First Part of the ESP Menu (Courtesy of OrCAD Inc.)

startup. Prefix Options sets the directory path of the location of the image file. An image file is a snapshot of the environment from where the **suspend to system** command is initiated. This is done so that the software can return back to the environment. Redirection Options is the place where the redirection file name is written. Color Tables provides the option to choose the color for icon buttons, text, border, etc.

4-6. CONFIGURATION OF USER BUTTONS

There are four button-like icons, User 1 through User 4. These icons can invoke utilities and other software tools while you are working with OrCAD. Each of these icons can be preprogrammed differently so that it can invoke your desired software. These four buttons are located in SDT_MENU or PCB_MENU. Follow these steps to program the buttons:

1. Place the mouse pointer inside one of the User icons, and click the left button of the mouse. A Transit Menu, shown in Figure 4-10, will appear at the upper left corner of the screen.

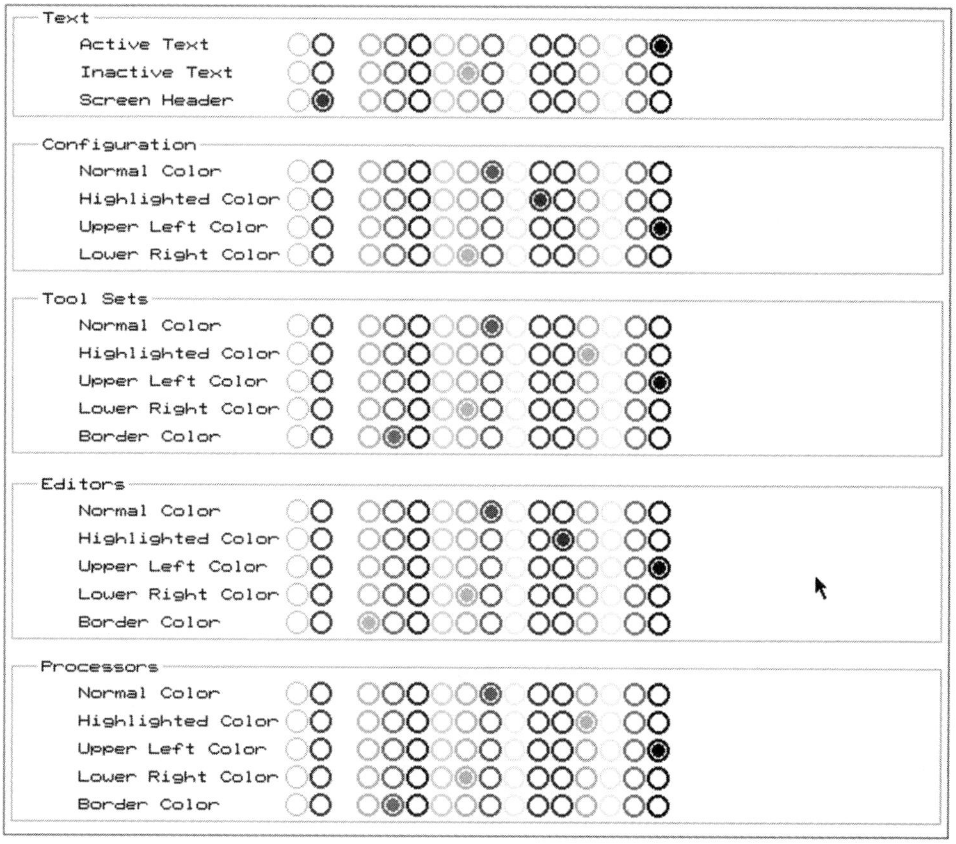

Figure 4–8. First Part of the Color Options Table of the ESP Menu (Courtesy of OrCAD Inc.)

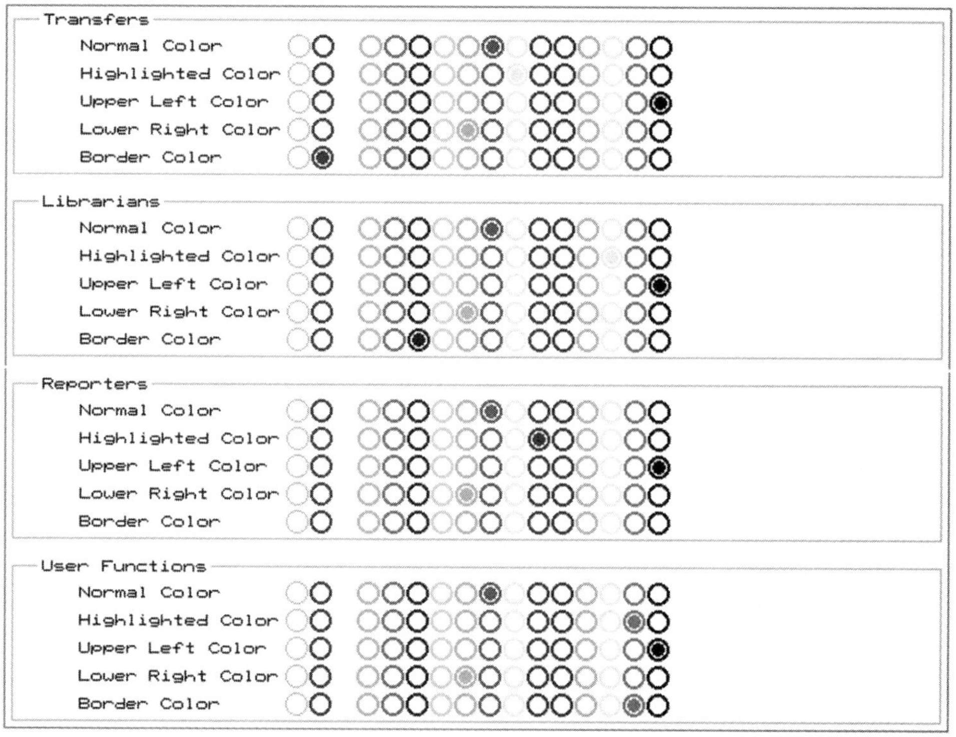

Figure 4–9. Second Part of the Color Options Table of the ESP Menu (Courtesy of OrCAD Inc.)

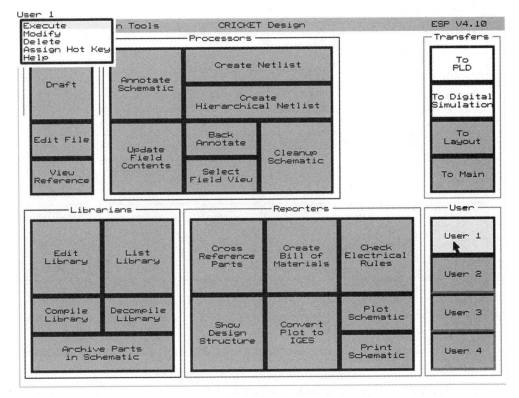

Figure 4–10. Transit Menu for User 1 through User 4 Buttons (Courtesy of OrCAD Inc.)

2. Select **Modify** from the menu. The configuration screen shown in Figure 4–11 will appear.

3. Click the left mouse button in each of these dialogue boxes and write appropriate information.

4. Left-click the mouse on the OK icon to save the changes in the User icon configuration.

5. Click on the User icon you just programmed. This time, select **Execute** from the Transit Menu. If you have written the command correctly in the Command dialogue box, OrCAD will transfer the control to the desired software.

Chapter 4 / Electronic Design Automation Environment

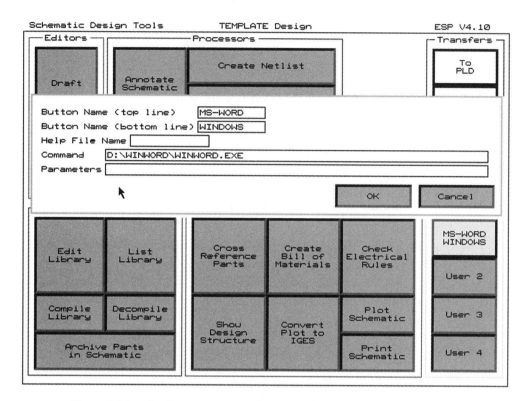

Figure 4-11. Configuration Screen for User 1 through User 4 Buttons (Courtesy of OrCAD Inc.)

QUESTIONS

1. What is the most appropriate design environment for PCB in today's world?

2. What do you need to type to invoke the OrCAD ESP menu from the MYORCAD subdirectory?

3. When starting a new project, which one of the following functions should be used?

   a. Design Management Tools
   b. Schematic Design Tools
   c. PCB Layout Tools
   d. None of the above

4. After a new design has been created using OrCAD's Design Management Tools, which one of the following design options is used next?

   a. Design Management Tools
   b. Schematic Design Tools
   c. PCB Layout Tools
   d. None of the above

5. Why is it recommended that you create a new name for each design?

6. The Transfers icon allows you to do what?

7. The Users icon allows you to do what?

8. What are the three major steps in designing a PC board layout?

9. How does the Transit Menu make the software tools flexible?

10. How does Local Configuration make the software tools flexible?

11. Left click on the OK icon means _____.

12. Left click on the Cancel icon means _____.

13. How can you enter text response into a dialogue box?

14. Normally, what color of the option button indicates the active condition?

15. What is the difference between a single arrow and a double arrow at the DOS prompt?

16. How do you invoke the configuration menu of the ESP environment?

File Organization

5

FILE ORGANIZATION AND DESIGN MANAGEMENT TOOLS

5-1. DESIGN MANAGEMENT TOOL SET

Since the OrCAD software tools are for a complete EDA environment, there are tools available within the environment for managing the files for a project. For the purpose of design and fabrication of electronic circuits, it may be necessary to transfer information between tool sets. Thus, a systematic management of files for a design under a single data base is essential. The Design Management Tools fulfills this need. This tool set has a menu with several layers. The tool set can be accessed from the ESP_MENU by clicking the left button of the mouse on the Design Management Tools icon.

Typically, a design will have many files with different names and extensions. As many as 150 separate designs can be supported by the Design Management tool set, and each of these designs can have as many as 1,600 files. A description of a few words can be added to each design and file name. This added information facilitates locating a particular design or file.

Objectives

After completing this chapter, you should be able to

1. Create, load, delete, copy, back up, and restore designs
2. Edit ASCII files and rename, copy, and delete files
3. Suspend to system and return to OrCAD environment
4. Organize design files for a project

5-2. MANAGING DESIGNS

Design Management Tools has two groups of tools to manage the design and its files, Design View and File View. Design View allows the management of designs and their files. These design managing tools and their functions are described below. File View allows the management of individual files belonging to a design. File managing tools and their functions will be described in the next section. Figure 5–1 and Figure 5–2 show the Design View and File View window display, respectively. These two windows can be invoked by left-clicking the mouse on the respective round button at the top of the window display. The green button means that the option is active. Note that green is the default color setting. This color setting can be changed from the ESP configuration.

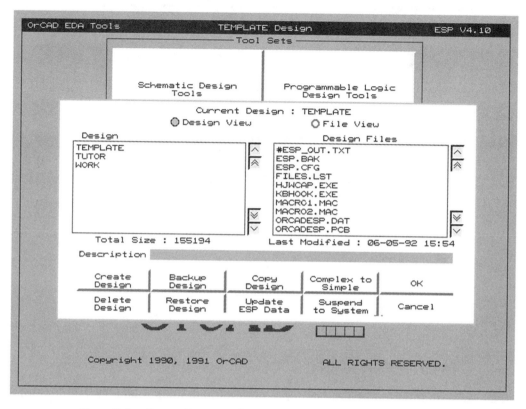

Figure 5–1. Design View Window Display (DM_MENU) (Courtesy of OrCAD Inc.)

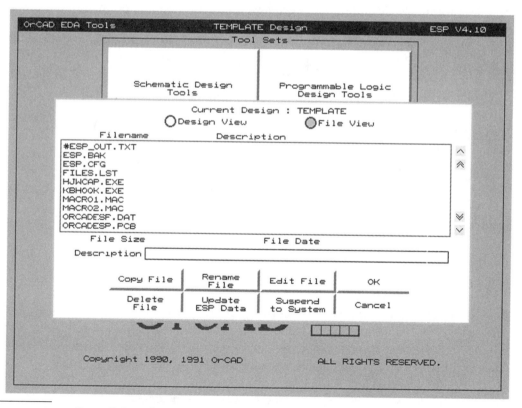

Figure 5–2. File View Window Display (DM_MENU1) (Courtesy of OrCAD Inc.)

5-2.1 Create Design

If the design is new, no file exists for the project. **Create Design** is used to create a new design name. While in the DM_MENU window display and under the Design View option, a design name can be created by using the **Create Design** tool. A left click on the icon will bring up the Create Design window display shown in Figure 5–3. Next, the mouse prompt needs to be placed and left-clicked inside the New Design Name dialogue box. This action will change the mouse arrow to dash-type prompt. The dash mouse prompt is in the insert mode, and the name of the new design needs to be entered. After providing the new design name, left-click the mouse once again, and the mouse will return to arrow prompt. Left-clicking the mouse once more on the OK icon creates the new design name. This action also copies all the files of the TEMPLATE subdirectory into a new design subdirectory. The subdirectory name will be the new design name you just provided. This subdirectory is created under OrCAD. The new design name will appear in the left list box of the DM_MENU, where all the design names are displayed.

5-2.2 Load Existing Design

While in the DM_MENU window display, a left click of the mouse on the desired design name located in the left list box will highlight the design name. This action will also display in the right list box of the window all

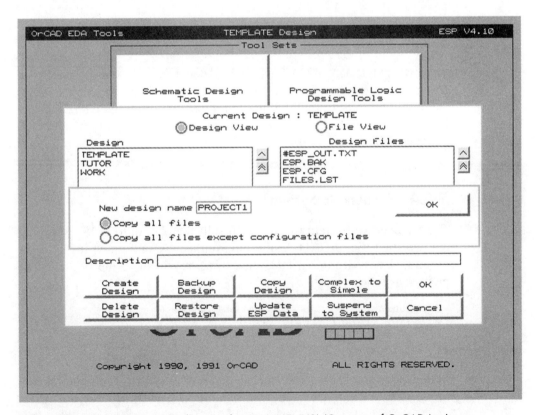

Figure 5–3. Create Design Window Display (DM_MENU2) (Courtesy of OrCAD Inc.)

filenames belonging to the design. At this point, a left click on the OK icon will replace the TEMPLATE Design environment with the selected design environment name. Control will be shifted to the ESP_MENU, and the new design name will appear at the top of the screen, replacing TEMPLATE. TEMPLATE Design is the default design environment name.

Whenever OrCAD is invoked for the first time, TEMPLATE Design will appear at the top of the ESP_MENU window display. Files in this design are the master copies of the design environment. Whenever a new design is created using the tool, all files belonging to TEMPLATE Design are copied to the new design environment. The tool uses these files to create a new design environment. Any changes made in the TEMPLATE files become part of the new design. Therefore, changes and additions for the TEMPLATE Design must be made carefully. Files in this design are the master copy of the Design and File Management tool set.

The changes and additions to files are made from the File View window, whereas the management of designs is done from the Design View window. Schematic and board files belonging to a design cannot be viewed by using this tool set. For example, schematic files with the .SCH extension and board files with the .BRD extension can be viewed only by using SDT tool set and PCB tool set, respectively. The files with these extensions are loaded by selecting them from the local configuration of the respective tool. However, ASCII files can be viewed by using the **Edit File** tool under the File View window display.

5-2.3 Delete Design

An unwanted design and its files can be deleted from the data base from the Design View window display. The deletion can be performed by selecting the design by left-clicking the mouse prompt on the design name and then left-clicking on the **Delete Design** icon. Figure 5–4 shows the Delete Design Window.

5-2.4 Suspend to System

The **Suspend to System** function will shift the control to DOS. To prompt the user that the OrCAD design environment is suspended in the background, the environment adds an extra arrowhead to the regular DOS prompt, as follows:

```
        subdirectory ↰        ↱ design subdirectory
C:\MYORCAD\ORCAD\PW_SUPP>>
        ↳ subdirectory                ↳ extra arrowhead to indicate
                                        suspended OrCAD environment
```

In this mode any DOS command can be used. In order to shift control back to the OrCAD EDA environment, enter **EXIT** at the double arrowhead

5–2. Managing Designs

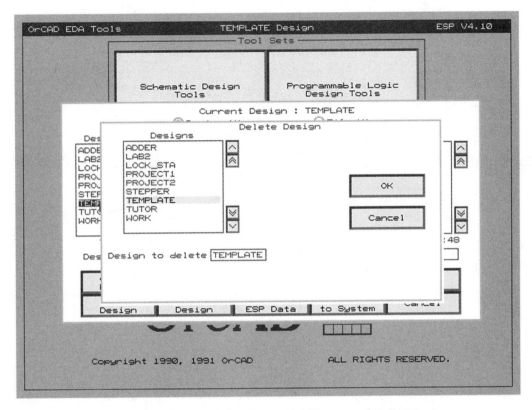

Figure 5–4. Delete Design Window Display (DEL_DGN) (Courtesy of OrCAD Inc.)

prompt. Thus, by typing as follows one can return back to the OrCAD design environment:

C:\MYORCAD\ORCAD\PW_SUPP>>EXIT [Enter ↵]

Exit from DOS system will return control to the place of the design environment, from which the **Suspend to System** command was initiated.

5–2.5 Copy Design

This tool copies all files belonging to a design to a new design name. To do this, select the design to be copied from the Design View window and then left-click on the **Copy Design** icon. The Copy Design window shown in Figure 5–5 will appear on the monitor screen with the source design as the selected design name. The name of the new design needs to be provided in the dialogue box next to Destination Design. Finally, you should left-click the mouse on the OK icon to perform the operation of copying all files from the old design name to the new design name. The new design name will appear in the left list box of the DM_MENU window display. However, left-clicking on the **Cancel** icon will abort the operation.

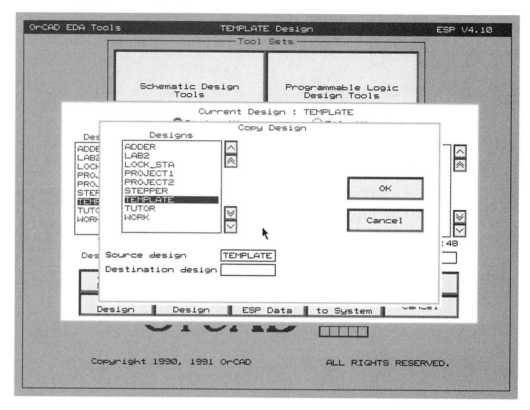

Figure 5-5. Copy Design Window Display (COPY_DGN) (Courtesy of OrCAD Inc.)

5-2.6 Backup Design

This tool backs up files belonging to a design onto floppy disks or the hard disk. In the Backup Design window display shown in Figure 5-6, the destination disk type can be selected by left-clicking the mouse on the button beside the desired disk size and density. However, when the destination prefix is provided in the respective dialogue box, depending on the disk size and density available on the selected drive destination disk type, the select button will automatically move to the proper location. Selecting the design to be backed up will put the design name in the respective dialogue box. At the same time, the size of the design in bytes will also appear in the line below it. A destination prefix such as A:\ for a floppy also needs to be provided in the destination prefix dialogue box. At this point, left-clicking of the mouse on the OK icon will start the backup operation. During the backup process, a few more windows with messages may appear on the screen. These windows are self-explanatory and can be easily acted upon.

5-2.7 Restore Design

This tool restores a design that was backed up using the **Backup Design** tool. Left-clicking of the mouse on the **Restore Design** icon from the Design View window will bring on screen the Restore Design window

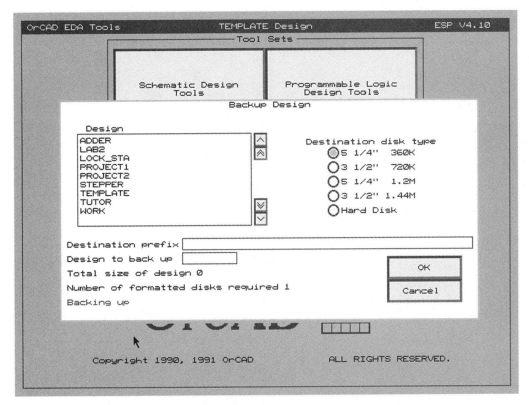

Figure 5-6. Backup Design Window Display (BACK_DGN) (Courtesy of OrCAD Inc.)

shown in Figure 5-7. A source prefix such as A:\ or the complete path name for a floppy needs to be provided in the respective dialogue box. Once the source prefix is provided, the tool will fetch the backup files from the source and put them in the Design to Restore list box. Generally, the source of backup designs is floppies placed in the A or B drive. Now the design to be restored needs to be selected from the list, and the design name to restore to must be provided in the dialogue box next to Design To Restore To. At this point, a left click of the mouse on the OK icon will begin the restore process.

5-2.8 Update ESP Data

This tool saves the current environment settings and local configuration options. However, the design environment will also prompt for saving the configuration when another tool set is opened.

5-2.9 Complex to Simple

This tool converts a complex hierarchical design structure to a simple one. Hierarchical design structure is used to represent circuits, parts of which are repetitive. In this kind of circuit structure the repetitive circuit is drawn only

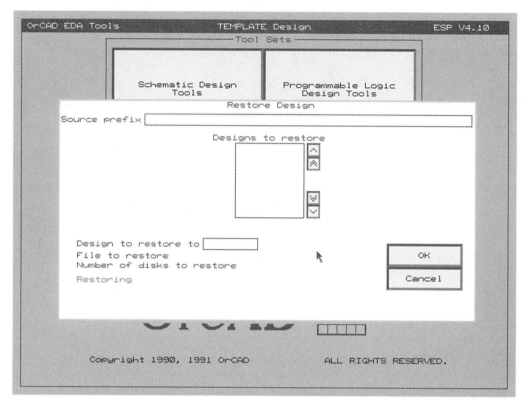

Figure 5-7. Restore Design Window Display (REST_DGN) (Courtesy of OrCAD Inc.)

once. In the actual main circuit the repetitive part is represented by a sheet symbol. However, this representation cannot be used by the PC Board Layout Tools to create a physical component map. The PC Board Layout Tools need a map that has one-to-one correspondence between schematic symbol and physical component. This tool enables you to convert a complex hierarchical design into a one-to-one correspondence map. For example, a full adder circuit can be represented by two half adder sheet symbols, both referring to the same half adder circuit. This arrangement is not a one-to-one correspondence. To make a one-to-one correspondence, two half adder circuits are necessary. The following procedure will convert a complex hierarchical design into a simple file whose symbols have one-to-one correspondence:

1. Invoke DM_MENU and left-click on the **Complex to Simple** icon. A window will appear.
2. Select the name of the design you want to convert from the list box. The name will appear in the Source Design dialogue box.
3. Provide another name for the design in the Destination Design dialogue box.

4. Left-click on the OK icon. The software will convert the source and store the simple design in the design name list box.

5. If annotation was done before the execution of Complex to Simple, you need to unannotate your entire design and execute the Annotate again.

6. Use this simple file to generate a netlist for the PC Board Layout Tools.

Figure 5–8a is a full adder circuit with two half adders represented by sheet symbols. Both of these sheet symbols refer to the one half adder schematic shown in Figure 5–8b. This kind of file structure is called a complex hierarchical structure because one schematic is referenced by multiple sheets. When you annotate the full adder schematic, it provides unique reference designators to each component of the schematic. However, the full adder sheet symbols still refer to one half adder schematic. For the purpose of PCB Layout Tools, this will not work because PCB needs the physical existence of each half adder. If you execute the **Complex to Simple** tool, it will create another half adder circuit. Now the two sheet symbols will be able to refer to two different half adder circuits. The tool will also change the sheet name and the filename of the sheets as follows:

Sheet Name:
    hadder_A   change to   **hadder_A_1**
                                                  added to extension to the sheet name

    hadder_B   change to   **hadder_B_2**

Filename:
    HALFADD.SCH   will not change   **HALFADD.SCH**
                                                  no added extension

                                                  added extension to the filename

    HALFADD.SCH   change to   **HALFADDA.SCH**

Although the tool created two separate half adders, one for each sheet symbol, it will not change the reference designators. To provide a unique reference designator to each component of the entire full adder schematic, you must annotate it unconditionally or first unannotate and then annotate it again. Figure 5–8c shows the full adder schematic, Figure 5–8d shows the first half adder referenced by the hadder_A_1 sheet symbol, and Figure 5–8e shows the second half adder referenced by the hadder_B_2 sheet symbol, after the entire design has been run through the **Complex to Simple** tool.

However, if you invoke the half adder schematics from the sheets one by one before the above annotation operation, you will find that the parts of the half adder schematics do not have unique reference designators. If you draw the simple full adder schematic and execute the **Complex to Simple** tool, you will see this process more clearly.

104   Chapter 5 / File Organization and Design Management Tools

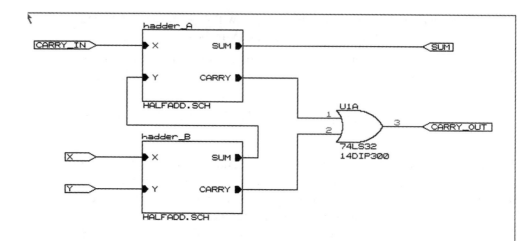

Figure 5-8a. Root Schematic of the Full Adder before Being Processed by the Complex to Simple Tool

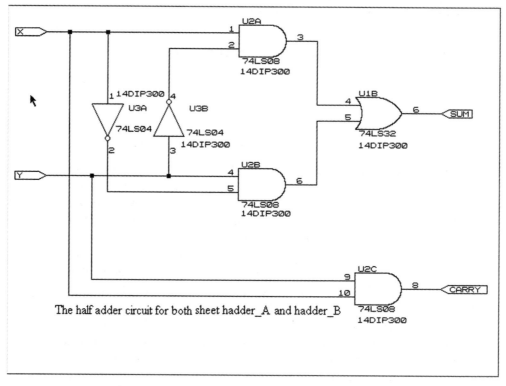

Figure 5-8b. Schematic of the Half Adder Referenced by Both Hadder_A and Hadder_B

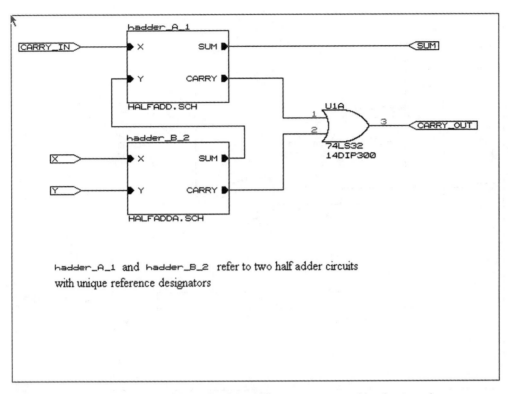

Figure 5–8c. Root Schematic of the Full Adder after Being Processed by the Complex to Simple Tool

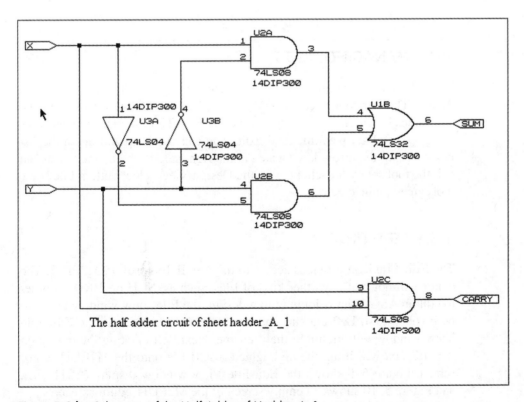

Figure 5–8d. Schematic of the Half Adder of Hadder_A_1

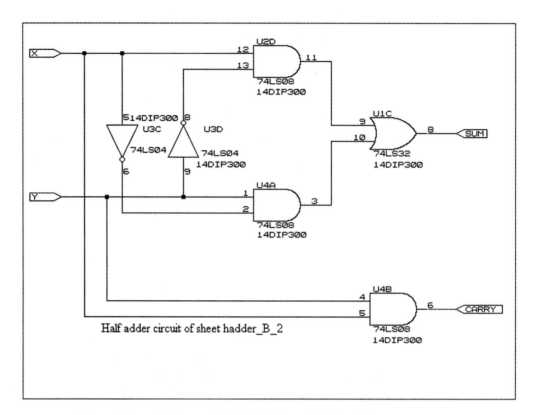

Figure 5–8e. Schematic of the Half Adder of Hadder_B_2

5-3. MANAGING FILES

The File View is another set of tools for managing files belonging to a design. The window is called DM_MENU1. The window displays all the file names belonging to a design in the list box. The file management tools are **Copy, Delete, Rename,** and **Edit.** You can scroll up and down the list box by using the arrow key located at the right side of the list box. You can exit the tool set by left-clicking on the Design View select button. File View tools are explained below.

5-3.1 EDIT File

The **Edit File** icon provides access to the ASCII Text Editor, M2EDIT. The editor will not edit any other kind of files, such as .SCH or .BRD. In order to edit an ASCII file belonging to a design, the File View window needs to be invoked first. Left-clicking the mouse on the file name in the **Edit File View** window will highlight the filename. Next, left-clicking the mouse on the OK icon will bring the highlighted ASCII file onto the M2EDIT editor screen. Figure 5–9 shows the Edit File View window display (EDIT_FL), and Figure 5–10 shows an empty screen of the M2EDIT editor screen.

5-2. Managing Designs

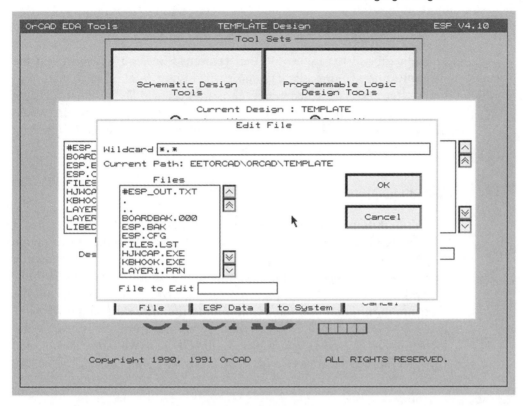

Figure 5-9. Edit File View Window Display (EDIT_FL) (Courtesy of OrCAD Inc.)

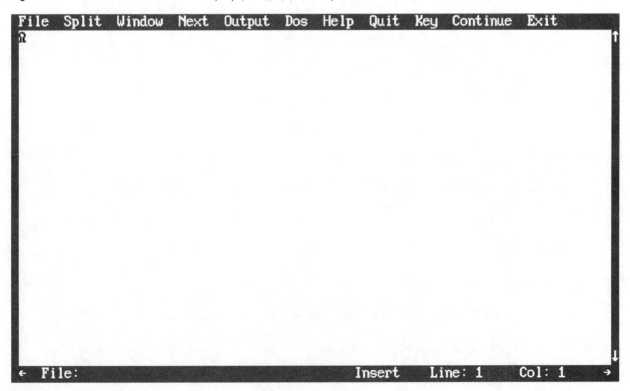

Figure 5-10. M2EDIT Edit Screen Display (Courtesy of OrCAD Inc.)

5-3.2 RENAME File

Left-clicking of the mouse on the **Rename File** icon will bring up the **Rename File** window display shown in Figure 5–11. The display will prompt for original filename and the new filename. The original filename must be selected from the list box, and the new file name must be provided in the dialogue box. At this point, a left click of the mouse on the OK icon will replace the old filename with the new name.

5-3.3 COPY File

The **Copy File** tool copies a file to another directory or design, or renames and copies a file. The tool can be invoked by left-clicking the mouse on the **Copy File** icon in the File View window. The **Copy File** window shown in Figure 5–12 will appear on the screen, with file names in the list box. Left-clicking of the mouse on the desired filename in the list box will put the filename in the dialogue box of the source file. The destination filename along with the path name, if any, must be provided in the destination dialogue box. At this point, a left click on the OK icon will copy the selected file to the appropriate place.

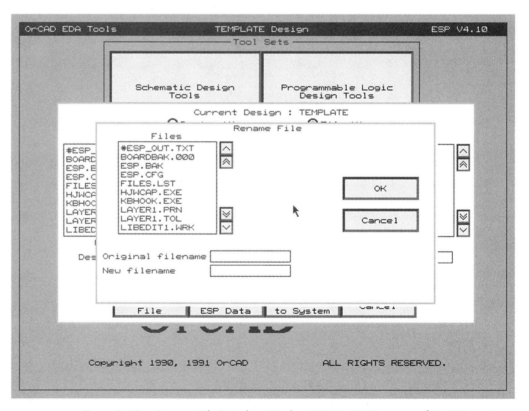

Figure 5–11. Rename File Window Display (RENM_FL) (Courtesy of OrCAD Inc.)

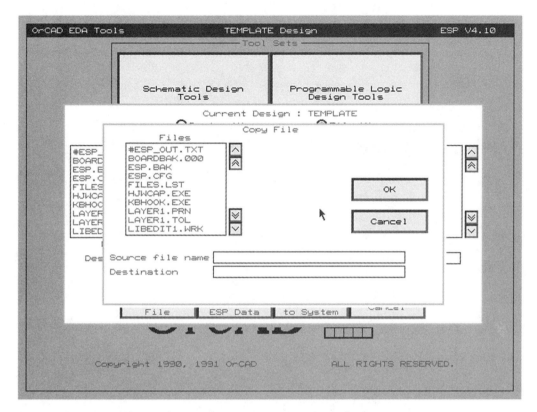

Figure 5-12. Copy File Window Display (COPY_FL) (Courtesy of OrCAD Inc.)

5-3.4 DELETE File

The **Delete File** icon will delete selected files from the File View window. In order to delete a file belonging to a design, the file needs to be selected by left-clicking the mouse on the filename. This operation will highlight the filename. Second, left-clicking of the mouse on the **Delete File** icon will remove the file from the design. Figure 5–13 shows the **Delete File** window display.

5-3.5 Suspend to System

Left-clicking of the mouse on the **Suspend to System** icon will transfer control to the DOS system by suspending the EDA environment in the background. This function works the same as the other **Suspend to System** function.

5-4. ORGANIZATION OF DESIGN FILES FOR PROJECTS

The Design Management Tools are used to create a new design environment. Files belonging to a design environment are created by related tools. For example, schematic files with the .SCH extension are created by the Schematic

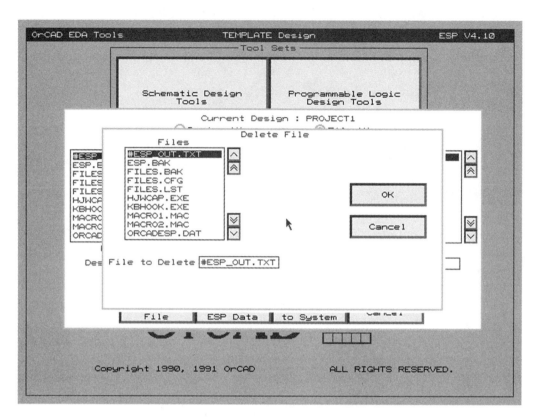

Figure 5-13. Delete File Window Display (DEL_FL) (Courtesy of OrCAD Inc.)

Design tool set, PC board artwork files with the .BRD extension are created by the Printed Circuit Board tool set, and all ASCII files are created by the M2EDIT ASCII editor provided with OrCAD tool sets. M2EDIT editor can be accessed from the Design Management, SDT, and PCB tool sets.

The following step-by-step procedure will create a new design name. The design environment of the four projects described in Chapter 2 will be created here.

Step 1: Boot the microcomputer system. If this is done correctly, a C:\> prompt will appear on the screen.

Step 2: At C:\> prompt, type cd\myorcad and press [Enter↵]. Response from the DOS system will be C:\MYORCAD>.

Step 3: Invoke the OrCAD ESP_MENU window by typing orcad. The response to this action will be the display of the ESP_MENU window on the computer monitor screen. At the top of the ESP_MENU, the default design name TEMPLATE will appear with the window display.

Step 4: Invoke the Design Management tool set by left-clicking the mouse on the **Design Management Tools** icon. The response to this action will be the display of DM_MENU. On the left list box of DM_MENU, the design name TEMPLATE will be highlighted. This highlighting means

5-4. Organization of Design Files for Projects

TEMPLATE is selected. Files belonging to TEMPLATE Design will be seen in the right list box. Always keep TEMPLATE selected before creating a new design name. You want to copy the files of the TEMPLATE design every time you create a new design. At least two design names, TEMPLATE and TUTOR, will be in the list box. Both of these designs come with OrCAD software.

Step 5: Left-clicking the mouse on **Create Design** will invoke DM_MENU2. A left click of the mouse inside the dialogue box next to New Design Name will change the mouse from arrow to dash. The dash mouse prompt is the insert mode.

Step 6: Provide the new design name. The number of characters allowed for the name will automatically be limited by the size of the dialogue box. After providing the name, left-click the mouse again to get back the mouse arrow prompt. Then left-click the mouse on the OK icon located straight to the right of the new design name dialogue box. DM Tools will create a new design environment, and the design name will appear in the list box where all the design names are located.

Using the above procedure, any design environment can be created. The File View window can be invoked by left-clicking the mouse on the **File View** button. This window will let you check the details of each file belonging to a design.

The design name of the projects described in Chapter 2 are created by using the above procedure. Project number, title, and design name of each project are given below:

Project No.: 01-1995
Project Title: The Deluxe Logic Probe
Design Name: LOG_PRO

Project No.: 02-1995
Project Title: The Electronic Cricket
Design Name: CRICKET

Project No.: 03-1995
Project Title: The Infrared Object Counter
Design Name: IR_COUNT

Project No.: 04-1995
Project Title: The Mini Stereo Amplifier
Design Name: AMPLIFIE

Figure 5–14 shows the Design View window display, with design names and their short project titles in the left list box. At this time the designer first needs to set the design environment for a particular project and then invoke the Schematic Design tool to start drawing schematics for that project. Generally, the design environment of each project is separated to enable the easy management of files. Files can easily be transferred among designs.

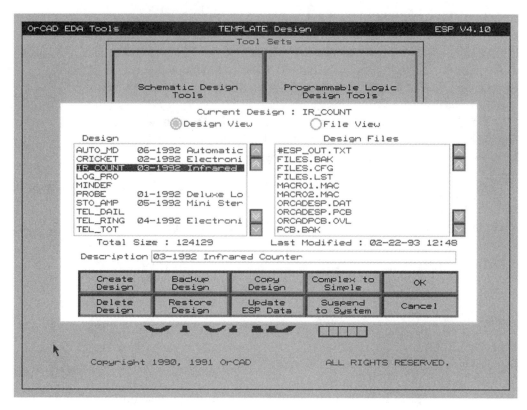

Figure 5-14. Design View Window Display (Courtesy of OrCAD Inc.)

QUESTIONS

1. What is the difference between a design and a file?

2. What is the maximum number of designs that the OrCAD Design Management tool will support?

3. What is the maximum number of files a design will support?

4. Is it possible to see through the files belonging to a design from the Design View window?

5. Is it possible to delete a file from the Design View window?

6. If you want to create a design, which design files should you copy into your new design directory?

7. How will you know which design environment is active?

8. If you delete a design, will it also delete all the files that belong to the design?

9. When you Suspend to System from the Design Management tool, what indicator will tell you that the OrCAD design environment is in the background?

10. Is there any tool available to back up and restore a design on a floppy?

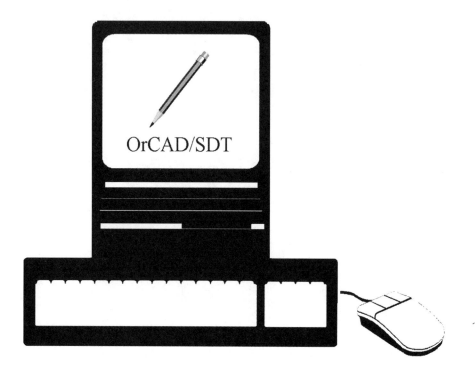

OrCAD/SDT Tool Set

6

SCHEMATIC DESIGN TOOL SET

6-1. WHAT IS AN ELECTRONIC CIRCUIT SCHEMATIC?

The term *electronic circuit schematic* means the scheme of an electronic circuit and how it works together. The schematic of an electronic circuit is composed of symbols that represent electrical, electronic, and electromechanical parts, and the lines that represent conducting paths for interconnecting components. The schematic describes interconnection among various electronic components and how they work with one another or with other electronic circuits to perform an electronic function. The components are assigned reference designators to distinguish among similar symbols.

To understand the operation of such a circuit schematic, one must first be able to identify the electronic symbols used in the schematic. Second, from the interconnections, one must identify the electronic circuit formed by the parts, and then its behavior can be understood. It is very helpful to learn to interpret a circuit schematic. An electronic circuit schematic is a powerful means of communicating electronic information.

Objectives

After completing this chapter, you should be able to

1. Identify various graphic symbols used by OrCAD to represent electronic parts and their function on a schematic
2. Identify the file structure required for a design
3. Describe the various SDT tools and their functions
4. Configure the SDT tool set
5. Describe the functions and use all the commands under SDT tool set

6-2. SCHEMATIC GRAPHIC SYMBOLS USED BY OrCAD

Schematics are drawn by using a variety of graphic symbols. Each of these symbols has different meaning to the software. Any number of them could be added to a circuit schematic. The following paragraphs describe the graphic symbols and how they are used to draw electronic circuit schematics.

6-2.1 Parts

Parts are graphic symbols used for such electronic devices as resistors, capacitors, diodes, transistors, and integrated circuits (IC), etc. Since there

116   Chaper 6 / SCHEMATIC DESIGN TOOL SET

are many different types of ICs, they cannot always easily be distinguished by their symbols. They are usually identified by their unique part number. However, designers familiar with electronic parts can identify them by their number and types of inputs and outputs. Symbols for various schematic parts are shown in Figure 6–1.

6–2.2 Wires

Wires are graphic objects used to make connection among pins of parts, power, and ground. They actually represent electrical connection between parts. Figure 6–2 shows wire connection between output pin 3 of 74LS08 and pin 1 of 74LS32, and between output pin 3 of 74LS32 and input pin 2 of 74LS74.

6–2.3 Buses

Buses are graphics objects almost like wires, but a little wider, to help distinguish them from a single wire. They are used to represent a number of wires carrying almost the same types of electrical signals, such as a microprocessor's 16 bit address bus or an 8 bit data bus. A bus is used to represent these signals on a schematic. Figure 6–3 shows a 10 bit address and 8 bit data bus of a 10H8 memory module.

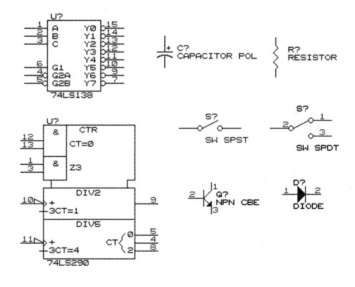

Figure 6–1. Symbols for Various Schematic Parts

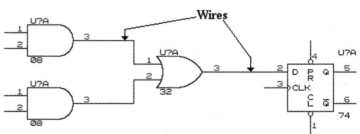

Figure 6–2. Wires Connecting Schematic Parts

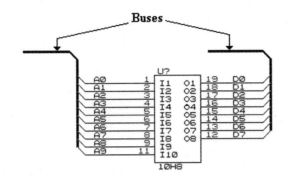

Figure 6–3. Bus Representation

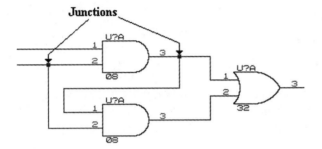

Figure 6–4. Valid Wire Junctions

6–2.4 Junctions

Junctions are graphic symbols that indicate a connection between two or more wires. When two wires cross each other, OrCAD treats them as unconnected wires without a junction box at the point of intersection. When two wires overlap each other, they are also treated as having no connection. However, when two wires meet with each other at a point, they are considered connected without a junction box. When more than two wires meet at a point, a junction is necessary, or they are considered unconnected. Figure 6–4 shows an example of valid wire junctions. More about valid wire connections will be described in the schematic design process section.

6–2.5 Power and Ground

Power and Ground are graphic symbols that are used to connect power and ground, respectively, to electronic devices. Generally, for an OrCAD library part, power and ground pins are hidden from the drawing. To connect to these hidden pins of power and ground from the outside world, many methods are used. Figure 6–5 shows three examples of power and ground connections. Since power and ground symbols are global in scope, these connections mean pins that are labeled VSS and VCC are connected respectively to GND and +5V of the power supply circuit. On some worksheets within a design, the power supply can be isolated by connecting it to module ports. The sheet on which it appears means that all the VCC pins on that sheet are connected to a module port called BACKUP. The BACKUP is a module port in a special power supply. Before you define your power and

Figure 6–5. Various Types of Power and Ground Connections

Figure 6–6. Four Kinds of Module Ports

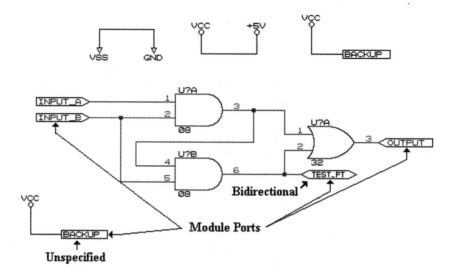

ground connections using this symbol, you must know how they are actually labeled by the device, such as VCC, VSS, or some other label. One way to find that information is by browsing through the same part in the library.

6–2.6 Module Ports

Module ports are graphic symbols used to represent input, output, and bidirectional electronic signals of a schematic. They are also used to conduct signals from one schematic to another. There are four types of module ports available in OrCAD/SDT: input, output, bidirectional, and unspecified. Figure 6–6 shows the four different module ports.

6–2.7 Sheet Symbols

Sheet symbols are block-shaped symbols representing another schematic worksheet. Each sheet symbol in a circuit schematic represents a subsheet schematic belonging to the circuit it appears on. Figure 6–7 shows a part of a schematic with a sheet symbol. In this case the filename of the counter schematic is COUNTER.SCH, meaning that this is the filename for the schematic representing the sheet symbol.

6–2.8 Labels

Labels are identifiers placed on a schematic that can be electrically connected to another signal or signals without actually showing the connection on the schematic. A bus identifier is an example of that. Buses are identified on a schematic by using labels that are typically incremental in nature. By using labels for common connections such as power and ground, you can reduce the number of wires on a schematic. Figure 6–8 shows an example of such a connection.

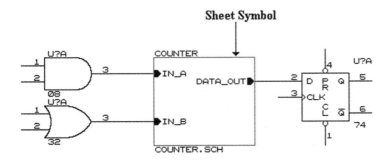

Figure 6–7. Sheet Symbol

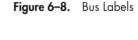

Figure 6–8. Bus Labels

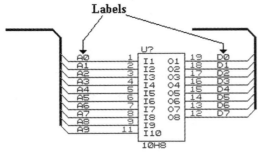

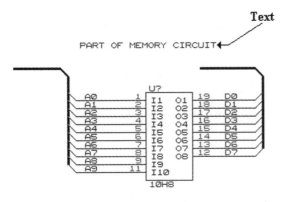

Figure 6–9. Text in a Circuit

6–2.9 Text

Text can be placed in a schematic worksheet. Text can be used to describe a certain function of the circuit operation, making the circuit schematic more readable and understandable. Text does not interfere with electrical operation of the circuit. Figure 6–9 shows an example of text in a schematic circuit.

6–2.10 Title Block

A title block is used to label the schematic worksheet for the purpose of proper documentation. It contains such information as name, address, drawing title, drawing number, worksheet size, date, and version. Figure 6–10 shows two kinds of title blocks.

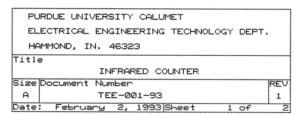

Regular title block

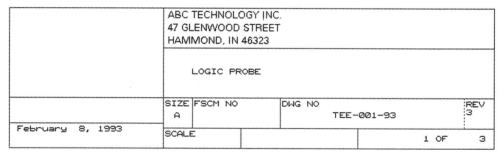

IEEE standard title block

Figure 6-10. Title Blocks

6-2.11 Custom Symbols

Custom symbols of reasonable size, shape, and complexity can be created by the Schematic Design tool set. Once they are created, they should be saved in an appropriate library such that they can retrieved when necessary for a schematic. Custom symbols should be saved in a custom library and must never be stored in an OrCAD-provided library. Figure 6–11 shows several custom symbols.

6-3. STRUCTURING SCHEMATIC FILES

Some electronic circuits are small enough to be drawn entirely on a single schematic worksheet. However, there are also electronic designs that may not fit even on the largest worksheet available with OrCAD/SDT. In the latter case the electronic design must be partitioned into two or more worksheets. Even if a design can fit on one sheet, there may be many practical reasons for partitioning it into several worksheets. These reasons could be numerous, but some common ones are (a) to organize an electronic design by its functional parts and (b) to organize such that several people can work on the design at the same time. Whatever the reasons, there are two ways that multiple sheet designs are handled in OrCAD/SDT: Flat File Design and Hierarchical File Design. (The procedure of creating a flat or hierarchical file structure for a design will be discussed later in this book, along with practical design examples.)

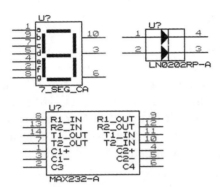

Figure 6-11. Custom Symbols

6-3.1 Flat File Design

A comparatively simple electronic circuit no more than five sheets in size is generally structured as a flat file design. In this type of structure, design files are electronically linked by module ports. Module ports having identical names are considered to be electrically connected. A group of schematics belonging to a design are connected together by listing them using a **LINK** command on the root schematic worksheet. List the filenames as text anywhere on the root worksheet as follows:

¦ LINK
¦ CIRCUIT1.SCH
¦ CIRCUIT2.SCH
¦ CIRCUIT3.SCH

The command symbol ("¦") must be placed before each schematic filename and the **LINK** command so that OrCAD treats them as commands instead of text. For the purpose of reference, a text note on each schematic worksheet may be added. Figure 6-12a shows a link pattern of flat file design structure. Figure 6-12b shows three flat file structure worksheets with the **LINK** command on the root worksheet.

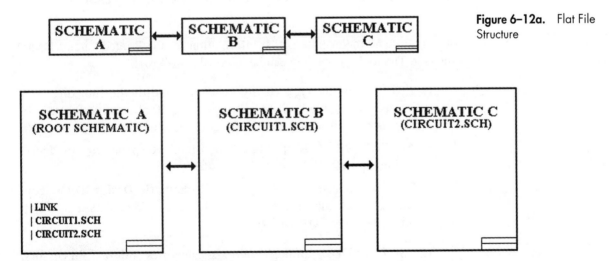

Figure 6-12a. Flat File Structure

Figure 6-12b. Flat File Structure Worksheets with LINK Command

6–3.2 Hierarchical File Design

For a large, repetitive, and complex electronic design problem, a hierarchical file structure is generally used. In this type of file structure, schematics are arranged in the form of an upside-down tree. This layered arrangement is created by sheet symbols. By using a sheet symbol a schematic is placed within another schematic. Two types of hierarchical file structures can be created. They are the simple hierarchical design file structure and the complex hierarchical design file structure.

Simple hierarchical. A simple hierarchical design structure interconnects schematic sheets in an inverted tree-like pattern vertically and horizontally. One root sheet contains symbols representing other worksheets called subsheets. The file structure is such that there is one-on-one correspondence between sheet symbols and the schematic diagrams they reference. This type of file structure is called *simple hierarchy*. Figure 6–13a shows the linking pattern of a simple hierarchical file structure of an arbitrary design example. Figure 6–13b shows a real example of a simple hierarchical design.

Complex hierarchical. A design in which a logical form is such that a particular schematic must be used in several places and thus two or more sheet symbols reference a single worksheet is called a *complex hierarchical design file structure*. These are more sensible for a designer to create because repetitive circuitries are drawn only once. Figure 6–14a shows the linking pattern of a complex hierarchical file structure of an arbitrary design example. Figure 6–14b shows a real example of a complex hierarchical design.

6–4. DESCRIPTION OF THE SCHEMATIC DESIGN TOOL SET

The Schematic Design tool set is primarily used to design electronic circuit schematics and to generate netlist information for the Printed Circuit Board tool set. The tool set window can be invoked as follows:

1. At C\> prompt, type cd myorcad or name of the root directory instead of myorcad where all the OrCAD tool sets are located. In response to this action, C:\MYORCAD> will appear on the screen.
2. At C:\MYORCAD> prompt, type orcad. Response to this action will bring up ESP_MENU.
3. Place the mouse prompt inside the **Schematic Design Tools** icon, and left-click the mouse or press the [Enter ⏎] key. This action will bring about the SDT_MENU.

The SDT_MENU is arranged into six categories: Editors, Processors, Librarians, Reporters, Transfers, and User. Each of these subdivisions has several tools. To invoke any of these tools, perform the following operation:

6-4. Description of the Schematic Design Tool Set

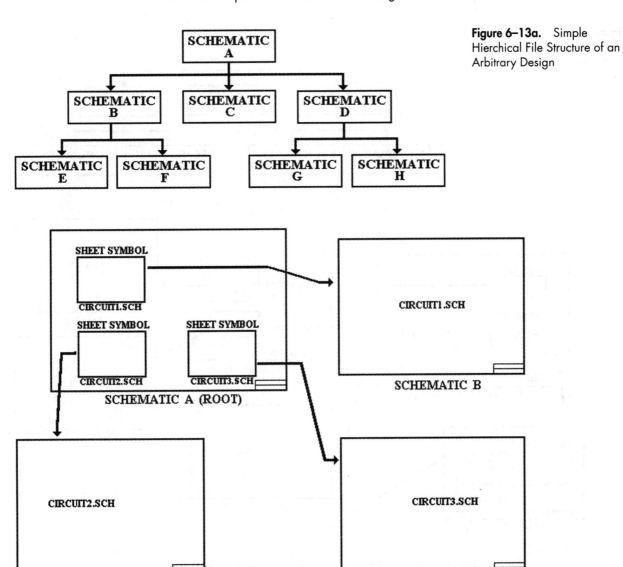

Figure 6-13a. Simple Hierchical File Structure of an Arbitrary Design

Figure 6-13b. Simple Hierchical File Structure of a Real Design

1. Place the arrow prompt of the mouse on the icon button.
2. Left-click the mouse or press the ⌈Enter ↵⌋ key.
3. Select **Execute** from the Transit Menu and left-click the mouse once again. The above action will either execute the respective operation of the tool or bring about another menu screen to choose from.

With the exception of **Edit File, View Reference Material, To Main,** and **User 1** through **User 4,** all tools have local configuration. The local configuration associated with each of the tools will alter the behavior of the same tool in many different ways. More details of each of these tools and how they are used to design and process schematics will be discussed later in this chapter. The functions of each tool on the SDT screen will be discussed next to give designers a clear idea about which tool to use under what circumstances.

Figure 6–14a. Complex Hierarchical File Structure of an Arbitrary Design

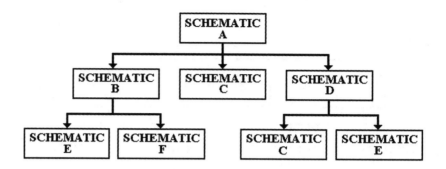

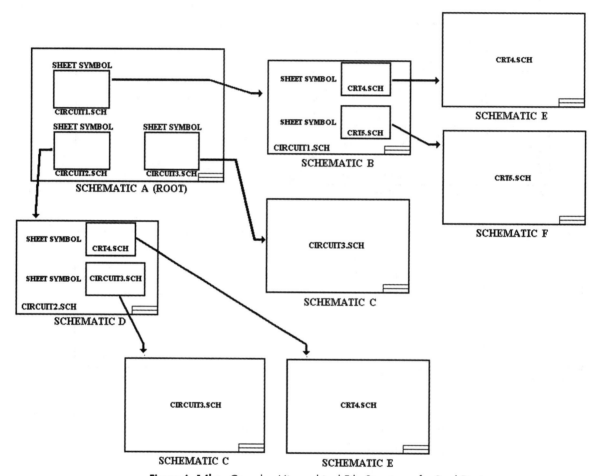

Figure 6–14b. Complex Hierarchical File Structure of a Real Design

6-4.1 Editors

Editors has three tools: **Draft, Edit File,** and **View Reference Material.** Draft is used to draw, edit, and view circuit schematics. **Draft** has a menu called SDT_MAIN. Most of these tools have several layers of commands. Figure 6–15 shows the command layers for **PLACE**. A tool such as **GET** is used to fetch schematic parts from the library, and **DELETE** is used to

6-4. Description of the Schematic Design Tool Set   125

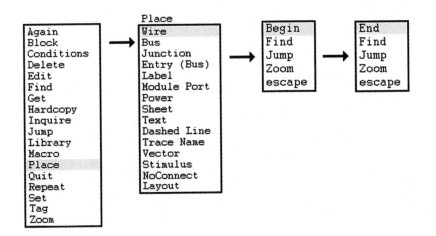

Figure 6-15. Command Layers for PLACE

delete parts, wires, and junction boxes on a schematic. More details about Draft tools will be apparent when you use it to draw schematics.

Edit File tool is a general purpose, full-screen text editor. OrCAD calls this M2EDIT. The editor can be accessed from many different places in OrCAD tool sets.

View Reference Material tool will allow the designer to view reference material regarding the tools provided by OrCAD.

6-4.2 Processors

Processor has seven tools: **Annotate Schematic, Update Field Content, Create Netlist, Create Hierarchical Netlist, Back Annotate, Select Field View,** and **Cleanup Schematic.**

Annotate and unannotate schematic. Annotate Schematic automatically assigns device numbers and reference designators to parts in a schematic. Often times, two, four, or six similar electronic devices are packaged in a single DIP (Dual-In-Line Package) chip. Four AND gates are packaged in a single 74LS08 TTL IC chip. When an AND gate is retrieved from the TTL library, only one gate comes on the screen. Depending on which gate of the 74LS08 chip is selected, the reference designator for the gate could be U?A or U?B, etc. Similarly, when a resistor or a capacitor is retrieved, it comes as R? or C?, respectively. Let us suppose that there are eight TTL AND gates and twelve TTL NOT gates retrieved from the library for a circuit. The default reference designator for all of them is U?A. The designer could edit each part and assign reference designators for the AND gates as U1A, U1B, U1C, U1D, and U2A through U2D, and for NOT gates as U3A through U3F and U4A through U4F. But editing each part is a tedious process. The **Annotate Schematic** tool will automatically do this for a circuit. Figure 6–16a shows a schematic before processing by Annotate Schematic, and Figure 6–16b shows the same schematic after processing by Annotate

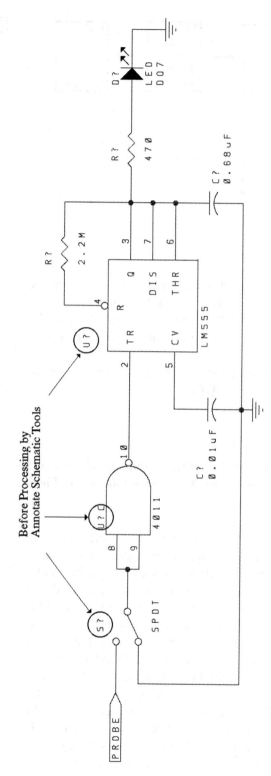

Figure 6-16a. Schematic before Processing by Annotate Tools

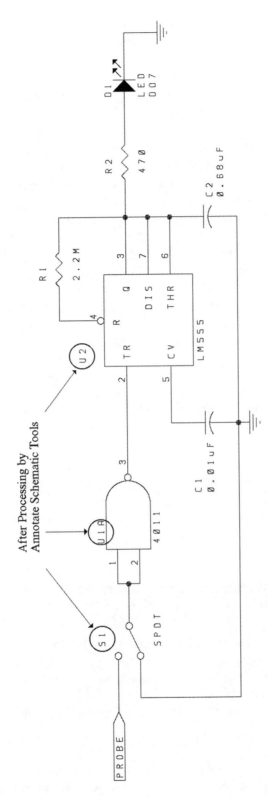

Figure 6-16b. Schematic after Processing by Annotate Tools

Schematic. The tool will annotate all the parts of a schematic and provide a unique reference designator to each of them.

Under the local configuration of the Annotate Schematic tool, there is a processing option called **Unannotate.** When this option is selected for a schematic, components are returned back to their original reference designator, such as R?, C?, or U?A.

Update field content. Each part (or symbol) in a schematic diagram has ten Part Fields. The first two Part Fields contain information about reference and value of a part. The other eight could contain user-defined information such as types of packaging, power/voltage/current rating, and any other pertinent information about the part. The packaging information is essential and is used by the PC Board Layout Tools. There are two different ways this information can be inserted in Part Fields. One way is editing each part separately, and another is to use a tool called **Update Field Content**.

The first method, editing each part and inserting the desired information in the Part Fields, is a slow and inefficient way to do this if your schematic contains a large number of parts. The schematic shown in Figure 6–17a when stuffed with packaging information will look like the schematic shown in Figure 6–17b. Each part symbol of the schematic of Figure 6–17a has only the part reference and the associated part value. After passing the schematic through the **Update Field Content** tool, the packaging ID, such as RC07, RC07, 14DIP300, or D07, is affixed to each part symbol.

Create netlist. This tool generates a file that contains connectivity and packaging information of electronic parts of a schematic. Netlist in OrCAD/PCB format is the interface between OrCAD/SDT and OrCAD/PCB tool sets. By using this tool, a netlist file can also be generated in various industry-accepted formats. Creation of a netlist file is an essential step toward the PC board artwork design process. Figure 6–17c shows the netlist file of the schematic shown in Figure 6–17b. The procedure for generating a netlist file will be described later.

Create hierarchical netlist. This tool is used to produce a hierarchical formatted netlist for a simple hierarchical file structure. Complex hierarchical design must be converted to simple hierarchical before this tool can be used. Complex hierarchical can be transformed into simple hierarchical by using the **Complex to Simple** tool available with the Design Management tool set.

Back annotate. This tool scans a design or a single sheet schematic and updates part reference designators. To run this tool, a text file containing a list of old and new designators is required. Let us suppose the reference designators R1, R2, and U2 of the schematic of Figure 6–17b need to be

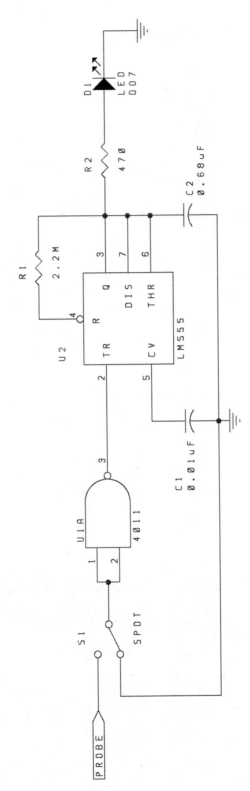

Figure 6-17a. Schematic before Processing by Update Field Content

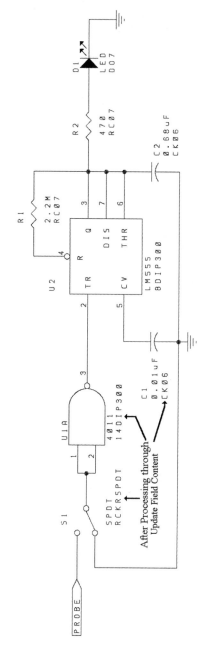

Figure 6-17b. Schematic after Processing by Update Field Content Tools

6-4. Description of the Schematic Design Tool Set

Figure 6-17c. Nestlist of Figure 6-17b

```
({OrCAD/PCB II Netlist Format
DELUXE LOGIC PROBE

Time Stamp - }
 (C563B41F CK06 C1 0.01UF)
  ( 1 N00006)
  (2 GND)
)
 (C563B41E CK06 C2 0.68UF
  (1  N00002)
  (2 GND)
)
 (C563B425 RC07 R1 2.2M
  (1 N00001)
  (2 N00002)
)
 (C5639E11 14DIP300 U1 4011
  (1 N00003)
  (2 N00003)
  (3 N00004)
  (4 ?1)
  (5 ?2)
  (6 ?3)
  (7 VSS)
  (8 ?4)
  (9 ?5)
  (10 ?6)
  (11 ?7)
  (12 ?8)
  (13 ?9)
  (14 VDD)
)
 (C563B42C RC07 R2 470
  (1 N00002)
  (2 N00005)
)
 (C563B42B DO7 D1 LED
  (1 N00005)
  (2 GND)
)
 (C5639E10 8DIP300 U2 LM555
  (1 GND)
  (2 N00004)
  (3 N00002)
  (4 N00001)
  (5 N00006)
  (6 N00002)
  (7 MOD_PORT)
  (8 VCC)
)
 (C563B42E RCKRSPDT S1 SPDT
  (1 GND)
  (2 N00003)
  (3 PROBE)
)
)
```

changed to R7, R6, and U4, respectively. The **Back Annotate** tool will reassign the reference designators if an ASCII file with the following information is provided to the tool:

R3 R7

R2 R6

U2 U4

Figure 6–18 shows the schematic after processing by the **Back Annotate** tool.

Select field view. By using **Select Field View**, the visibility of a part's specified field attributes can be changed. This means that reference designator, part value, or any part field designator in a schematic can be made invisible or visible. Only one type of attribute of a schematic can be processed at one time. If two of the attributes of parts in a schematic need to be made invisible, the schematic needs to be processed twice. Making the attributes of parts invisible never means deleting the attributes. The attribute of the part can be made visible by processing the schematic once again for visibility.

Cleanup schematic. The **Cleanup Schematic** tool scans an entire design or a single worksheet and removes the duplicate or overlapping wires, buses, and junctions. However, the tool is not capable of checking objects overlapping part leads, wires overlapping buses, or wire bus entries overlapping bus entries.

6–4.3 Librarians

Librarians has five tools: **Edit Library, List Library, Compile Library, Decompile Library,** and **Archive Parts in Schematic.**

Edit library. **Edit Library** is a graphic editor tool. This tool is used to create new schematic parts and to edit existing parts in the OrCAD schematic part library. Two types of activity are not recommended unless the user is absolutely sure about the consequence of the act. These are modification of existing library parts and storing them under the same name in the library, and modification of library parts and storing them in one of the schematic part libraries provided by OrCAD. The first action will modify original schematic parts of OrCAD library. The second action may cause you to lose the effort spent in modifying and creating a new part by loading the library update from OrCAD. Therefore, it is always recommended to store any new and modified part into a custom library.

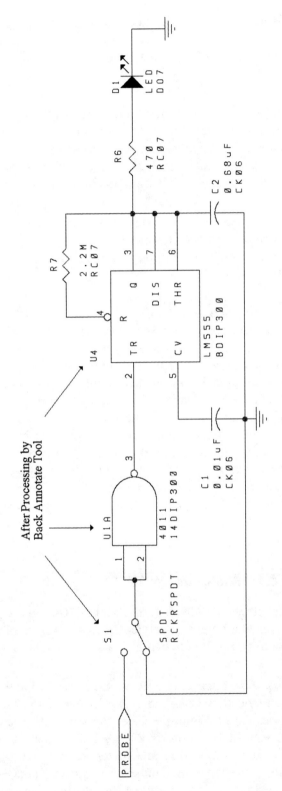

Figure 6–18. Schematic after Processing by the Back Annotate Tool

List library. The **List Library** tool will list the names of the schematic parts of a selected library. The list can be seen on a text screen by first writing the list in a text file and then viewing the file by using the **Edit File** tool.

Compile library. The **Compile Library** tool compiles a library source file. The uncompiled library source of a part is a text file. The schematic editor can fetch only those parts whose compiled library file is available in the database. Tools available with Editor, Processor, Reporter, and Transfer can use only the compiled version of the library file.

Decompile library. **Decompile Library** will construct the library source code of a schematic part. The source of a library part is a text file created with a text editor. Compiling and decompiling are two opposite activities.

Archive parts in schematic. **Archive Parts** scans a design or a single worksheet schematic, takes all the schematic symbols, and stores them in a single library file dedicated to the schematic. With the proper path name, this file can be sent to the current design directory. By providing the proper name and extension only, this library file could be used to display the schematic. For a large schematic file where parts are fetched from many different libraries, all of the libraries must be configured to display the schematic. This situation can be avoided by archiving the parts of a schematic into a single library. This tool is especially useful for systems where memory has to be used very efficiently. Also, the Schematic Design file along with the archived library file can be used by an SD tool residing in another microcomputer system to display the schematic without loading any of the libraries.

6–4.4 Reporters

Reporters has seven tools: **Cross Reference Parts, Create Bill of Materials, Check Electrical Rules, Show Design Structure, Convert Plot to IGES, Plot Schematic,** and **Print Schematic.**

Cross reference parts. **Cross Reference Parts** scans through specified schematic files and generates a listing of the location of each part. The tool can scan a multiple-sheet file structure or a one-sheet file structure. It is especially useful for a large and multisheet design. When this tool is used to scan a design, it will report parts with identical part reference, parts with the wrong number of parts per package, and different part values within the same package. In a large design, these problems are difficult to detect.

Create bill of materials. **Create Bill of Materials** generates a list of all parts used in a design or in a single-sheet schematic. Special information for a part can also be included in the list by configuring the **Key Fields** located under the **Configure Schematic Design Tools** screen.

Check electrical rules. **Check Electrical Rules** scans the design, checks, the validity of the basic electrical rules, and reports the problems back into a destination file. Electrical problems generally reported are unused inputs of parts, two part pins defined as outputs connected together, power and ground connected together, etc. Conditions to be checked are specified in the Electrical Rules Matrix located in the **Configure Schematic Design Tools** screen. Electrical errors reported must be corrected before proceeding further with the design process. However, a warning can be ignored.

Show design structure. **Show Design Structure** scans the schematic of a design and generates a report. The report shows sheet names, their associated worksheet filenames, and the date of their creation. If the structure of the design in question is hierarchical, the Descent into Sheet Path Parts processing option needs to be selected. Without this, the tool will consider the sheet path as a part, and complete structural evaluation will not be done.

Convert plot to IGES. **Convert Plot** translates a plot file of a single worksheet schematic or a complete design into Initial Graphic Exchange Specification (IGES) data format. IGES is an application-independent data format in text form.

Plot schematic. The **Plot Schematic** tool can be used to plot an entire design, including the schematic associated with each sheet symbol. The tool allows proper scaling of the schematic. The plotter tool outputs vector commands to the output device. A vector is a series of points with a specific function defined.

Print schematic. The **Print Schematic** tool is used to print one copy of the entire design. If a schematic is referenced more than once by sheet symbols, only one of them will be printed. The printer tool outputs raster commands to the output device. A raster is an array of dots.

6–5. CONFIGURATION OF THE SCHEMATIC DESIGN TOOL SET

The configuration screen of the Schematic Design tool set can be accessed from several places, one of which is Draft's Transit Menu. This configuration process allows the designer to set many global configuration parameters for the tool set. The configuration screen is quite long and cannot be seen at one time. The mouse or **PageUp** and **PageDown** keys are used to travel through the configuration screen. Generally, the mouse is more convenient to use. The configuration screen is divided into several chunks: **Driver Options, Printer/Plotter Output Option, Library Option, Worksheet Option, Macro Options, Hierarchy Options, Color and Pen Plotter Table, Template Table, Key Fields,** and **Check Electrical Rules Matrix.** Figures 6–19a through Figure 6–19g show the configuration screen for Schematic Design Tools.

6–5.1 Driver Options

Driver Prefix is the directory path or disk drive where Schematic Design Tools finds and loads the display, printer, and plotter drive. Driver Prefix is set during the installation process; therefore, if everything is running, do not disturb it needlessly. Display, printer, and plotter driver can be searched using the up/down arrow keys, selected, and then configured using the Enter ↵ key.

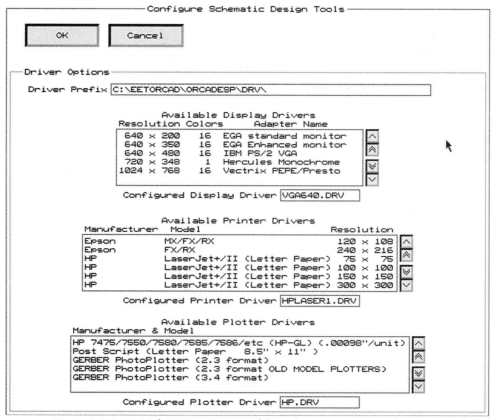

Figure 6–19a. Configuration Screens of the Schematic Design Tools (Courtesy of OrCAD, Inc.)

b.

```
┌─Printer/Plotter Output Options──────────────────────────────────────┐
│ Printer Port   ●LPT1: ○LPT2: ○LPT3: ○COM1: ○COM2: ○COM3: ○COM4:     │
│                    Baud Rate        ○No Parity    ○8 Data Bits      │
│                   ○300   ○4800      ○Odd Parity   ○7 Data Bits      │
│                   ○1200  ○9600      ○Even Parity                    │
│                   ○2400  ○19200                   ○1 Stop Bit       │
│                                                   ○2 Stop Bits      │
│                                                                     │
│ Plotter Port   ○LPT1: ○LPT2: ○LPT3: ●COM1: ○COM2: ○COM3: ○COM4:     │
│                    Baud Rate        ●No Parity    ●8 Data Bits      │
│                   ○300   ○4800      ○Odd Parity   ○7 Data Bits      │
│                   ○1200  ●9600      ○Even Parity                    │
│                   ○2400  ○19200                   ●1 Stop Bit       │
│                                                   ○2 Stop Bits      │
└─────────────────────────────────────────────────────────────────────┘
┌─Library Options─────────────────────────────────────────────────────┐
│  Library Prefix  C:\EETORCAD\ORCADESP\SDT\LIBRARY\*.LIB             │
│                                                                     │
│       Available          ● Insert a Library      Configured         │
│       Libraries          ○ Remove a Library      Libraries          │
│      ┌─────────────┐                            ┌─────────────┐     │
│      │ ALTERA_M.LIB│         ┌───────────┐      │ ANALOG.LIB  │     │
│      │ ALTERA_P.LIB│         │ > Insert >│      │ DEVICE.LIB  │     │
│      │ AMD.LIB     │         └───────────┘      │             │     │
│      │ ANALOG.LIB  │                            │             │     │
│      │ ANALOG1.LIB │         ┌───────────┐      │             │     │
│      │ ANALOG2.LIB │         │ < Remove <│      │             │     │
│      │ ANALOG3.LIB │         └───────────┘      │             │     │
│      │ ANALOG4.LIB │                            │             │     │
│      │ ANLG_DEV.LIB│                            │             │     │
│      │ ASSEMBLY.LIB│                            │             │     │
│      └─────────────┘                            └─────────────┘     │
│                                                                     │
│ Name Table Location ●Main memory   Symbolic Data Location ○EMS memory│
│                     ○EMS memory                           ●Disk      │
│                     ○Disk                                            │
│ Active library size 64                                               │
└─────────────────────────────────────────────────────────────────────┘
```

c.

```
┌─WorkSheet Options───────────────────────────────────────────────────┐
│    ☐ ANSI title block                                               │
│    ☐ ANSI grid references                                           │
│    ☐ Use alternate worksheet prefix                                 │
│         Worksheet Prefix  [                              ]          │
│    Default worksheet file extension  SCH                            │
│    Sheet size          A                                            │
│    Document number    [                                  ]          │
│    Revision           [   ]                                         │
│    Title              [                                  ]          │
│    Organization name  [                                  ]          │
│    Organization address[                                 ]          │
│                       [                                  ]          │
│                       [                                  ]          │
│                       [                                  ]          │
└─────────────────────────────────────────────────────────────────────┘
┌─Macro Options───────────────────────────────────────────────────────┐
│ Macro Buffer Size 8192                                              │
│ Draft Macro File   [                                     ]          │
│ Draft Initial Macro[                    ]                           │
│ Edit Library Macro File    [                             ]          │
│ Edit Library Initial Macro [                 ]                      │
└─────────────────────────────────────────────────────────────────────┘
┌─Hierarchy Options───────────────────────────────────────────────────┐
│ Hierarchy Buffer Size 1024                                          │
└─────────────────────────────────────────────────────────────────────┘
```

d.

Color and Pen Plotter Table

Item	Pen	Width	Speed
Part Body	1	.010	DEF
Pin Number	1	.010	DEF
Pin Name	1	.010	DEF
Part Reference	1	.010	DEF
Part Value	1	.010	DEF
1st Part Field	1	.010	DEF
2nd Part Field	1	.010	DEF
3rd Part Field	1	.010	DEF
4th Part Field	1	.010	DEF
5th Part Field	1	.010	DEF
6th Part Field	1	.010	DEF
7th Part Field	1	.010	DEF
8th Part Field	1	.010	DEF
Wire	1	.010	DEF
Bus	1	.010	DEF
Junction	1	.010	DEF
Power Object	1	.010	DEF
Power Text	1	.010	DEF
Sheet Body	1	.010	DEF
Sheet Name	1	.010	DEF
Sheet Net	1	.010	DEF
Module Port	1	.010	DEF
Module Text	1	.010	DEF
Label	1	.010	DEF
Comment Text	1	.010	DEF
Dashed Line	1	.010	DEF

e.

Item	Pen	Width	Speed
Title Block	1	.010	DEF
Title Text	1	.010	DEF
Command Prompt	1	.010	DEF
Grid Dots	1	.010	DEF
Trace Object	99		
Test Vector Obj	99		
Stimulus Object	99		
Error Object	99		
No Connect Obj.	1	.010	DEF
Layout Object	99		
Sheet Filename	1	.010	DEF

Template Table

Units: ● Inches ○ Millimeters

	A	B	C	D	E
Horizontal	9.700	15.200	20.200	32.200	42.200
Vertical	7.200	9.700	15.200	20.200	32.200
Pin-to-Pin	.100	.100	.100	.100	.100
Pin Number	.060	.060	.060	.060	.060
Pin Name	.060	.060	.060	.060	.060
Part Reference	.060	.060	.060	.060	.060
Part Value	.060	.060	.060	.060	.060
1st Part Field	.060	.060	.060	.060	.060
2nd Part Field	.060	.060	.060	.060	.060
3rd Part Field	.060	.060	.060	.060	.060
4th Part Field	.060	.060	.060	.060	.060
5th Part Field	.060	.060	.060	.060	.060
6th Part Field	.060	.060	.060	.060	.060
7th Part Field	.060	.060	.060	.060	.060
8th Part Field	.060	.060	.060	.060	.060

f.

Power Text	.060	.060	.060	.060	.060
Sheet Name	.060	.060	.060	.060	.060
Sheet Net	.060	.060	.060	.060	.060
Module Text	.060	.060	.060	.060	.060
Label	.060	.060	.060	.060	.060
Comment Text	.060	.060	.060	.060	.060
Title Block	.060	.060	.060	.060	.060
Border Text	.060	.060	.060	.060	.060
X Border Width	.100	.100	.100	.100	.100
Y Border Width	.100	.100	.100	.100	.100
Plot X Offset	.000	.000	.000	.000	.000
Plot Y Offset	.000	.000	.000	.000	.000
Roll Form Size	.000	.000	.000	.000	.000
Spacing Ratio	1.333	1.333	1.333	1.333	1.333

―Key Fields――――――
Annotate Schematic
    Part Value Combine   [   ]

Update Field Contents
    Combine for Value   [   ]
    Combine for Field 1 [V   ]
    Combine for Field 2 [   ]
    Combine for Field 3 [   ]
    Combine for Field 4 [   ]
    Combine for Field 5 [   ]
    Combine for Field 6 [   ]
    Combine for Field 7 [   ]
    Combine for Field 8 [   ]

g.

Create Netlist
    Part Value Combine   [V   ]
    Module Value Combine [1   ]

Create Bill of Materials
    Part Value Combine   [V   ]
    Include File Combine [   ]

Extract PLD
    PLD Part Combine   [   ]
    PLD Type Combine   [   ]

―Check Electrical Rules Matrix―――

[Set to Defaults]

		in	i/o	out	oc	pas	hiz	oe	pwr	mI	mO	mB	mU	sI	sO	sB	sU	NC
Input Pin	in												W				W	W
Input/Output Pin	i/o		W				W	W	W				W				W	
Output Pin	out			W	E		E	E	E				W			E	W	
Open Collector Pin	oc			E			W	E	E				W			E	W	
Passive Pin	pas																	
High Impedence Pin	hiz			E	W			E	E				W			E	W	
Open Emitter Pin	oe			E	W		E	W	E				W			E	W	
Power Pin or Object	pwr		W	E	E		E	E					W			E		
Input Module Port	mI		W	E	E		E	E	E				W			E	W	W
Output Module Port	mO		W				E	E					W			E	W	
Bidirectional Module Port	mB		E				E	E					W				W	
Unspecified Module Port	mU	W	W	W	W		W	W	W	W	W	W		W	W	W	W	
Input Sheet Net	sI												W				W	W
Output Sheet Net	sO		W	E	E		E	E	E				W			E	W	
Bidirectional Sheet Net	sB		E				E	E					W				W	
Unspecified Sheet Net	sU	W	W	W	W		W	W	W	W	W	W		W	W	W		
Unconnected	NC	W					W						W				W	

6-5.2 Printer/Plotter Output Option

Output ports for printer and plotter are selected from this part of the configuration screen. LPT1:, LPT2:, and LPT3: are parallel ports generally used for printers. COM1:, COM2:, COM3:, and COM4: are generally used for plotters. In some microcomputer systems BIOS may restrict the usage of COM3: and COM4:.

6-5.3 Library Option

Library Prefix is the dialogue box where the directory path for the location of the schematic parts library is entered. The **Library Option** has two list boxes. **Available Library,** located on the left, contains all the schematic parts libraries available with the OrCAD/SDT tool set, including any custom libraries created by the designer. The list box on the right contains all the **Configured Library.** While drawing, schematic parts can be fetched only from the configured library. The number of configured libraries depends on the system's RAM (main memory). However, a configured library can also be located on a disk or in **Extended Memory (EMS).** A library has two parts: **Name Table,** the names of library parts, and **Symbolic Data,** symbol information of library parts. For a configured library a number of combinations for the location of **Name Table** and **Symbolic Data** can be selected from this part of the screen. The order in which libraries are located in the configured area is important. Whenever a part is retrieved from a configured library, the tool searches the part starting with the library located at the top of the list box and travels to the bottom to find the part. The first part it finds with a matching name is fetched to the draft screen. Different parts with duplicate names in the same or different libraries will cause disagreeable problems.

6-5.4 Worksheet Options

The designer can choose to draw a schematic on worksheet size A through E (American Standard) or A4 through A0 (International Standard). Using this part of the configuration screen, the desirable worksheet size and title block format can be selected.

6-5.5 Macro Options

A macro is a series of commands that executes automatically at the touch of a single key or key combination. Macros are series of keystrokes performed by one keystroke. While you are designing a schematic, macro usage can speed up the design process.

6-5.6 Hierarchy Options

Designs having many worksheets and a hierarchical file structure with large hierarchy buffer size must be large enough to hold hierarchical worksheets and their path names.

6-5.7 Color and Pen Plotter Table

The **Color and Pen Plotter Table** allows the selection of the plotter pen color, pen number, pen size, and pen speed. Colors for part body, pin number, part value, wire, junction, bus, text, and title block can be selected separately from here. There are sixteen different pen colors that can be selected just by putting the pen number in the box. 0 in this box will pause the plotter, and 99 will inhibit the pen from plotting the object. Pen width determines the number of strokes needed to draw a part. Thus, putting a smaller value in the width box and providing a thicker pen on the actual plotter will create thick lines. Pen speed may be important, depending on the type of plotter and quality of ink in the plotter pens.

6-5.8 Template Table

By configuring the **Template Table** parameters, global changes in size and various other parameters of a schematic part, title block, distance between grid dots, or page border can be achieved. Left-clicking inside the entry box will allow the alteration of these values. Do not change the Template Table values randomly; keep them at their default values until you feel comfortable using the software. Change these values only if you are sure about the effect of these changes.

6-5.9 Key Fields

Key Fields configuration is not as straightforward as the others. Electronic circuit designers must understand how Key Fields should be configured and the effect of each different configuration on their outcome.

Every schematic library part has ten **Part Fields.** A part field is a slot for holding text or data associated with that part. These fields can be edited after a part is placed in a schematic. The first field in the group is reserved for holding the part's reference designator. The second field is reserved for the value of the part. Eight other Part Fields are used to hold the designer's optional data related to the parts. For the purpose of printed circuit board design, packaging information of each schematic part is necessary. The first Part Field of each schematic part is usually used to hold the packaging information of the part. Other Part Fields can be used to store useful information regarding the part, such as voltage rating, power rating, and vendor's name. Visibility of each of these eight Part Fields can be controlled by configuring the **Select Field View** tool.

A maximum of 127 characters can be stored in each field. Sixteen Key Fields of the SDT configuration screen are divided into five groups. **Annotate Schematic** uses one, **Update Field Contents** uses nine, **Create Netlist** uses two, **Create Bill of Materials** uses two, and **Extract PLD** uses two. These Key Fields can contain characters **R** and **V,** and numbers **1** through **8.** Character **R** represents values found in the part's **Reference** field, character **V** represents values found in the part's **Part Value** field, and numerals **1** through **8** represent values found in the part's **Part Fields 1** through **8,** respectively. Any combination of these characters, including blank spaces, commas, or any other punctuation, constitutes a text string formed by the characters found in Part Fields.

Example. Consider a schematic file, LAB1.SCH, with three parts: a resistor, a capacitor, and an op-amp. The resistor, R1, has 10K and RC01 in its **Part Value** field and **1st Part Field**, respectively. The capacitor, C1, has 8 µf and RC02 in its Part Value field and 1st Part Field, respectively. The op-amp has 741 and 8DIP300 in its Part Value field and 1st Part Field, respectively. During configuration of Schematic Design Tools, **Combine for Field 2** dialogue box of Key Fields located under **Update Field Contents** was stuffed with two characters, **V** and **1**, separated by a comma (V,I). An ASCII file, LAB1.STF, containing the following information was created:

'10K,RC01'   '1/4 Watts'

'8µf,RC02'   '30WV'

'741,8DIP300'   'Gain 10^3'

When the Update Field Contents tool is executed, it will first form a different text string for each part of the schematic using characters found in the Part Value field and 1st Part Field. Commas are also part of the string characters. Text strings for each of these parts are matched with the text string found in LAB1.STF. The destination of these matched strings is the **2nd Part Field** of each part. This means the 2nd Part Field of 10K resistor will receive 1/4 Watts, 8µf capacitor will receive 30WV, and op-amp 741 will receive Gain 103 because they correspond to Combine for Field 2 of the Key Fields. More about these configurations will be provided later.

6-5.10 Check Electrical Rules Matrix

The configuration of **Check Electrical Rules Matrix** determines the rules to use to check the schematic. A test condition is determined by the intersection of a row and a column. The intersection point of the row and column can either be empty, contain a **W**, or contain an **E**. An empty intersecting point represents a valid connection between the units. A **W** at the intersecting point will generate a warning signal when the units are connected. An **E** represents an error. Any of the intersections can be toggled among empty, **W,** and **E.** Some of the design rules are interrelated; a change in one may change the other also. Do not change them randomly without knowing their behavior thoroughly.

There are default values of electrical rules. To return to default values, left-click on Set to Defaults. In order to understand the matrix, consider a few intersecting points on the matrix. According to the default values of the matrix, when two input pins are connected together, it is considered as a valid connection. When two output pins are connected together, it is considered as an invalid connection and will generate an error.

6-6. COMMON FUNCTIONS OF THE DRAFT TOOL

The schematic editor is the key tool necessary for designing and drafting a schematic of an electronic circuit. This schematic editor is located in SDT_MENU under an icon named **Draft.** In order to invoke this editor screen, simply place the arrow prompt of the mouse inside the **Draft** icon and left-click the mouse. It may be necessary to select an existing schematic file name from the Transit Menu of the Draft editor in order to invoke the schematic. The Draft schematic editor has eighteen commands in the main menu. Many of the commands have submenus and are arranged in the form of several layers of command scheme. These submenus appear along the top, or upper left corner, of the screen while a command is being executed. Figure 6–20 shows commands of the SDT_MAIN menu and their first-level submenus. Commands are described in alphabetical order. Any command in SDT_MAIN menu (which is also the **Draft** menu) can be executed either by entering the capitalized letter of the command from the keyboard or left-clicking the mouse on the command.

6-6.1 AGAIN

AGAIN repeats the last main menu command executed. For example, if the last command executed is **GET,** the command may be repeated by selecting **AGAIN.** Since it is the first command in the menu, it saves time. When a command needs to be repeated several times, instead of going through the main menu and selecting the command, AGAIN can be used. This command does not have any submenus.

6-6.2 BLOCK

The **BLOCK** command is used to manipulate specific areas of a schematic worksheet. By selecting this command, a specified part of a schematic can be moved around the worksheet, duplicated, imported, or exported. A specified part of the schematic can be moved with its wires, which are connected to the components outside the area, by stretching the wires like rubber bands or isolating the part of the schematic by cutting the wires. Figure 6–21 shows the **BLOCK** submenu.

BLOCK Move moves the selected objects from one location to another location of the worksheet. When this command is selected from the submenu, it will prompt to specify the block to be moved. To select the desired part of the schematic, a box needs to be drawn around it. Place the mouse

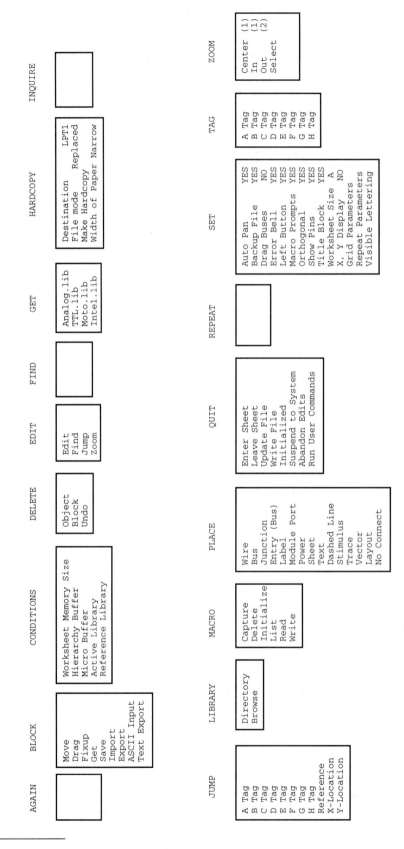

Figure 6-20. SDT_MAIN Command Structure

6-6. Common Functions of the Draft Tool

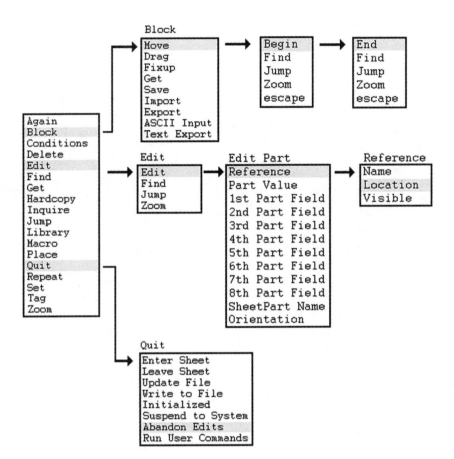

Figure 6-21. Submenu Command Layers for BLOCK

pointer on the corner of the box to be started. Select **Begin** and drag the mouse to create the box. The direction and length of your drag will determine the size and orientation of the box. Left-click the mouse once and select **End** to complete the selection of the part of the circuit to be moved. Now as the mouse pointer is moved, the part of the circuit in the box will move with it. Move the circuit to the desired location and select **Place** to put the part of the schematic at the new location. Figure 6-22 shows a circuit, part of which has been moved using the **BLOCK Move** command.

BLOCK Drag works the same way as BLOCK Move. However, while BLOCK Drag is used to move a part of the circuit schematic from one part of the worksheet to another, the objects will maintain their original connectivity. While moving, the wires or buses will appear to be stretching as the selected block is moved from one place to another. This effect is called rubber banding. Figure 6-23 shows a circuit, part of which is dragged using the BLOCK Drag command. Comparing Figure 6-22 and Figure 6-23 reveals that the **BLOCK Drag** command moved the selected objects with their original connections whereas **BLOCK Move** moved the selected objects without rubber-banding their original connections.

The **BLOCK Fixup** command is used to fix up wires and buses that became non-orthogonal due to **BLOCK Drag.** In order to add new wire or bus segments to make connections orthogonal, select **BLOCK Fixup** from

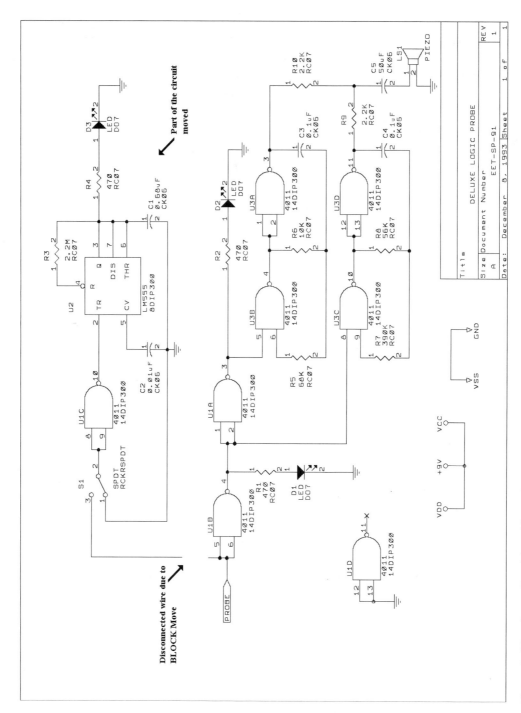

Figure 6-22. Circuit Part Moved by BLOCK Move Instruction

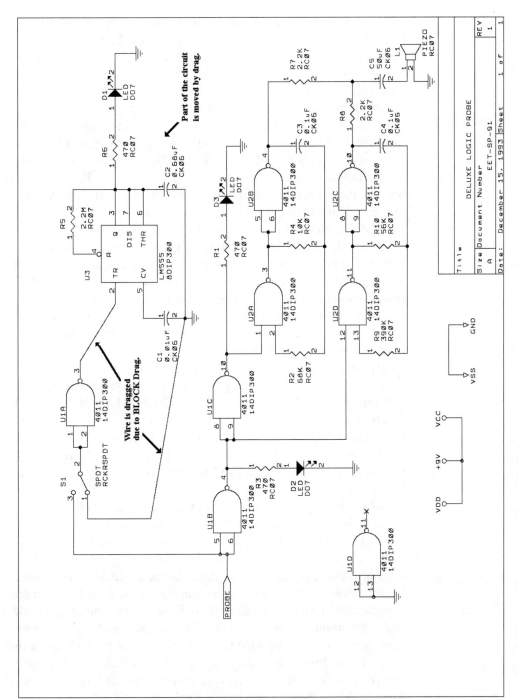

Figure 6–23. Circuit Parts Moved by DRAG Move Instruction

the submenu. Place the mouse arrow prompt on the end of the wire or the bus that needs to be made orthogonal, and select **Pick.** To move all wires or buses meeting at the junction, select **Drag All** or else **Pick One** to move only one wire or bus at a time. The process of fixup is not difficult to perform and can be learned easily by practicing on a circuit.

BLOCK Get retrieves part of the circuit previously saved in a buffer using **BLOCK Save** and places it on the worksheet. Selecting **BLOCK Get** from the BLOCK submenu will retrieve the part of circuit with a pointer attached to it. This part of the circuit can be placed in a desired location on the worksheet by moving the mouse and selecting the **Place** command. The box and the part of the circuit remain on the screen, and the desired number of copies of it can be placed on the worksheet.

BLOCK Save stores a copy of the selected part of the circuit in a buffer. There is only one buffer. Therefore, when a part of a circuit is saved in the buffer, it remains until something else is saved again by using **BLOCK Save.** In order to save a part of a circuit in the buffer, select **BLOCK Save** from the submenu and select **Begin** to draw a box around the part of the circuit to be saved in the buffer. Select **End** when the box is large enough to encompass all the desired parts. The saved part of the circuit can be retrieved by using **BLOCK Get.**

BLOCK Export saves a copy of a part of a circuit in a file. This command can be used to export a part of a circuit to another worksheet or to a different part of the same worksheet. Select **BLOCK Export** from the **BLOCK** submenu, select **Begin** to start creating a box around the part of the circuit to be exported, and then select **End** when the box encloses all of the desired objects. Finally, at the prompted message (Export Filename?), provide the path and filename to which part of the circuit will be exported.

BLOCK Import retrieves part of the circuit that was exported to a file by using **BLOCK Export.** Select **BLOCK Import,** provide path and filename to be imported at the prompted message (File to Import?), position the pointer at the desired place on the worksheet, and select **Place** to place the parts of the circuit on the worksheet. Using BLOCK Import and BLOCK Export, several parts of a circuit schematic can be exported from one worksheet and imported to another. This is very convenient in an electronic design environment. Instead of drawing the same part of a circuit in another worksheet, it can be imported from a worksheet where it already exists.

BLOCK Text Export saves a copy of a selected text in a text file. The text file can be edited with an ASCII editor. Select **BLOCK Text Export** from the **BLOCK** submenu, select **Begin** to create a box around the text to be exported, select **End** when the box encompasses all of the desired text, and at the prompt (Text Export Filename?), provide the path and filename to which it is to be exported.

BLOCK ASCII Import is used to retrieve text that was exported by the **BLOCK Text Export** command. Select **BLOCK ASCII Import** from the **BLOCK** submenu, enter path and filename of the ASCII file to be imported at the prompt (ASCII File to Import?), and select **Place** to position the text at the desired place.

6-6.3 CONDITIONS

The **CONDITIONS** command displays the condition of the computer memory and the memory available for the worksheet, hierarchy buffer, and macro buffer. When the command is selected from the SDT_MAIN screen, a status window showing the memory status is displayed on the screen. The memory allocation cannot be changed from this screen, but can be reassigned from the SDT configuration menu. The **CONDITIONS** command is used only to review the status of the memory allocation. Figure 6–24 shows the memory allocation by the software that is displayed when the CONDITION command is invoked. Reallocation of the system memory can be performed from the SDT configuration.

6-6.4 DELETE

DELETE and its subordinate commands erase objects or blocks of objects. In case of an accidental erasure, the subordinate command **Undo** will restore the deleted object back to its original position. The only time Undo will recover the object after you have deleted it is when Undo is the next immediate command executed after DELETE. **DELETE** has three subordinate commands: **Object, Block,** and **Undo.** Figure 6–25 shows the submenu for DELETE.

DELETE object. **DELETE Object** erases an object on the worksheet from which the command is executed. To delete an object from the worksheet, place the mouse pointer on the object to be deleted, select **DELETE** from the SDT_MAIN menu, select **Object** from the subordinate menu, and select **Delete** from the menu bar at the top of the screen. When an object is deleted from a worksheet, sometimes a few dots are left behind at the location of the object. To get rid of these, click the left button of the mouse once.

DELETE block. **DELETE Block** deletes a block of objects on the worksheet. Select **DELETE** from the SDT_MAIN menu, select **Block** from the subordinate menu, select **Begin**, position the mouse prompt, and draw the box around the part of the circuit to be deleted. Finally, select **End** to delete the objects inside the box.

```
DELETE
┌────────┐
│ Object │
│ Block  │
│ Undo   │
└────────┘
```

Figure 6–25. DELETE Submenu

Figure 6–24. System Memory Allocation

```
Press Enter to Continue ■
CONDITIONS:
                       Location   Allocated      Used   Available
Worksheet Memory Size  Main       ----------     365    163360
Hierarchy Buffer       Main       1024           0      1024
Macro Buffer           Main       8192           128    8064
Active Library         Main       157124         55124  102000
Reference Library
  Name Table           Main       65536          32000  33536
  Symbol Information   EMS        320000         256000 64000
```

DELETE undo. **DELETE Undo** restores the last deletion. This command restores accidentally deleted objects or blocks of objects in the worksheet. If an object is deleted accidentally, select **DELETE Undo**. The object deleted using the last **DELETE Object** or **DELETE Block** command will be restored.

6-6.5 EDIT

EDIT is a versatile command. The command can be used for the following purposes:

1. Modify labels, text, module ports, power objects, sheets, part reference designators, part value, part field, and title block.
2. Change pin names and numbers on devices with multiple parts-per-package.
3. Move part reference designators, part value, and part fields to new locations on the worksheet.
4. Make part reference designators, part values, or part fields visible or invisible.
5. Change the style of parts, power objects, module ports, labels, and text.
6. Change the orientation of parts, text, and labels.
7. Change the size of text and labels.
8. Change sheets, sheet names, sheet nets, sheet net types, and sheet filenames.
9. Add sheet nets or delete sheet nets.

To edit an object on the worksheet, place the mouse pointer on the object to be edited and select **EDIT** from the SDT_MAIN menu located at the upper left corner on the worksheet. Since EDIT is a versatile command, depending on the object, it will display an appropriate subordinate menu. The EDIT submenu is shown in Figure 6–26. If the SDT_MAIN menu is not present at the corner, right-click the mouse a couple of times, left-click it once, and the menu should appear.

Common keys used with EDIT command when editing text

1. Positioning of the cursor can be done by the left arrow ←, right arrow →, **Home**, and **End** keys, or the mouse. In some cases a mouse is too sensitive; therefore, you may need to use the arrows and other keys to perform the operation.
2. Erasing of characters is done with **Backspace** or **Delete** key.
3. New characters are added with the alphanumeric keys.

EDIT labels. To edit a label on the worksheet, place the mouse pointer on the label and select **EDIT** from the SDT_MAIN menu. The **EDIT Label**

6-6. Common Functions of the Draft Tool

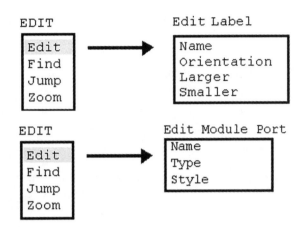

Figure 6-26. EDIT Label Submenu

Figure 6-27. EDIT Module Port Submenu

submenu shown in Figure 6-26 will appear at the upper left corner of the screen. Make the appropriate selection.

EDIT label name. If the **Name** option is selected, the current label name will appear at top of the screen. Common keys used with **EDIT** command (described earlier) for text editing can be used to modify the name to the desired one.

EDIT label orientation. If the **EDIT Label Orientation** option is selected, a menu with two options, **Horizontal** and **Vertical**, will appear at the upper left corner of the screen. If Horizontal is selected, selected text will be horizontally placed on the worksheet. If Vertical is selected, selected text will be placed vertically on the worksheet.

EDIT label larger. If the **EDIT Label Larger** option is selected, character size of the label will be larger. This option may be selected more than once to adjust the size of the label character.

EDIT label smaller. The **EDIT Label Smaller** option will make the character size of the label smaller. Like Larger, this option can also be selected more than once to make the label character a desired size. For both of these options, there is a limit to the size of the character.

EDIT module ports. To edit a module port in a schematic, place the mouse pointer within the boundary of the module port symbol and select **EDIT** from the SDT_MAIN menu. The **EDIT Module Port** submenu shown in Figure 6-27 will appear on the screen with its three options. Next, you need to make the appropriate selection from the submenu for changing the desired attributes of the selected module port.

EDIT module port name. If the **EDIT Module Port Name** option is selected, the existing module port name will appear at the top of the screen. Common keys used with the **EDIT** command for text editing can be used to modify the name to the desired one. After entering the desired name, press

⌜Enter ⏎⌟, and the new module port name will appear inside the module port of the schematic.

EDIT module port type. The **EDIT Module Port Type** option will allow the user to change the type of module port. If this option is selected, the submenu shown in Figure 6–28 will appear at the upper left corner of the screen. Selection of any of the four options will change the existing module port to the new type.

EDIT module port style. The **EDIT Module Port Style** option will change the style of the module port regardless of its type. This means that the bidirectional module port style can be used to represent an input module port. However, this kind of style change may be very confusing, so it is not recommended to change the default type unless there is a pressing need. Figure 6–29 shows the submenu for the **EDIT Module Port Style** option.

EDIT power symbol. Three characteristics of a power or a ground symbol can be edited: **Name, Type,** and **Orientation.** To edit any of the attributes of power and ground symbols, place the mouse pointer on a power or ground object and select **EDIT** from SDT_MAIN menu. The **EDIT Power or Ground** submenu shown in Figure 6–30 will appear in the upper left corner of the screen.

EDIT power or ground name. If the **EDIT Power or Ground Name** option is selected from the **EDIT Power** submenu, the existing name of the

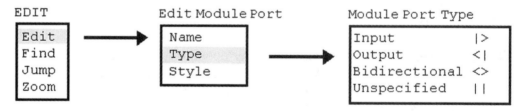

Figure 6–28. EDIT Module Port Type Submenu

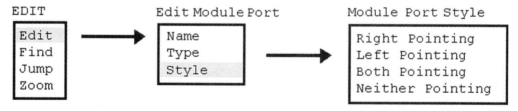

Figure 6–29. EDIT Module Port Style Submenu

6–6. Common Functions of the Draft Tool

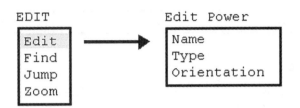

Figure 6–30. EDIT Power Submenu

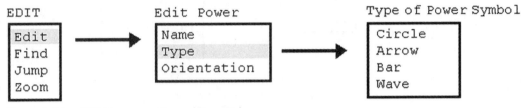

Figure 6–31. EDIT Power or Ground Type Submenu

power or the ground symbol will be displayed. At this time a new name can be entered if so desired.

EDIT power or ground type. If the **EDIT Power or Ground Type** option is selected, the submenu shown in Figure 6–31 will appear with the following four options: **Circle, Arrow, Bar, Wave.** Although **Circle** and **Arrow** are the types commonly used for power and ground symbols, the other types can also be used.

EDIT power or ground orientation. Oftentimes, power and ground symbols of different orientations are necessary to place the appropriate connection onto a schematic. If the **EDIT Power or Ground Orientation** option is selected, a submenu with four choices will appear on the screen: **Top, Bottom, Left,** and **Right.**

EDIT sheet. Sheet symbols are used in hierarchical designs to represent another schematic worksheet. On a sheet symbol, only inputs, outputs, worksheet name, and schematic filename are shown. New inputs and outputs can be added to the sheet symbols or, in other words, to the schematic. To edit a sheet symbol, place the mouse pointer inside it and left-click the mouse. The menu bar shown in Figure 6–32a will appear at the top of the screen. At this time you may use capitalized letters from the options to execute any of the commands, or click the mouse again to change this to the step menu that is shown in Figure 6–32b. The mouse pointer movement will be restricted inside the sheet symbol.

EDIT sheet add-net. To add a new net (i.e., add an input/output connection) to a sheet symbol, move the mouse pointer to the edge of the sheet symbol where you want to place your net, and select **Add-Net** from the sub-

```
Add-Net   Delete   Edit   Name   Filename   Size   Zoom
```

Figure 6–32a. Menu Bar for EDIT Sheet

Figure 6–32b. EDIT Sheet Pull-Down Submenu

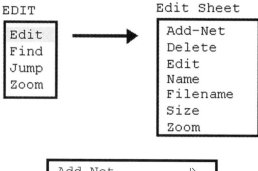

Figure 6–32c. EDIT Add-Net Netname Submenu

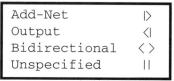

menu. The software will display the prompt **Net Name?** Enter the desired net name, and another menu, shown in Figure 6–32c, will appear. Select the appropriate type from the menu, and the net will be placed on the sheet.

EDIT sheet delete. To delete a net name from the sheet symbol, place the mouse pointer on the net name and click the **Delete** option.

EDIT sheet edit. Both type and name of a net on the sheet symbol can be edited. Place the pointer and left-click the mouse on the net whose name or type is to be changed. A submenu will appear with two options: **Name** and **Type.** Name and type can be respecified by using these options.

EDIT sheet name. The **EDIT Sheet Name** option will allow the name of the sheet symbol to change. The name of the sheet symbol usually should provide the type of circuit within the symbol, such as Power Supply or Memory. If this option is selected, the existing name will be displayed, and at this time a different name can be edited in.

EDIT sheet filename. The filename represents the actual schematic filename (.SCH filename) under which the schematic file is stored. Existing schematic filename can be altered by selecting the **EDIT Sheet Filename** option. The software automatically changes the name of the actual schematic file.

EDIT sheet size. Often it may be necessary to adjust the size of the sheet symbol. Select the **EDIT Sheet Size** option, and the mouse pointer will move to the lower right corner of the sheet. Move the mouse to adjust the size, and, when finished, select **End.**

6–6. Common Functions of the Draft Tool

EDIT part. Reference designator, value, and 1 through 8 part field of a schematic part can be edited, moved, and made visible. A specific device can be selected from a package having multiple parts. Also, orientation of a symbol can be changed. To edit a part, place the mouse pointer inside the part to be edited and left-click the mouse. Select **EDIT** from SDT_MAIN, and the submenu shown in Figure 6–33a will appear on the screen. If the device is generally available as a multiple part package, Figure 6–33b will appear; otherwise Figure 6–33a will appear.

EDIT part reference. If **Reference** is selected for the **EDIT Part** submenu, the Reference submenu shown in Figure 6–33c will appear with three options. The Reference submenu will allow the modification of the **Name, Location,** and **Visibility** of a reference designator.

If **Name** is selected from the Reference menu options, the existing name of the reference designator will appear. At this time it can be modified or a new name can be entered by using the common text editing keys.

If **Location** is selected, the location of the reference designator will be highlighted, and at this time the highlighted box can be moved to any position with the mouse pointer. When the highlighted box is at the desired position, left-click the mouse, select **Place** from the submenu appearing at the upper left corner of the screen, and left-click the mouse again. The reference designator will position itself at the new place. Remember that when the highlighted box is moved, the reference designator will not move with it, but with the left click of the mouse it will move to the new location.

If **Visible** is chosen, **Yes** and **No** options will appear. If **Yes** is chosen, the reference designator will remain visible in the schematic. If **No** is chosen, the designator will remain invisible in the schematic. If a part designator is in invisible status, that never means that the designator is nonexistent. For all processing purposes, that designator is there, and the software is

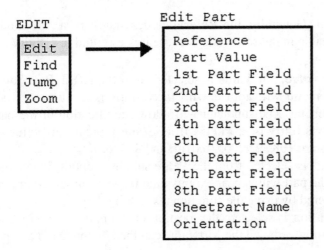

Figure 6–33a. EDIT Part Submenu for Parts with Single Device per Package

Figure 6–33b. EDIT Part Submenu for Parts with Multiple Devices per Package

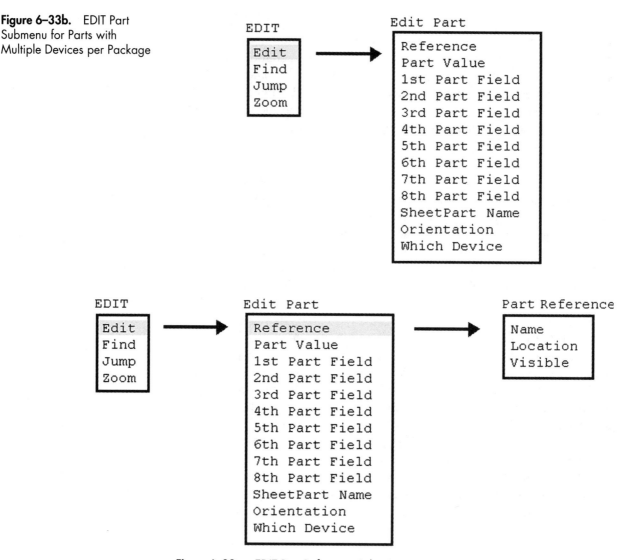

Figure 6–33c. EDIT Part Reference Submenu

aware of it. Generally, the **Reference** designator of a part should not be made invisible in the schematic because it is essential to identify a part.

EDIT part value. If **Value** is selected from the **EDIT Part** submenu, the **Value** submenu will appear with three options. Figure 6–34 shows the **Value** submenu. The submenu will allow modification of the name, location, and visibility of a **Part Value.** Some typical part values found in schematics are 10K, 2N2005, 0.01uf, and 80386.

If **Name** is selected from the **Value** submenu options, the existing **Part Value** of the part will appear. At this time it can be modified or a new name can be entered by using the common text editing keys.

If **Location** is selected, the location of the part value on the schematic will be highlighted, and the part value can then be moved to any position by using the mouse pointer. When the highlighted box is at the desired position, left-click the mouse, select **Place** from the menu as it appears at the upper left corner of the screen, and left-click the mouse again. The part value of the part will position itself at the new location. Remember that when the highlighted box is moved, the part value will not move with it, but with the left click of the mouse it will move.

If **Visible** is chosen, **Yes** and **No** options will appear. If **Yes** is chosen, the part value will remain visible in the schematic. If **No** is chosen, the part value will remain invisible in the schematic. Generally, the **Part Value** of a part should not be made invisible in the schematic, because it is essential information for a part.

EDIT 1st part field through 8th part field. There are eight part fields associated with each part. These part fields can be used for various purposes, such as part packaging information, vendor information, tolerance, voltage rating, and many others. For printed circuit board fabrication, **1st Part Field** is generally used for storing the packaging information of a part.

If 1st Part Field is selected for the **EDIT Part** submenu, the **1st Part Field** submenu will appear with three options. Figure 6–35 shows the **1st Part Field** submenu. The 1st Part Field submenu will allow modification of the name, location, and visibility of the contents of the 1st Part Field. The procedure is same as for **Reference** designator and **Part Value** edit.

If **Name** is selected from the **1st Part Field** submenu options, the existing content of it will appear. At this time it can be modified or a new name can be entered, using the common text editing keys.

If **Location** is selected, location of the first part field on the schematic will be highlighted, and at this time the highlighted box can be moved to any position with the mouse pointer. When the highlighted box is at the

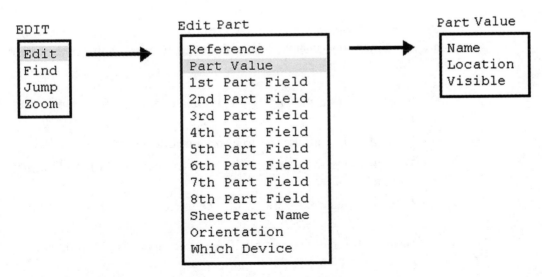

Figure 6–34. EDIT Part Value Submenu

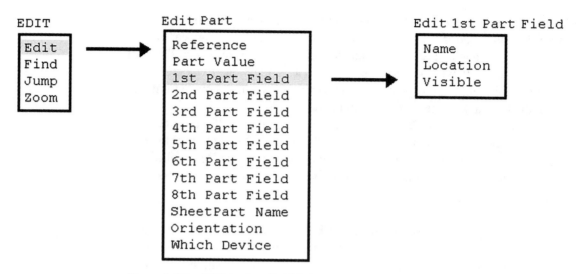

Figure 6–35. EDIT 1st Part Field Submenu

desired position, left-click the mouse, select **Place** from the menu as it appears at the upper left corner of the screen, and left-click the mouse again. The content of the part field of the part will position itself at the new location. Remember that when the highlighted box is moved, its content will not move with it, but with the left click of the mouse the content will also move to the new location.

If **Visible** is chosen, **Yes** and **No** options will appear. If **Yes** is chosen, the content of the first part field will remain visible in the schematic. If **No** is chosen, the content will remain invisible in the schematic.

The content of any of the **Part Field (1st Part Field** through **8th Part Field)** can be similarly edited by using the above procedure.

EDIT part orientation. A schematic part can be oriented to various forms on a two-dimensional plane. If **Orientation** is selected, a submenu shown in Figure 6–36 will appear.

- **Rotate** will turn the part by 90° counterclockwise from the current position.
- **Normal** will return the part to its original orientation.
- **Up** will rotate the part once 90° counterclockwise from its normal position.
- **Over** will rotate the part by 180° counterclockwise from its normal position.
- **Down** will rotate the part by 270° counterclockwise from its normal position.
- **Mirror** will display a mirror image of a part.

6–6. Common Functions of the Draft Tool

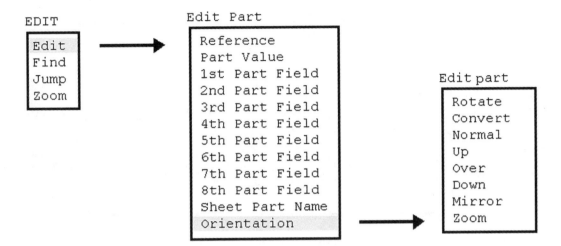

Figure 6-36. EDIT Part Orientation

EDIT which device. For many electronic devices, more than one device is packaged in a single DIP. For example, a 74LS08 has four NAND gates packaged in a single 14 pin DIP. The **EDIT Which Device** option will allow the selection of any device from the package. When you select this option, the software will let you know how many devices there are in one package and display a sequence of letters (A, B, C . . .). Move the mouse pointer and click on the desired device. Notice that the pin numbers are different on the schematic symbol, if you select device A versus device B.

EDIT convert. The **EDIT Convert** option of the **EDIT** submenu will display the logic symbol in alternative form. For example, the 74LS00 is an AND gate and is normally displayed by using an AND gate symbol. However, if the symbol is converted, it will be displayed using a NOR gate symbol. By using the **Normal** and **Convert** command, you can go back and forth between these symbols. Figure 6–37a and Figure 6–37b are, respectively, normal and converted symbols of a few TTL logic gates.

EDIT title block. Information in the title block, located at the lower right hand corner of a worksheet, can be edited. Left-clicking of the mouse button after placing the pointer inside the block will bring up the **EDIT Title Block** menu. The submenu shown in Figure 6–38 will allow editing of the information in various fields.

In order to input appropriate information in a field, the respective field needs to be selected from the submenu, and the current information can be changed by using common edit keys. For example, if the **Title of Sheet** option is selected, the current title of the sheet written in the title block will be prompted as follows:

160 Chapter 6 / Schematic Design Tool Set

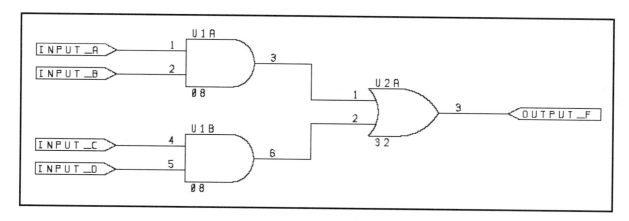

Figure 6–37a. A Circuit Using Normal Logic Symbol

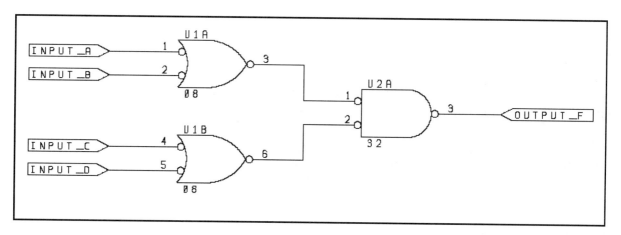

Figure 6–37b. A Circuit Using Alternative Logic Symbol

Figure 6–38. EDIT Title Block Submenu

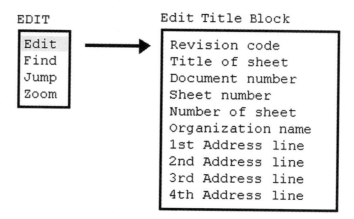

Title of Sheet? XXXXXXXXXXXXXXXX
(where XXXX is the current title of the sheet in the field)

Using this same technique, any other field of the **Title Block** can be edited.

6-6.6 FIND

The **FIND** command is used to locate a text string anywhere in a schematic worksheet. A search string can be any number of characters identifying any of the following items: module ports, labels, reference designators, part values, data stored in a part field, sheet symbol names, power objects, and text.

If **FIND** is selected from the SDT_MAIN menu, it will prompt for the character string. **FIND** is not case sensitive, so the character string can be entered in upper or lower case. After entering the string, press the the ⏎ key or left-click the mouse to place the pointer near it. If more than one such string exists on the schematic, the pointer will point to the closest one first. Selecting **FIND** or **AGAIN** will bring the string up again on the screen, at which time a new string can be entered, or pressing the ⏎ key or left-clicking the mouse at the previous string will put the pointer close to the new string or near the next one of the previous string.

For example, if all 10K resistors are to be replaced by 12K resistors, this means they need to be located one after the other on the schematic. This can be done easily by using the FIND command.

6-6.7 GET

The **GET** command is used to retrieve schematic parts from the symbol library and place them in the worksheet in desired form. The command can retrieve parts only from those libraries that are configured in the SDT configuration menu. **GET** searches through the libraries if a part name is specified at the prompt. A part can also be retrieved by scanning through the library. Figure 6–39a shows the GET submenu in step form, and Figure 6–39b shows the GET submenu in bar form.

GET Place will place the schematic part on the worksheet. Until the **Esc** key is pressed, the part will remain selected and moves along with the mouse pointer on the worksheet. This option will allow placing of duplicates of a part on the worksheet without going to the SDT_MAIN menu. The **Esc** key or a right-click of the mouse will return the control to the SDT_MAIN menu.

The **GET Rotate** command will rotate the part counterclockwise once by 90°.

The **GET Convert** command of the GET submenu will display the logic symbol in alternative form. For example, 74LS00 is a NAND gate and

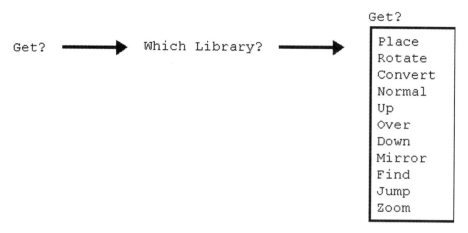

Figure 6-39a. GET Submenu

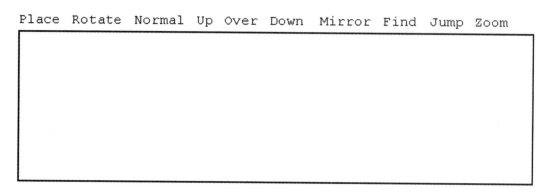

Figure 6-39b. Submenu Bar for GET Command

is normally displayed using the NAND gate symbol. However, if the symbol is converted, it will be displayed using a negated OR gate symbol. By using the **Normal** and **Convert** commands, one can go back and forth between these symbols. Figure 6-37a and Figure 6-37b show normal and converted symbols of a few TTL logic gates.

The **GET Normal** command will display the symbol in its original form—i.e., the form in which it was retrieved from the library. The command will also return the mirrored, converted, and rotated part to its original form.

The **GET Up** command will display the part by rotating it once counterclockwise by 90°.

The **GET Over** command will display the part by rotating it once counterclockwise by 180°.

The **GET Down** command will display the part by turning it once by 270° counterclockwise.

The **GET Mirror** command will display the mirror image of the normal view of the part.

```
Hardcopy
┌─────────────────────────────────┐
│ Destination         LPT1        │
│ Filemode            Replaced    │
│ Make Hardcopy                   │
│ Width of Paper      Narrow      │
└─────────────────────────────────┘
```

Figure 6–40. HARDCOPY Submenu

6–6.8 HARDCOPY

The **HARDCOPY** command is used for fast draft copy of a schematic. Figure 6–40 shows the **HARDCOPY** submenu. The command can be used to create a binary file for a schematic, which can be printed by using the DOS copy command.

6–6.9 INQUIRE

INQUIRE is used to display text associated with stimulus, trace, vector, and layout directives. Each of these categories has a different symbol. They are used in the schematic to provide a special directive, algorithmic function, or signal value to other tool sets. For example, a special route width for a connection or a special distance between two specific components can be provided to PC Board Layout Tools and other tool sets of the EDA environment. The symbol associated with these objects is displayed only in the schematic; the text is hidden. The INQUIRE command is used to display this text.

6–6.10 JUMP

The **JUMP** command quickly positions the mouse pointer in the specified location. A location on a worksheet can be specified in three different ways:

- Tag
- Text Reference
- X-Y Location

Figure 6–41 shows the **JUMP** submenu. The command is very useful when a schematic designer needs to jump back and forth into different locations on the worksheet.

The **JUMP A through H Tag** command will reposition the mouse pointer in any of eight different (A through H) previously specified locations on the worksheet, using the **Tag** command. For example, select **JUMP** from the SDT_MAIN menu, select **B Tag** from the **JUMP** submenu, and the mouse pointer will reposition to the location stored under B Tag.

The **JUMP Reference** command will move the mouse pointer to a specified grid reference location. A schematic worksheet can have a visible alphabetic Y-axis and numeric X-axis grid reference. The command will move the

```
JUMP
┌───────────────┐
│ A tag         │
│ B tag         │
│ C tag         │
│ D tag         │
│ E tag         │
│ F tag         │
│ G tag         │
│ H tag         │
│ Reference     │
│ X Location    │
│ Y Location    │
└───────────────┘
```

Figure 6–41. JUMP Submenu

pointer to a specified X-Y location. For common references the X-axis is divided into eight equal divisions, 1 through 8, and the Y-axis is divided into four equal divisions, A through D. To move the pointer to a specified X-Y grid position, select **JUMP** from SDT_MAIN, select **Reference** from the **JUMP** submenu, select the Y-axis alphabetic value, and then select the X-axis numeric value. The pointer will jump to the specified grid reference point.

JUMP X-location. To move the pointer to any desired X location, select **JUMP** from the SDT_MAIN menu, select **X-Location** from **JUMP** submenu, and provide a number without any sign. The pointer will move to the specified X grid coordinate position. However, if a positive or a negative number is provided, the command will move the pointer by that many grid points right or left from its present position. A positive number will move the pointer to the right, and a negative number will move the pointer to the left.

JUMP Y-Location is similar to **JUMP X-Location** except that it will move the pointer in the Y direction.

6-6.11 LIBRARY

The **LIBRARY** command displays a list of the configured libraries, a list of their parts, and schematic symbols of parts from those libraries. Using this command, you can display on the screen, write to a file, or print a list of parts belonging to a library. Also, you can use this command to view a specific symbol or all schematic symbols of a library in sequential order. Figure 6–42a shows the **LIBRARY** submenu.

The **LIBRARY Directory** command is used to display, print, or write to a file the list of libraries and list of parts of a library. Figure 6–42b shows the submenu.

The **LIBRARY Browse** command is used to view a specific symbol or all schematic symbols of a configured library in sequential order. The command will scan a selected library and can display all parts symbols on the screen sequentially in either forward or reverse order.

6-6.12 MACRO

The **MACRO** tool is used to capture the many keystrokes of a complex command and replace them with one or two keystrokes. Thus, a schematic designer can perform repetitive and complex tasks with a reduced number of keystrokes. Figure 6–43 shows the **MACRO** submenu. Macros can also be written by using syntax of macro definitions provided by OrCAD. To be used, macros must be stored in the macro buffer. The **MACRO Read** command can be used to read macros from a file to the macro buffer. **MACRO Write** can be used to save a macro or macros from the macro buffer to a file. A macro can be created in two different ways: using the MACRO capture command and writing a macro text file using macro syntax.

```
LIBRARY
┌─────────────┐
│ Directory   │
│ Browse      │
└─────────────┘
```

Figure 6–42a. LIBRARY Submenu

```
Which Library?
┌─────────────┐
│ PCBDEV.LIB  │
│ TTL.LIB     │
│ INTEL.LIB   │
│ CMOS.LIB    │
│ MEMORY.LIB  │
└─────────────┘
```

Figure 42b. LIBRARY Directory Submenu

```
MACRO
┌─────────────┐
│ Capture     │
│ Delete      │
│ Initialize  │
│ List        │
│ Read        │
│ Write       │
└─────────────┘
```

Figure 6–43. MACRO Submenu

MACRO capture. The **MACRO Capture** command is used to capture macros, with the keystrokes made to perform a certain task and replace it with a fewer number of keystrokes. To capture, select **MACRO** from SDT_MAIN menu. The **MACRO** submenu will appear on the screen. Select **Capture, Draft** to prompt for the key or key combination that you plan to use to invoke the macro. Notice that the key or key combination will appear on the prompt line. Once this is done, press the ⎡Enter ⏎⎤ key, and **Draft** will display the macro at the upper left corner indicating that it is in the capture mode. In this mode any keystrokes or sequence of keystrokes that are made will be captured. These keystrokes are saved in the macro buffer under the given name of the macro. This macro can be permanently saved in a file by using the **MACRO Write** command. However, if it is not saved it will be erased when you get out of the SDT_MAIN screen (or **Draft**).

MACRO Delete is used to delete a Macro.

MACRO Initialize will erase all macros in the buffer. However, this command will not erase macros saved in a file.

MACRO List will list all key combinations assigned to macros.

MACRO Read will load a macro file into macro buffer memory. Once it is loaded, it can be accessed by **Draft.** However, the macro file can be loaded automatically when SDT_MAIN menu is invoked by configuring the **MACRO Option** area of the SDT configuration menu.

MACRO Write saves macros from the buffer into a file.

6–6.13 PLACE

The **PLACE** command will place wires, buses, junction boxes, labels, text, module ports, power, and sheet symbols on the worksheet. The command will place almost anything except schematic parts that are brought from the symbol library. PLACE has a submenu, which is shown in Figure 6–44a. To invoke the **PLACE** submenu, click the mouse on PLACE from SDT_MAIN.

PLACE Wire will let you draw wires to connect components on the worksheet. To draw a wire, select and left-click the mouse on **Wire** in the **PLACE** submenu. The **Wire** submenu shown in Figure 6–44b will appear across the top of the screen. Before beginning to draw a wire segment, position the tip of the mouse pointer on the place where the wire segment is supposed to begin, and then select **Begin** by pressing the letter **B** for Begin on the keyboard. If you drag the mouse pointer, a wire will be drawn on the worksheet. When you finish drawing the wire, press **N** for **New,** and **Draft** will remain in the wire draw mode. At this point **Draft** will let you draw a new wire segment. It is possible to make only one 90° turn while drawing a wire segment. To draw a wire that needs to make more than one 90° turn, a new wire segment needs to be drawn for each turn. To draw a new wire segment, press the letter **N,** position the pointer on the appropriate place, and drag the mouse to draw the new wire segment. To end the wire draw mode, press **E** for **End,** and Draft

```
PLACE
┌─────────────┐
│ Wire        │
│ Bus         │
│ Junction    │
│ Entry (Bus) │
│ Label       │
│ Module Port │
│ Power       │
│ Sheet       │
│ Text        │
│ Dashed Line │
│ Trace Name  │
│ Vector      │
│ Stimulus    │
│ NoConnect   │
│ Layout      │
└─────────────┘
```

Figure 6–44a. PLACE Submenu

Figure 6-44b. PLACE Wire Submenu

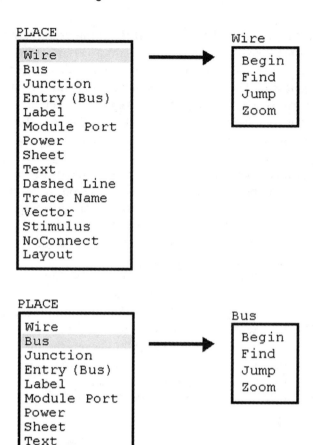

Figure 6-44c. PLACE Bus Submenu

will return to the SDT_MAIN menu. Several wire segments, meaning a wire that has several 90° turns, can also be drawn by successively left-clicking the mouse on **Begin** or pressing the letter **B** for **Begin** and dragging the mouse pointer in the desired direction.

PLACE Bus is a command similar to **PLACE Wire** except that it will place a **Bus** segment. **Begin, End,** and **New** of the **Bus** submenu work in the same way as the Wire submenu. Figure 6-44c shows the **Bus** submenu.

PLACE Junction will place a junction box on the worksheet. When wires and buses cross each other, a connection is distinguished from a crossover by a junction box. If more than two buses or wires connect to a common node, a junction box must be placed to that point to connect them. When two wires or buses cross each other, they are considered by **Draft** as no connection unless a junction box is placed at their point of intersection. However, if two wires or buses just meet at a point without one running over the other, it is considered as a connection without a junction box. This feature of **Draft**

must be followed carefully, especially while drawing wires and buses from the individual pins of a component. Always drag a wire up to or from a pigtail of a schematic symbol and do not overlap them. If you overlap the wire and pigtail of the symbol, **Draft** considers them as no connection. (The pigtail of a symbol is the little piece of wire that is hanging out from the body of the part for each pin of a schematic symbol. When a schematic symbol is fetched from a library, pigtails for each symbol come with it.)

To place a junction box, select **Junction** from the **PLACE** submenu, position the tip of the mouse pointer at the point where the junction box is desired, and select **Place** from the Junction submenu or press the letter **P** on the keyboard to place a junction on the worksheet.

The **PLACE Entry (Bus)** command is used to place bus entries. Bus entries are individual wire connections between buses, a bus and a pigtail, or a bus and wires. Since one segment of a bus contains several wires, a bus will have a number of entry wires. Junction boxes are not required to connect a bus entry to a bus.

The **PLACE Entry (Bus) Wire** command is used to place wire entries to a bus. This command is used for both entry wire or exit wire to a bus.

The **PLACE Entry Bus** command is used to join one bus to another bus or when a bus makes a turn.

The **PLACE Label** command is used to place an identifier on the worksheet. Labels are not ordinary text or comments; they have specific meaning for other tools, such as Create Netlist. Buses generally have labels, such as an address bus with A0, A1, etc., and a data bus with D0, D1, etc.

The **PLACE Module Port** command is used to put four different types of module ports on a worksheet. Module ports are used to connect signal input or output from one worksheet to another worksheet. Module ports having the same names are considered the same node or electrically connected. Signals not going out from a worksheet are usually marked with labels and not module ports. Labels with the same name are also considered to be electrically connected. To invoke the command, select PLACE Module Port. The command will prompt for the name of the module port. Once the module port name is provided, **Draft** will display the Module Port submenu for four different types and their default style. Once the type is selected, another submenu will appear across the top, and **Draft** will bring the module port on to the screen. Move the mouse pointer to place the module port at the desired point. Select **Place,** to place the module port on the worksheet. The style of a module port is independent of its type. This means that an input module port can either be right pointing or left pointing; it is only a matter of personal choice. However, it is strongly recommended to use the default style.

The **PLACE Power** command is used to place power symbols on a worksheet. To invoke a power object, select **Power** from the **PLACE** submenu. The power symbol will appear on the screen, ready to be positioned by the mouse. Select **Place** to place the power symbol on the worksheet. **Orientation, Value,** and **Type** can be changed before doing so. The value of a power object should be labeled as +5V, VSS, VEE, −12VDC, or any other appropriate string.

```
PLACE Sheet
┌─────────────┐
│ Add-net     │
│ Delete      │
│ Edit        │
│ Name        │
│ Filename    │
│ Size        │
└─────────────┘
```

Figure 6–45. PLACE Sheet Submenu

```
Quit Filename.SCH
┌─────────────────────┐
│ Enter Sheet         │
│ Leave Sheet         │
│ Update File         │
│ Write to File       │
│ Initialize          │
│ Suspend to System   │
│ Abandon Edits       │
│ Run User Commands   │
└─────────────────────┘
```

Figure 6–46. QUIT Submenu

The **PLACE Sheet** command is used to place a sheet symbol. Sheet symbols are used to create a hierarchical design. A symbol will represent a worksheet that is connected to the rest of the schematic by module ports defined on the sheet symbol. Each sheet symbol has inputs, outputs, sheet name, and a schematic filename that represents the schematic within the sheet.

To place a sheet symbol on a schematic worksheet, click the mouse on PLACE from SDT_MAIN menu; then click the mouse again on **Sheet** from the **PLACE** submenu. **Begin** outline and then select **End** to complete it. Since a sheet has no meaning without a net, the software will automatically display the **PLACE Sheet** submenu shown in Figure 6–45.

If at a later time you need to add a net, edit a name, change the size of the sheet symbol, etc., you can invoke the above menu as follows: Place the mouse pointer on the sheet symbol and click on **EDIT** from the SDT_MAIN menu. If you want to enter into the schematic represented by the sheet, do as follows: Place the mouse pointer on the sheet symbol, click on QUIT from the SDT_MAIN, and then click the mouse on **Enter Sheet** from the **QUIT** submenu shown in Figure 6–46. You can leave the sheet and get back to the main schematic by clicking on QUIT and then on **Leave Sheet.**

PLACE Sheet Add-Net will let you add a netname to the sheet symbol. These nets are the connection between the main schematic and the schematic represented by the sheet symbol. A net can be input, output, bidirectional, or unspecified.

PLACE Sheet Delete will delete a sheet net. Place the pointer on the net to be deleted and click on **PLACE→ Sheet → Delete.**

PLACE Sheet Edit will let you edit the names and types of the inputs and outputs of a sheet.

PLACE Sheet Name will let you edit the name of a sheet symbol. The default sheet name is a question mark.

PLACE Sheet Filename will let you change the filename representing the sheet. This filename is the schematic file that is within the sheet symbol.

PLACE Sheet Size will let you change the size of the sheet symbol.

PLACE Text will let you place text information on a schematic. Using this command, a text of various size and orientation can be placed on the schematic.

PLACE Dashed Line will let you place a dashed line on the worksheet. The dashed line is useful to encompass certain areas of a schematic and label them with text.

PLACE Trace Name, PLACE Vector, and **PLACE Stimulus** are used with the OrCAD Digital Simulation Tools software and are beyond the scope of this book.

PLACE NoConnect is used to place a special symbol for no connection. The symbol on a pin of a device indicates to the software that the pin is intentionally left unconnected. All the pins of a device need to be connected, for if a pin remains unconnected, an error message may appear during Electrical Rules Check or during Netlist generation. Therefore, if a pin needs to remain unconnected, put the **NoConnect** symbol at the pin.

PLACE Layout is used to place a special marker that gives the PC Board Layout Tools specific directives or conditions for routing the board.

These directives are passed on to the board designer by the schematic designer. For every directive a symbol like this ↻ is placed on the schematic to indicate the conditions. To view the content of a directive, use the **INQUIRE** command from the PCB_MAIN. A few examples of layout directives are shown below:

Width (0.010 in.)
Isolation (0.025 in.)
Via Through (0.035 in.)

Each one of these conditions will have a layout directive symbol in the schematic file.

6-6.14 QUIT

The **QUIT** command is used to enter and leave a hierarchical worksheet; load, update, and write to files; clear the worksheet; suspend to DOS; and abandon EDIT.

QUIT Enter Sheet is used to view a schematic represented by a sheet symbol in a hierarchical design.

QUIT Leave Sheet is used to leave the schematic belonging to a sheet symbol.

QUIT Update File writes the latest version of the worksheet under the existing filename.

QUIT Write to File allows you to save the schematic under a new filename.

QUIT Initialize lets you either load a new schematic file or erase everything from the worksheet.

QUIT Suspend to System lets you temporarily leave the schematic worksheet to go to DOS mode. Here you can execute DOS commands normally. To remind the user that OrCAD/SDT is in the background, the DOS prompt has an additional arrowhead. You may return to OrCAD/SDT by typing **EXIT** at the double arrowhead prompt.

QUIT Abandon Edits will let you exit from the worksheet without saving the changes to the worksheet.

QUIT Run User Commands will let you suspend to the operating system and will issue the command **DRAFTUSR.**

6-6.15 REPEAT

REPEAT is used to duplicate the latest entered object, label, or text string and place it on the worksheet.

6-6.16 SET

SET is used to control various options of the **Draft** tools. The set has a submenu that is shown in Figure 6–47. For most of these options, you need to provide a **YES** or **NO** answer. For example, if you need to draw buses and

Figure 6–47.
SET Submenu

```
SET
Auto Pan         YES
Backup File      YES
Drag Buses       NO
Error Bell       YES
Left Button      NO
Macro Prompts    YES
Orthodonal       YES
Show Pins        YES
Title Block      YES
Worksheet Size   A
X,Y Display      NO
Grid Parameters
Repeat Parameters
Visible Lettering
```

wire at any angle, left-click the mouse on **SET,** then on **Orthogonal,** and select **NO.** Another example, if you want to keep the standard title block, left-click the mouse on **SET,** select **Title Block,** and then select **YES.**

6-6.17 TAG

TAG is used to jump from one part of the schematic to another part by tags. Eight tags, A through H, can be set on the board, and the mouse pointer will jump from one part of the board to a tag location if you specify the desired tag. The tags are invisible and are not saved with the worksheet. While working on a large schematic, being able to jump from one part to another specified point is very useful.

6-6.18 ZOOM

ZOOM is used to view the worksheet in detail. It is like a magnifying glass. **Zoom In, Out, Center,** and **Select** will conveniently magnify the schematic to a variety of levels.

REVIEW QUESTIONS

1. What does an electronic schematic describe?
2. Does an electronic schematic show physical or logical connections of the components?
3. Wires, buses, and junction boxes—are these symbols also?
4. What is the difference between a text and a label on a schematic?

5. Can you describe the content of a title block anywhere other than on the schematic worksheet?

6. How is the ability to structure schematic files useful to a designer?

7. How many different types of editor are available within the OrCAD/SDT tool set?

8. Why is the Annotate tool useful when you are designing a large schematic?

9. After you have annotated a schematic, can you change it? If so, what is the easiest way to do that?

10. Each schematic part has eight Part Fields in addition to Reference Designator and Part Value fields. What should you usually put in the first one?

11. Can you change the location of a Reference Designator, a Part Value, or any of the Part Fields?

12. Can you make any of these fields invisible in your schematic?

13. What is the netlist file for a schematic?

14. Why should one archive part of a schematic?

15. How can you specify electrical rules for your schematic?

16. How are Key Fields used?

17. Can you delete a schematic symbol by using the Delete command?

18. How do you edit a reference designator?

19. What are the different types of module ports available in OrCAD/SDT?

20. If the schematic you are designing is repetitive in nature, what symbol will reduce the size of it on the worksheet?

21. What command can you use to represent a normal logic symbol to an alternative one?

22. What command can you use to find a specific reference designator in a large schematic?

23. In a normal mode of SDT, the lines you can draw are orthogonal. What should you do to draw a non-orthogonal line?

24. What command sequence should you go through to place a 7408 logic gate on a schematic worksheet?

25. What command sequence should you go through to place a sheet symbol on a schematic worksheet?

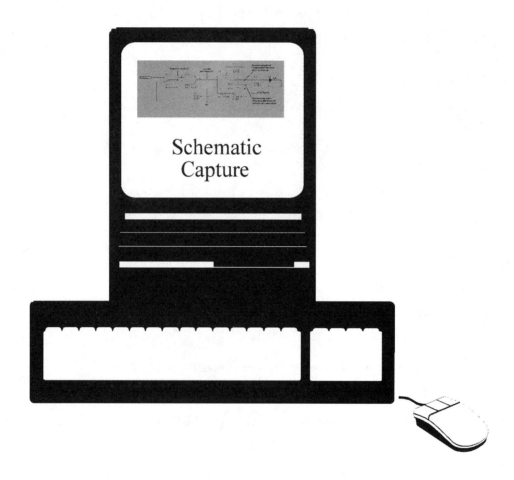

Schematic Capture

7

COMPUTER-AIDED SCHEMATIC CAPTURE

7-1. DRAFTING AND DESIGNING SCHEMATICS

When an electronic project is undertaken, a hand-drawn schematic should already be available. The decision has to be made about the number of drawings required, appropriate sheet size for each drawing, scale required, on-board and off-board components, etc. After all these decisions are made, schematic design sessions can begin. This is the vital part of the project because mistakes made here most likely will be carried over to the end. As a part of the schematic capture process, the SDT tool generates a netlist file, a list of parts, and other valuable information for documentation. The netlist file is used by the PC Board Layout Tools for master artwork design. The step-by-step schematic design process is discussed in the rest of this chapter.

7-2. ORGANIZING FILE AND SET DESIGN ENVIRONMENT

Design Management Tools are used to organize the complete PC board design effort. The tools are used to organize files belonging to a design, protect the design with backup copies, and add descriptions to design files so that they can be easily located in a particular design. Using the procedure described in Chapter 5, the design environment for all of the four examples has been created. All that is required to draw the necessary schematic is to load one design environment at a time and create the associated schematics for the design.

7-2.1 Create a New Design

If the design name for the desired project has not been created before, you can create it using the following procedure:

Objectives

After completing this chapter, you should be able to

1. Create a new design or load an existing design
2. Fetch schematic parts and begin drawing a schematic using the SDT Tool Set
3. Decide on the number and size of drawings required for an electronic project
4. Decide on the parts needed to be mounted off board
5. Configure a Schematic Design tool for a specific electronic project
6. Edit reference designator, part value, and part fields of a schematic symbol
7. Place label, text, and comments into a schematic
8. Save and update a schematic file
9. Verify the electrical rules of a schematic by using rule checker
10. Annotate, unannotate, and back annotate a schematic file

Objectives (continued)

11. Create an ASCII file using M2EDIT editor
12. Update the content of the various fields of a part
13. Archive parts of a schematic into a single library
14. Create the netlist of a schematic
15. Create the Bill of Materials for a design
16. Print or plot a hardcopy of a schematic using utilities available in OrCAD

1. Invoke DM_MENU from ESP_MENU. Make sure that the design environment is set to TEMPLATE Design. Many of the files will be copied from TEMPLATE Design to your new design directory. Then left-click on the Create Design icon.
2. Left-click inside the dialogue box and type the desired design name.
3. Left-click on the OK icon. The design name will be created and will appear in the left list box where all the design names are displayed.

7-2.2 Load an Existing Design

If the design name already exists and it needs to be invoked, use the following steps:

1. Invoke DM_MENU from ESP_MENU, locate the design name in the left list box, and left-click on it.
2. Left-click on the OK icon. The design name TEMPLATE at the top of the screen will be replaced by the new design name you just selected.

7-2.3 Set the Design Environment

To set a desired design environment for the logic probe, for example, follow the procedure described below:

1. Invoke the OrCAD ESP_MENU window by typing orcad at the DOS prompt of MYORCAD subdirectory as: C:\MYORCAD>orcad. The ESP_MENU window will appear on the screen. Notice at the top of the ESP_MENU that the default design name, TEMPLATE, will appear with the window.
2. Invoke the Design Management tool set by left-clicking the mouse on that icon. This action will bring the DM_MENU window up on the display terminal. For example, to invoke a design name called LOG_PRO, locate the name in the list on the left side of the window. If necessary, scroll through the list by clicking the mouse on the upper arrow or lower arrow located at the side of the left list box. Once the name appears inside the list box, left-click the mouse on the design name. The design name will be highlighted.
3. Left-click the mouse on the OK icon. The design name TEMPLATE at the top of the window will be replaced by the new design name, LOG_PRO. The design environment for the logic probe is now set. To begin the schematic drafting session, the Schematic Design tool needs to be invoked and configured.
4. Exit from the Design Management tool with a left click on the **Cancel** icon. The ESP_MENU window will appear.

7-3. DESIGNING AND DRAFTING A CIRCUIT SCHEMATIC

Designing and drafting a schematic involves several steps. In this section steps such as decisions about number of drawings, their sizes, configuration of SDT tools for a design, fetch, place, and connect a part will be described.

7-3.1 Determination about Number of Drawings and Approximate Sheet Size for a Project

The number of pages of schematic required for a project depends on the size of the circuit diagram, which in turn depends on the number of components and size of each component. If the circuit is too big and you can logically divide it into several 8.5" × 11" schematics, you should do that. If these logical divisions are repetitive in nature, you may use either a simple or a complex hierarchical file structure for the project schematic. If the schematic is not repetitive, you may still divide the circuit into several sheets, but use a flat file structure.

Schematic sheet size often depends on the amount of details you require on the schematic, your printer capacity, and in many cases, on personal preference. However, 8.5" × 11" schematics are easy to handle, put in a binder, and transport conveniently from one place to another.

7-3.2 Determination about Parts to Be Mounted Off Board

The parts that are mounted off board depend on several factors. You do not have to make this decision at this time. The final decision about off-board parts can be made during the PCB design process. However, a rough idea about off-board parts would help in designing the schematic more efficiently. The following criteria will help you decide about off-board parts for your circuits:

- Needs large or heavy components, such as power transformer or speaker
- Needs special heat sink, such as power transistor, power SCRs, or power regulators
- Needs to be accessed by user and display indicator, such as volume control potentiometer, switches, indicator light, or indicating meters

Other than these, most of the electronic parts should be mounted on the PCB.

7-3.3 Configure Schematic Design Tool

Configuring some part of the SDT tool is useful before starting on the schematic drawing. However, you can go back and forth between SDT configuration and the schematic worksheet at any time during the worksheet session. Configuration parameters of the SDT tools are only global for the particular design. Therefore, parameters configured for one design will not be available to the other design, unless you configure it in the same way. This means that each design has its own SDT configuration that is saved and invoked with the design. The following basic procedure can be used to configure Schematic Design Tools for any particular design; however, a few options, such as Worksheet Option and Library Option would be different for each design.

1. Invoke SDT_MENU from ESP_MENU and left-click the mouse on the **Draft** icon located at the upper left corner of the screen.

2. Scroll through the transit menu and left-click the mouse on the **Configure Schematic Design Tool.** The configuration screen will appear. Many options have default values, or you may have already chosen the option during loading of the software. However, configurations that are specific to a design need to be performed at this time.

3. Scroll down to **Library Option** and click on the **Insert a Library** option bubble. Library Option will be in insert mode. Scroll through the **Available Libraries** list box located on the left side, and left-click the mouse on the CMOS. CMOS will be highlighted. Left-click on the Insert icon. Notice that CMOS will appear on the right list box, identified as **Configured Libraries.** Now only schematic parts from the CMOS library can be fetched onto the worksheet. In this way the desired library can be configured on the right list box so that the schematic symbol can be fetched from it. While in this option, click on the Main memory bubble for Name Table Location and Disk bubble for Symbolic Data Location, if they are not selected. (Libraries to be configured for each project will be provided with SDT configuration for the projects.)

4. Go to **Worksheet Options,** and use SCH for Default Worksheet File Extension and 'A' for **Sheet Size.** Enter **01-1995** for Document Number and **Deluxe Logic Probe** as the title of the project.

5. At this time, use the default value for Macro, Hierarchy, Color and Pen Plotter Table, and Template Table Options. You may want to be a little more familiar with the software before changing any of the values in these options.

6. Go to Key Fields, place the mouse pointer inside the dialogue box of Combine for Field 1, and left-click the mouse. The prompt will go to insert mode. Type **V** in the dialogue box. What you are telling the software by this is to compare the content of the value field of each part while updating the Part Field 1 for it.

7. For Create Netlist, insert **V** in the dialogue box of Part Value Combine and **I** in the dialogue box of Module Value Combine.

8. For Create Bill of Materials, insert **V** in the dialogue box of Part Value Combine.

9. At this time, use the default setting of the Check Electrical Rule Matrix.

7-3. Designing and Drafting a Circuit Schematic

7-3.4 SDT Configuration for the Projects

SDT configuration for each of the four projects will not greatly differ. Therefore, using the above procedure, enter the appropriate SDT configuration parameter for each of the projects described below. Configure only the options mentioned here, and leave the ones not mentioned at their default values.

Project 01-1995, Deluxe Logic Probe; computer filename: LOG_PRO.

Library Options: Load the following libraries in the **Configured Libraries** list box by selecting the **Insert a Library** bubble and using the **Insert** icon: CMOS, PCBDEV, ANALOG3.

Worksheet Options: Enter the following parameters in the appropriate dialogue boxes:

- Sheet Size: A
- Document Number: 01-1995
- Revision: 001
- Title: Deluxe Logic Probe
- Organization Name: Put your name, course number, and course title in this dialogue box.
- Organization Address: Put your department name and name of your school in this dialogue box.

If you do not wish to enter this information now, you will be able to do so later in the worksheet title block while working on the schematic.

Key Fields: Enter the following parameters in the appropriate dialogue boxes:

Update Field Contents—Combine for Part Field 1:V

Create Netlist—Part Value Combine: V

—Module Value Combine: 1

Create Bill of Materials—Part Value Combine: V

Project 02-1995, Electronic Cricket; computer filename: CRICKET

Library Options: Load the following libraries in the **Configured Libraries** list box by selecting **Insert a Library** and using the **Insert** icon: CMOS, PCBDEV, ANALOG.

Worksheet Options: Enter the following parameters in the appropriate dialogue boxes:

- Sheet Size: A
- Document Number: 02-1995
- Revision: 001
- Title: Electronic Cricket

- Organization Name: Put your name, course number, and course title in this dialogue box.
- Organization Address: Put your department name and name of your school in this dialogue box.

If you do not wish to enter this information now, you will be able to do so later in the worksheet title block while working on the schematic.

Key Fields: Enter the following parameters in the appropriate dialogue boxes:

Update Field Contents—Combine for Part Field 1: V

Create Netlist—Part Value Combine: V

—Module Value Combine: 1

Create Bill of Materials—Part Value Combine: V

Project 03-1995, Infrared Counter; computer filename: IR_COUNT.

Library Options: Load the following libraries in the Configured Libraries list box by selecting Insert a Library and using the Insert icon: TTL, PCBDEV, ANALOG3.

Worksheet Options: Enter the following parameters in the appropriate dialogue boxes:

- Sheet Size: A
- Document Number: 03-1995
- Revision: 001
- Title: Infrared Counter
- Organization Name: Put your name, course number, and course title in this dialogue box.
- Organization Address: Put your department name and name of your school in this dialogue box.

Key Fields: Enter the following parameters in the appropriate dialogue boxes:

Update Field Contents—Combine for Part Field 1: V

Create Netlist—Part Value Combine: V

—Module Value Combine: 1

Create Bill of Materials—Part Value Combine: V

Project 04-1995, Mini Stereo Amplifier; computer filename: AMPLIFIE.

Library Options: Load the following libraries in the **Configured Libraries** list box by selecting **Insert a Library** and using the **Insert** icon: CMOS, PCBDEV, ANALOG3.

Worksheet Options: Enter the following parameters in the appropriate dialogue boxes:

7–3. Designing and Drafting a Circuit Schematic 179

- Sheet Size: A
- Document Number: 04-1995
- Revision: 001
- Title: Mini Stereo Amplifier
- Organization Name: Put your name, course number, and course title in this dialogue box.
- Organization Address: Put your department name and name of your school in this dialogue box.

Key Fields: Enter the following parameters in the appropriate dialogue boxes:

Update Field Contents—Combine for Part Field 1: V

Create Netlist—Part Value Combine: V

—Module Value Combine: 1

Create Bill of Materials—Part Value Combine: V

7–3.5 Draft Local Configuration and Invoke Worksheet

To invoke a worksheet for a particular design, the following procedure may be used. If your design name is not the same as the schematic file name or if you have multiple schematics under the same design name, you need to use the procedure described below:

1. Set the design environment for the desired design by using the procedure described earlier.

2. Left-click the mouse on **Draft** icon once, and the Draft Transit Menu will appear. Click on **Local Configuration,** and the Local Configuration menu will appear. Click on the name of the schematic design you would like to work on. Exit the Local Configuration by clicking on the OK icon.

3. Left-click the mouse on the **Draft** icon in SDT_MENU, followed by a left click on **Execute** in Draft Transit Menu. A blank worksheet will appear on the computer screen for drafting a schematic. If you have already drawn a schematic for the design, it will appear on the screen instead of a blank sheet. A menu will also appear at the upper left corner of the screen along with the worksheet. With a right click of the mouse the menu will disappear, and the mouse pointer can be moved around the worksheet. A left click will bring the menu back, and the mouse pointer can be used only to scroll through the menu.

Unlike a step menu, a bar menu and a no-menu will not lock the mouse within the menu. You can move the mouse pointer around the circuit when a bar menu is displayed.

4. Left-click the mouse on the Schematic Design Tools icon, and the SDT_MENU window will appear on the screen. At this point the designer needs to make a rough estimate of how many schematic part symbols will be used for the desired project. This assessment is necessary because configuring all of the library can cause a memory shortage. Initial assessment of the parts in the circuit will tell you which of the schematic part libraries to load before going to the schematic drafting session. For example, initial assessment of parts for the Logic Probe Project will lead you to load the CMOS, ANALOG3, and PCB DEVICE libraries. If the desired parts are not available in the configured library, other libraries could be configured by invoking the SDT configuration screen again. If your schematic has polarized devices, such as diode transistor, electrolytic capacitor, or SCR, you must get these symbols from the PCB DEVICE library and not from the DEVICE library. PCB DEVICE is a special library needed for OrCAD/PCB to recognize the parts. For the projects in this book, do not configure the DEVICE library under the Library Options area of Configure Schematic Design Tools.

7-3.6 Capture Circuit Schematic

The designing and drawing of schematics start with fetching parts from libraries. The various libraries are where schematic symbols are stored. These symbols are fetched, placed, and later connected on the worksheet as a part of the process of drawing a circuit schematic. The symbols in libraries are arranged by their type, manufacturer, and characteristics—for example, TTL, CMOS, Motorola, Intel, Device, and Memory. Some special parts may not be available in an OrCAD-provided library, in which case the part needs to be created and stored in a custom library. Parts or schematic symbols on a worksheet can be fetched only from the configured libraries. The configuration is done under the **Library Option** of the Configure Schematic Design Tools screen.

Fetch a part or a schematic symbol

1. Scroll through the SDT_MAIN menu and left-click on **GET.** The software will ask you GET?, meaning GET What? You may enter the library part name, if known, at the prompt, or you may left-click the mouse once more to display the names of the configured libraries.

2. Looking at the library names, you will be able to determine under which library your part symbol can be found. If you do not know which library contains your part, you may have to guess to initiate a logical search. Left-click the mouse on that library, and the software will display the names of all parts belonging to that library.

3. Scroll up and down and find your part numbers one at a time, and left-click the mouse on each one. The part will appear on the work-

7-3. Designing and Drafting a Circuit Schematic

sheet along with a menu bar at the top for desired placement. The complete list of parts of a library cannot be displayed in the list box containing the part numbers located at the upper left corner of the screen, so you may need to scroll up and down to see the complete list. It may be necessary to get familiar with the names and symbols of the electronic parts because parts such as resistors and capacitors are listed in the library by their packaging number, not by their value, and parts such as op-amps and transistors are listed by their type, values, or manufacturer.

About schematic part symbols for PCB tools. Schematic part symbols that are available in the library by their generic name and not by their value, such as resistors, capacitors, diodes, and transistors, must be fetched from the PCB DEVICE library instead of the DEVICE library. This is important if the schematic is used later to produce a netlist for the PCB software tool. However, if you do not intend to use the schematic for artwork design, you may fetch the symbol from either DEVICE or PCB DEVICE library.

Place a part on a worksheet

1. Once a part is on a worksheet, it needs to be placed on it. By moving the mouse, the part you just fetched can be dragged to the desired place on the worksheet. You may rotate, mirror image, or convert before placing each part.

2. Left-click the mouse once to convert the bar menu to a list menu, where you can scroll through the options using the mouse or the arrow keys and select the desired one. Left-click the mouse on the desired option of the **GET** submenu. Most of the time you may need to click on the option called **Place.** Left-clicking of the mouse will place the part on the worksheet. If you want to place the same part at a different place on the worksheet, move the mouse to the desired place and left-click the mouse again on option **Place.**

3. If the orientation of the part needs to be changed, you may use such options as **Rotate, Normal, Up, Down,** and **Mirror,** depending on your need.

Once you place the parts, you can move them by using the **BLOCK-Move** command. You may download all the parts required for the schematic first from the library and then move them around on the worksheet by using this command. The other way is to get one part at a time and place them on the worksheet as you lay out the circuit. During schematic layout, parts may have to be moved several times. Therefore, do not waste your time placing them in a specific place on the worksheet.

Example: fetch parts for the logic probe schematic.

1. Select GET from SDT_MAIN menu and click on it. A prompt (GET?) will appear.

2. Click the mouse again, and a list of the configured libraries with a Which Library? prompt will appear.

3. Select CMOS from the library list and left-click the mouse. The software will list all of the library parts in the CMOS library. All the parts that are in the library cannot be displayed in the list box, so scroll to see the full list.

4. Select part 4011 (quad NAND Gate) and left-click on it. The part will appear on the screen.

5. The part can be moved to the desired location on the worksheet by using the mouse. The part image temporarily simplifies as a symbol outline and moves with the mouse. This simplification of the part symbol always happens before placement and during movement of the part from one point to another. When the part is at the desired location, left-click the mouse. A submenu bar will appear at the top. You may choose from the menu bar to **Place, Rotate, Convert,** etc. If you like to use your mouse, left-click it once more. The bar menu will convert to a list menu in which the mouse can be used for scrolling and selecting the options.

6. Choose the option **Place** from the menu to position the CMOS 4011 NAND gate.

7. The 4011 will still be on the mouse pointer. Place seven of these on the worksheet by selecting and clicking on the **Place** option.

8. Go to the SDT_MAIN menu by right-clicking the mouse twice and left-clicking it once.

9. Select and click on **GET,** and click again at the Get? prompt. A list of configured libraries will appear on the screen. Select ANALOG3 from the list, and click on it.

10. Scroll down the list of parts to find LM555. Click on it, and place it on the worksheet.

11. Go to SDT_MAIN and click on **GET** again. As the library list appears on the screen, select PCB DEVICE library and click on it. All the parts in the DEVICE library will be displayed.

12. Scroll through the parts list of the library and fetch one kind of part symbol at a time. Place 1 Buzzer, 3 Capacitors Non Pol, 1 Capacitor Pol, 3 LEDs, 10 Resistors, and 1 SPDT switch. You must get these part symbols only from PCB DEVICE library and not from any other DEVICE library. These other DEVICE library parts are not suitable for the PCB software tools set.

Connecting parts using wires or buses. The task of connecting parts with wires and buses to establish signal connection among the components needs to be done carefully. You cannot miss a connection when duplicating your design. You may need to position a part close to another part with which it has the most number of connections. If you place the parts too

7-3. Designing and Drafting a Circuit Schematic

close to each other, it will be difficult to run wires and bus connections. On the other hand, if you place them too far from each other, the result will not look nice and you may have to use more than one sheet of paper. Placing the component properly on a worksheet is a challenge.

A few important points about wire connection

1. If two wires cross each other, OrCAD considers them as no connection unless a junction box is placed.
2. If two wires are connected end to end, they are considered as a connection without a junction box. However, if two wires overlap each other, they are considered as no connection.
3. If a wire overlaps a component pin, it is considered no connection. However, if a wire and a component pin are connected end to end, they are considered as a connection.
4. When you place a wire to a component pin, be sure not to overlap the wire and the pigtails of the schematic part; instead, connect them end to end.

Figure 7-1 shows several examples where junction boxes are required to indicate connection and examples where junction boxes are not required.

Connection components by using wires

1. Invoke the SDT_MAIN menu, select **PLACE,** and click the mouse on it. The **PLACE** submenu will appear.
2. Select the **Wire** option from the PLACE submenu, and click the mouse on it. The **PLACE-Wire** submenu will appear.
3. Move the mouse pointer to a pin end of a component or to the end of a wire where you want to begin your wire, click the mouse once, and select **Begin** from the **PLACE-Wire** submenu.
4. Now drag the mouse pointer to trace a wire to the end of another component pin or to the end of a wire you have decided to connect. Click the mouse once and select **New** from the **PLACE-Wire** submenu.
5. The length of a wire is limited only by the worksheet size, so this means you can have wire of any size within the worksheet. However, when you begin a wire, it can make only one right angle turn before a new wire needs to be introduced. If a connection requires several right angle turns, a new wire needs to be introduced before every second right angle turn.
6. Now you can position your mouse pointer on another wire or component end for a new connection.
7. To end your wiring session, select **End** from the **PLACE-Wire** submenu. You can also end the session and come back to SDT_MAIN menu by right-clicking the mouse twice and left-clicking it once.

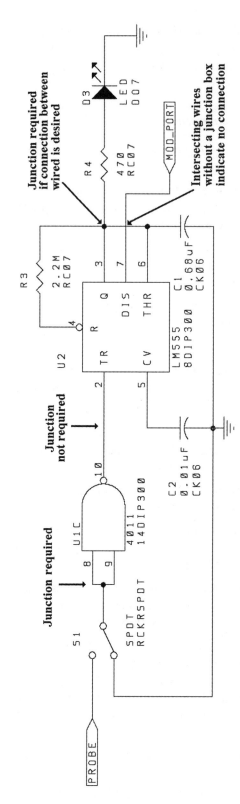

Figure 7-1. Required Junction Boxes

7-3. Designing and Drafting a Circuit Schematic

Connecting components by using buses. Tracing a wire from one component pin to another is not much different from connecting a set of component pins to a set of pins of another component. Buses are used in place of a number of similar wires. They are generally used to connect a number of similar pins of one component to the several other pins of another component. They are shown as thicker than a wire on the worksheet in order to distinguish them from wires. Use buses wherever possible to make the schematic more readable. The microprocessor address and data lines are connected to the address and data lines of memory chips by using buses. Instead of running 16 data and 32 address lines from the microprocessor through the worksheet to memory chips, buses can do this more conveniently. See the following procedure.

1. Invoke the SDT_MAIN menu, select the option **PLACE,** and click on it. The **PLACE** submenu will appear.

2. Select the option **Bus,** and the **PLACE-Bus** submenu will appear.

3. Move the mouse pointer to the point where you want to begin your **Bus.**

4. Select the option **Begin** from the **PLACE-Bus** submenu, and drag the mouse to trace a bus from one point to another. Figure 7–2 shows an example where a bus is connecting the address lines of a microcontroller to the data lines of a memory chip. Like a wire, a bus can make only one right angle turn. Therefore, successive **Begin** and **New** commands will let you make any number of turns.

5. To connect a bus to pins of a microprocessor's address lines, select the option **Entry (Bus)** from the **PLACE** submenu and click on it. The **PLACE-Entry (Bus)** submenu will appear.

6. First click on the option **Wire** from the **PLACE-Entry (Bus)** submenu and then click on the preferred direction for the entry wire. A little piece of wire will be available at the mouse pointer. Position the wire between one of the microprocessor pins and the bus in such a way that one end of the wire touches the bus and the other end touches the A0 pin of the microprocessor. Once the piece of wire is placed properly, click the mouse once and select the option **PLACE** from the **PLACE-Entry (Bus)** submenu.

7. In this way address pins (lines) can be connected to the address bus, and data pins can be connected to the data bus.

8. Now each entry wire to the bus needs to be labeled. Select the option **Label** from the **PLACE** submenu, provide the name of the first label at the prompt (use A0 for the address bus), and press Enter ⏎.

9. The label will be at the mouse pointer. Position the label at the first entry wire. If it is the A0 line of the address bus, place the label at the point where the A0 address line enters the bus.

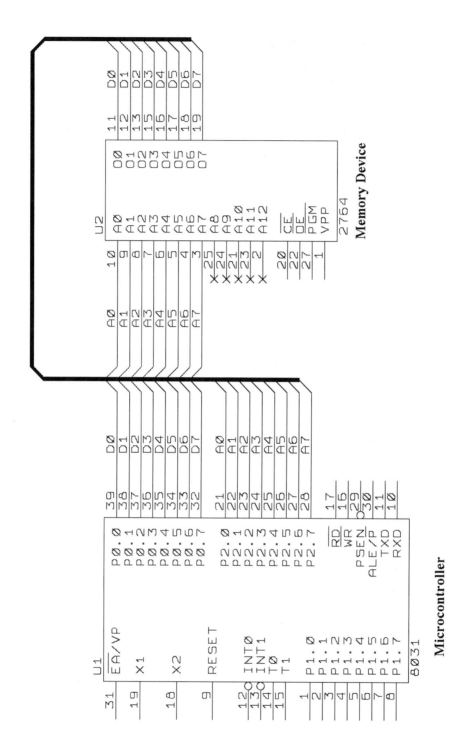

Figure 7-2. Bus Connecting Microcontroller and Memory Device

10. Select the option **Repeat** from the SDT_MAIN menu and click on it. The next bus entry will be labeled A1. Keep on left-clicking the mouse until all the bus entries are labeled. Figure 7–2 also shows an example of bus entry to a microcontroller. There is nothing called **Exit Bus;** it is just another orientation of the **Entry Bus.**

Placing a junction box at intersections. Two crossing wires do not represent a junction. To tell Schematic Design Tools that the crossing wires are connected, a junction box needs to be placed at the intersection of the wire. Also, if a wire end terminates at a wire in a T shape, it is also considered as no connection until a junction box is placed at the intersection. The following are the steps to place a junction box:

1. Left-click on **PLACE** from the SDT_MAIN menu.
2. Select **Junction** from the **PLACE** submenu and click on it. A bar menu will appear at the top.
3. Position the mouse pointer at the point where a junction box is desired.
4. Click on the mouse once and select **Place** from the **PLACE-Junction** submenu. A green junction box will appear at the tip of the mouse pointer.
5. A junction box can be placed at another intersection by placing the mouse pointer on it and selecting **Place** from the **PLACE-Junction** submenu.

7–3.7 Edit part fields content

Edit part fields. Each schematic part has ten reserved data fields to store information about the part:

- Reference
- Part Value
- 1st Part Value through 8th Part Value

The information that is generally stored is reference number, part value, packaging type, tolerance, manufacturer, and any other value the circuit designer may need pertaining to the part.

Reference values for different types of components are generally as follows:

- Resistor: R1, R2, etc.
- Capacitor: C1, C2, etc.
- Chip with multiple devices in a single package: U1A, U2B, etc.
- Chip with a single device in a package: U1, U2, etc.

- Diode: D1, D2, etc.
- Transistor: Q1, Q2, etc.

Part values for different types of components generally are as follows:

- Resistors: 10KΩ, 100Ω, etc.
- Capacitors: 50 µF, 10 µF, etc.
- Functional chips: 741 (Op-amp), 7408 (AND Gate), 4011 (CMOS NAND Gate), etc.

Part fields contain information such as packaging type, tolerance, and ratings:

- Resistors: RC07, RC01 (packaging type for resistors)
- Capacitors: CK01, CK07 (packaging type for capacitors)
- Single chip: 8DIP300, 14DIP300, 64DIP600 (DIP package for ICs)

Note: OrCAD provides a list of all types of packaging available for various components. You may also generate this packaging information from the PCB module library by using the procedure described in Appendix B.

When a schematic part is fetched from a library, the reference field is always unidentified, so the circuit designer needs to provide this information. If the part is a resistor, then it will have in its reference field an R?. If it is an IC chip with multiple devices, it will have U?A. If the schematic part is a resistor, a capacitor, a diode, or a transistor, the part value field does not come with any specific value. However, if it is an IC chip, it generally comes with a type number, such as 74LS08 or 74LS138 for TTL chips. Part fields 1 through 8 almost never have any information when a part is fetched from a library.

To be able to process correctly, the reference and part value fields must have appropriate information. Information in other fields depends on the type of processing. To generate a netlist for a printed circuit board, Part Field 1 must have packaging information. Each schematic part can be edited one at a time to provide information to part fields. However, there are tools within SDT that update this information much more efficiently.

Edit part fields one component at a time

1. Get the schematic whose part you want to edit on the worksheet.
2. Place the mouse pointer on the part you want to edit.
3. Select **EDIT** from the SDT_MAIN menu and click on it. The **EDIT-Part** submenu will appear.
4. Select **Reference** and click on it; a **Part-Reference** submenu will appear.
5. Select **Name** if you desire to edit the reference name, and click on it. If the part is a resistor, R? will appear at the top of the screen.

7-3. Designing and Drafting a Circuit Schematic

You can delete R? by using **Backspace** and then using other keys to add desired information to the field. However, it is strongly recommended to keep **R** as the first letter for all resistors, **C** for capacitors, and **U** for IC chips in order to be consistent.

6. Select **Location** if you want to change the location of the reference designator. The Reference designator will be highlighted and can be moved to the desired location by using the mouse pointer. Once the highlighted box is at the desired location, click once and select **Place** to move the reference designator to the desired location.

7. Select **Visible** if you want to change the visibility of the part reference. Often a schematic becomes unreadable due to too much text; therefore, it is a common practice to hide certain information from a schematic while displaying it. However, you never want to hide reference designators from the schematic, since these are necessary to identify the part.

8. Select **Yes** to keep the reference name visible or **No** if you want to keep the reference name invisible. For almost all processing purposes the reference name is there, but it is just not visible on the worksheet.

9. Select **Part Value** from the **EDIT-Part** submenu and click on it. The existing part value for the part will be displayed. If it is a resistor and you have not entered any value before, it will be RESISTANCE. If it is a single chip, such as a TTL chip, it will display its number—for example, 7408 or simply 08. Use **Backspace** to delete and use other keys to change the information of the field.

10. Select **1st Part Field** and click on it. Since the part field of a newly fetched part from a library is generally empty, nothing will be displayed. You can enter the packaging type of the component in this field. For an 8 pin DIP chip, enter 8DIP300, and for a 14 pin chip, enter 14DIP300, etc.

11. Select **Which Device** from the **EDIT-Part** submenu, and click on it. If the selected part has two devices per package, it will display the A and B device. For the first one you should choose A, and for the second similar device in the schematic you should choose B. This means that you have used both the devices of this part. If the part has four devices per package, you may choose any one of them or all, depending on your needs. Notice that as you select different devices of a part, the pin numbers of those devices change. This happens because input and output pins of each of these similar devices are different.

7-3.8 Placing Labels, Text, and Comments

Placing labels and comments. Labels on a schematic are not simple text; they are identifiers placed on a schematic that can physically connect signals together. For example, two wires ending with the same label are considered a connection without a physical wire link. Labels are generally used for common signals to many devices in the schematic, such as clock (CLK) signal, V_{CC} signal for transistor, and V_{DD} for MOSFETs. See the following procedure.

1. Get the schematic to which you want to add labels on the SDT_MAIN worksheet.
2. Select **PLACE** from the SDT_MAIN menu and click on it.
3. Select **Label** and click on it. A Label? prompt will appear. Enter the name of the label, such as CLK, if you are labeling a clock input/output, or V_{CC}, if you are labeling a common collector input/output power for a transistor. Suppose five other devices in the schematic have an input whose label is also CLK and five transistors have the collector input V_{CC}. You do not have to make a physical connection among all the CLK and all the V_{CC}. To OrCAD, the label represents an electrical connection.
4. Position the leftmost point of the label name close to the wire or bus you are labeling. Select **Place,** and the label name will appear at the desired position.

Placing comment text. It is often necessary to put a short descriptive text on a schematic so that other people can understand your schematic. The following procedure will show how to do this:

1. Get the schematic on the screen.
2. Select **PLACE** from the SDT_MAIN menu, and click on it. A submenu will appear.
3. Select **Text** and click on it. A Text? prompt will appear. Enter the desired text, left-click the mouse, and the **PLACE-Text** submenu shown in Figure 7–3 will appear. By choosing between **Larger** and **Smaller,** the entered text can be sized. By choosing the option **Orientation,** the entered text can be placed vertically or horizontally.
4. Move the text image to the desired place on the worksheet by using the mouse. Click on the mouse once and select **Place** to position the text on the worksheet.

7–3.9 Saving and Updating Schematic Files

Saving and updating schematic files at various time intervals must be thoroughly understood. After a long session of drawing schematics, if you cannot save your file, all your effort will be wasted. It is always recommended to update your schematic frequently (such as every 5 to 10 minutes) as a precaution against power failure or other unexpected events.

Figure 7–3. PLACE-Text Submenu

| Place | Orientation | Value | Larger | Smaller | Find | Jump |

Update is used if you have invoked a named worksheet with or without any components. **Save** or **Write to File** is used if you want to save the schematic in a different name or if you have invoked an unnamed worksheet. Remember that a design name may or may not be the same as the schematic filename. The default design name and schematic filename are the same. If you have more than one schematic file, you have to provide different filenames for each one. The following procedure will save a schematic under a new name:

1. From the SDT_MAIN menu, click on **QUIT,** and the submenu for **QUIT** will appear with the schematic filename at the top of the screen.

2. Click on **Write to File,** and the SDT software will prompt for a new filename. Respond to the prompt by providing a filename with SCH extension. The software will save the schematic file in the new filename. If you are running the design environment, the file will be saved under the design subdirectory. If, however, you desire to save the file on a floppy, provide the appropriate destination for the file. For example, to save the file called LOG_PRO under a subdirectory called LOGIC in floppy A, use the following in response to the prompt for a filename:

   A:\LOGIC\LOG_PRO.SCH

3. Right-click the mouse to get back to the SDT_MAIN menu to resume the schematic design process. (Caution: If you are using a computer that is shared with other people, always remember to save your work on a floppy to avoid possible loss.)

The following procedure will update a previously named worksheet:

1. From the SDT_MAIN menu, click on **QUIT.** The **QUIT** submenu will appear with the schematic filename.

2. Click on **Update File,** and the SDT software will save the file in the same destination from where it was invoked. If you have invoked the file from the floppy disk, it will save the file there. If it is from the **C-drive,** it will save the file there.

3. Right-click the mouse to get back to the SDT_MAIN menu to resume the schematic design process.

7-4. CHECK ELECTRICAL RULES

Before generating a netlist file for the PCB tool set, the schematic must be checked for any violations of basic electrical rules. This tool scans the electrical circuit of a design, checks the validity of the basic electrical rules, and reports the problems back into a destination file. Electrical problems generally reported are unused inputs of parts, two pins defined as outputs connected together, a power and a ground connected together, etc. Conditions to be checked are specified in the Check Electrical Rules Matrix, located in the Configure Schematic Design Tools screen. Electrical errors reported by the tool must be corrected before proceeding further with the design process. However, warnings may be corrected or can be omitted, depending on their type.

7-4.1 Using the Check Electrical Rules Matrix

Before running a circuit through rules check, it is imperative to get familiar with the matrix. The matrix shows the pins, module ports, and sheet net names in columns and rows. At the intersection of a row and a column, there is a little square box that is either empty, contains the letter **W** or **E**. An empty intersection represents valid connection between column and row items, a **W** represents a warning that you should avoid such a connection, and **E** represents an error, meaning an illegal connection that must be eliminated. To toggle an intersection among the three settings, click the mouse on the square until the desired settings appear. To get back to the default setting of the rule matrix, click the mouse pointer on **Set to Defaults.** To invoke the Check Electrical Rules Matrix, use the following procedure:

1. Invoke SDT_MENU from ESP_MENU.
2. Place the mouse pointer on the **Draft** icon and click once. The Schematic Tools Transit Menu will appear.
3. Select **Configure Schematic Tools** and click on it. The SDT configuration screen will appear. Scroll down to the end of the menu to the Check Electrical Rules Matrix.
4. Place the mouse pointer on an empty square and click on it a couple of times to see how it changes from empty to **W** to **E** and to empty again.

7-4.2 Running Check Electrical Rules for a Schematic

1. Invoke SDT_MENU and click on the **Check Electrical Rules** icon located within the **Reporters** tools. The Transit Menu will appear.
2. Select **Local Configuration** and click on it. The Local Configuration screen, shown in Figure 7-4, will appear. Provide the name of the schematic file whose electrical rules need to be checked. Also provide the destination filename. The destination file should have the same filename as the schematic file but with an ERC extension. The tool will write all the errors and warnings in this file.
3. Exit from the Local Configuration screen by clicking on the OK icon.
4. From the Transit Menu, select the **Execute** option and click on it. The tool will check the schematic against all the electrical rules described by

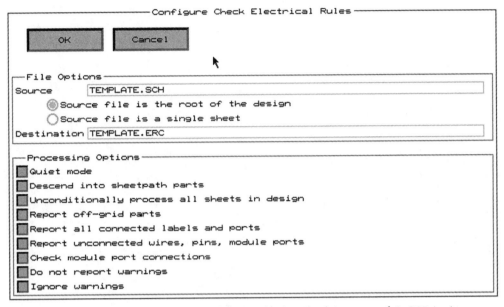

Figure 7-4. Local Configuration Screen of Electrical Rules Check (courtesy of OrCAD Inc.)

the matrix. Any error or warning will be reported to the ERC file. Errors are also designated by glowing circles in your original schematic. You may have more than one glowing circle for one error.

5. Invoke the schematic by clicking on the **Draft** icon. Examine the schematic for glowing circles. Your interpretation of the glowing circles will point you towards actual errors. Oftentimes these glowing circles do not point to the actual error but its effects. One missing junction box may cause several glowing circles. Therefore, logical interpretation is absolutely necessary. Warnings and errors are written in words, and their location is specified by the X-Y coordinate position in the ERC file. Figure 7–5a shows an example of a schematic file with two missing junction boxes, shown by arrows. Figure 7–5b shows the same circuit after processing through the Electrical Rules Check. Error circles, shown by arrows, either point directly to the error or point to the effect of an error.

7-5. ANNOTATE SCHEMATIC

7-5.1 Annotate

Annotate Schematic is a tool by which the schematic designer can assign device numbers and reference designators to parts in a new schematic. The reference designators need to be provided by the designer, depending on the number of IC chips, resistors, and switches the schematic has.

Sometimes two, four, six, or even more electronic devices are packaged in a single DIP chip. When such a device is fetched from a library, its refer-

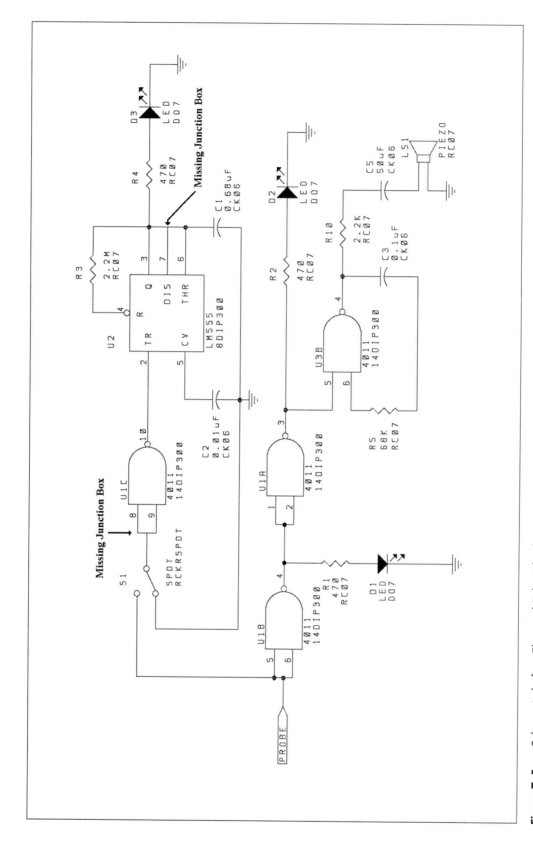

Figure 7-5a. Schematic before Electrical Rules Check

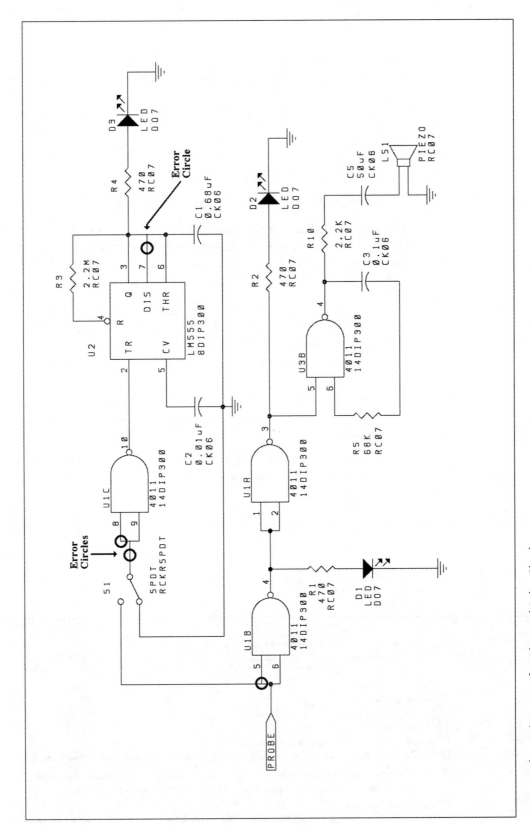

Figure 7-5b. Schematic after Electrical Rules Check

ence designator is marked as U?A. Even if ten of them are fetched, all will be marked as U?A. Therefore, when a schematic is drawn using devices, their reference designators need to be assigned as U1A, U1B, etc. U1A is the device A and U1B is the device B of IC chip U1. There are two ways that appropriate reference designators can be assigned to each part of a schematic:

- Edit the reference field of each part
- Use the **Annotate** tool

The first method is tedious and prone to error, especially for a schematic with many parts, whereas the second method is comparatively faster and not error prone. The first method, editing, was described earlier. The procedure for the second method is described below:

1. Invoke the schematic that needs to be annotated. If your design has more than one schematic sheet, invoke the one you need to annotate. Remember its name. If you want any of the parts to have reference designators of your choice, assign them now by editing the parts' reference field.

2. Invoke SDT_MENU and click on the **Annotate Schematic** icon, which is located under **Processors.** The **Annotate** Transit Menu will appear.

3. Click on **Local Configuration.** The Local Configuration menu of **Annotate Schematic** shown in Figure 7–6 will appear.

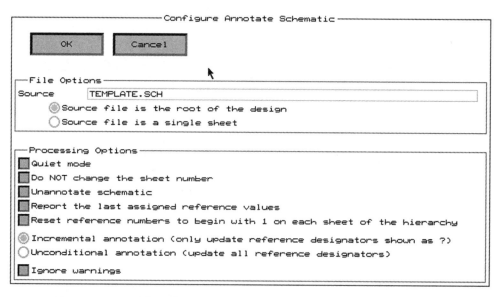

Figure 7–6. Local Configuration Screen of the Annotate Tool (Courtesy of OrCAD Inc.)

4. Provide the schematic filename you want to annotate as the source under the file option. If all the schematics of the design need to be annotated, then click on the first bubble (source file is the root of the design) or else the second bubble (source file is a single file). Under **Processing Options,** choose Incremental Annotation if you have assigned special reference designators to some parts. Otherwise, choose Unconditional Annotation.

5. Exit the Local Configuration menu by clicking on the OK icon.

6. Click on the **Annotate Schematic** icon again, and this time select **Execute** from the Transit Menu and click on it. Unless there is a problem, annotation will be done for the schematic or the design name you provided.

7. Invoke the schematic you just annotated for verification by first clicking on **Draft** icon and then selecting **Execute** from the Draft Transit Menu. Figure 7–7a and Figure 7–7b are the two schematics before and after processing by the Annotate Tool.

7–5.2 Unannotate

When a part is fetched from a library, its reference designator is a **?** instead of a number. For example, if a resistor is fetched, it will have R? in its reference designator field instead of R1 or R2, etc. After the schematic is annotated, the reference designator of each part will change to the appropriate value. Let us suppose that you do not like the way the software has done the annotation; you can go back to its unannotated condition. The following procedure will let you unannotate the schematic.

1. Invoke the Transit Menu for **Annotate Schematic** by clicking on the **Annotate Schematic** icon.

2. Select **Local Configuration** from the Transit Menu and click on it. The Local Configuration menu will appear.

3. Provide the name of the appropriate source file. Under **Processing Options,** click the mouse on the square button that says Unannotate Schematic. The button will change color to green, indicating that the option is active.

4. Return to SDT_MENU by clicking on the OK icon.

5. Invoke the Transit Menu of **Annotate Schematic** again, and select the option **Execute.**

6. Invoke the schematic you have just unannotated, using the **Draft** icon for verification.

7. Assign reference designators to those parts that you want to specify. After providing the reference designators of your choice, you may annotate the schematic again. However, first you should deselect the **Unannotate** option located in the Local Configuration screen.

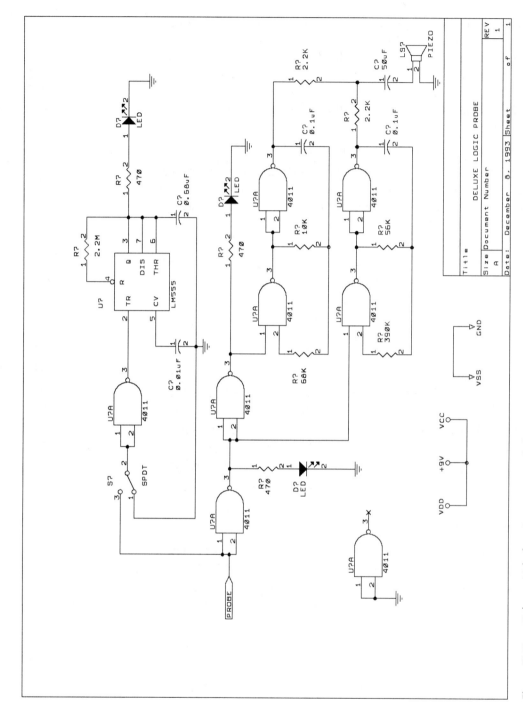

Figure 7-7a. Schematic before Processing with the Annotate Tool (*Source:* Delton T. Horn, *50 CMOS IC Projects*, 1988, p.64, used with permission from TAB Books Inc.)

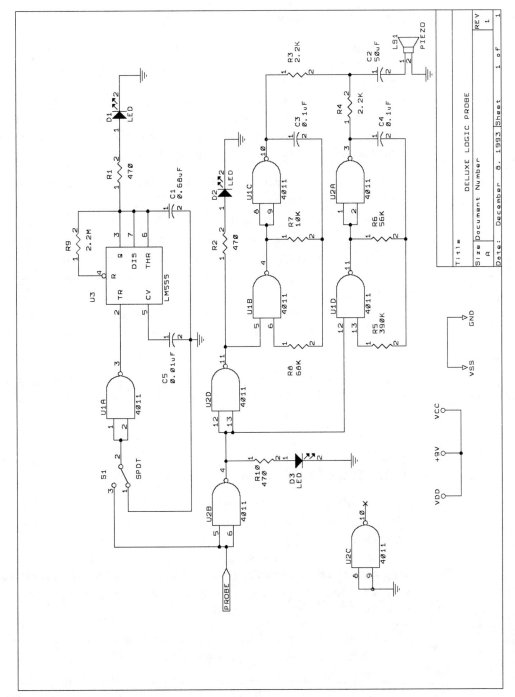

Figure 7-7b. Schematic After Processed by the Annotate Tool (*Source: Delton T. Horn, 50 CMOS IC Projects, 1988, p.64,* used with permission from TAB Books Inc.)

Chaper 7 / Computer-Aided Schematic Capture

7–5.3 Back Annotate

Back Annotate is a tool by which reference designators of parts of a schematic can be changed from old to new reference designators. To perform this, a text file must be provided that lists the old designator on the left and the new designator on the right. This is called the **Was/Is** file. The **Was/Is** file is used by the tool to modify the reference designators of the schematic. This tool is especially useful for modifying large numbers of reference designators. See the following procedure.

1. Invoke the Transit Menu of **Back Annotate** by clicking on the **Back Annotate** icon.
2. Select **Local Configuration** and click on it. The Local Configuration screen of **Back Annotate** will appear. Provide the appropriate schematic filename you want to back annotate. Under the **Was/Is** option, provide the text filename that contains the list of changes you want to make.
3. Invoke the Transit Menu for **Back Annotate** and select the option **Execute.** Invoke the schematic you have just back-annotated to verify it.

The schematic of Figure 7–7b when processed by the Back Annotate tool using the **Was/Is** file shown in Figure 7–8a will look like the schematic shown in Figure 7–8b.

```
Was/is File
'R1'   'R12'
'R2'   'R11'
'C1'   'C10'
'C3'   'C9'
'U1A'  'U1A'
'U1C'  'U1A'
```

Figure 7–8a Was/Is File for Back Annotation

7–6. SELECT FIELD VIEW

The **Select Field View** tool is used to change the visibility attribute of the specified field. Often a schematic contains a lot of pertinent information about parts, but displaying this information makes the schematic look crowded. Note that each schematic part has ten fields that can contain information. This tool can help hide some of this information from the display. The following procedure will make the contents of any field of a part invisible.

1. Invoke the **Select Field View** Transit Menu by clicking on the tool icon located under **Processors** in the SDT_MENU screen.
2. Select the **Local Configuration** option and click on it. The Local Configuration screen shown in Figure 7–9 will appear on the screen.
3. Under **File Options,** provide the name of the schematic file (with SCH extension) whose visible attribute needs to be changed.
4. Under **Processing Options,** click on the bubble beside the field name whose visibility attribute you want to modify. Only one field at a time can be modified. If you need to change the visible attribute of two fields of a schematic, the circuit needs to be processed twice.

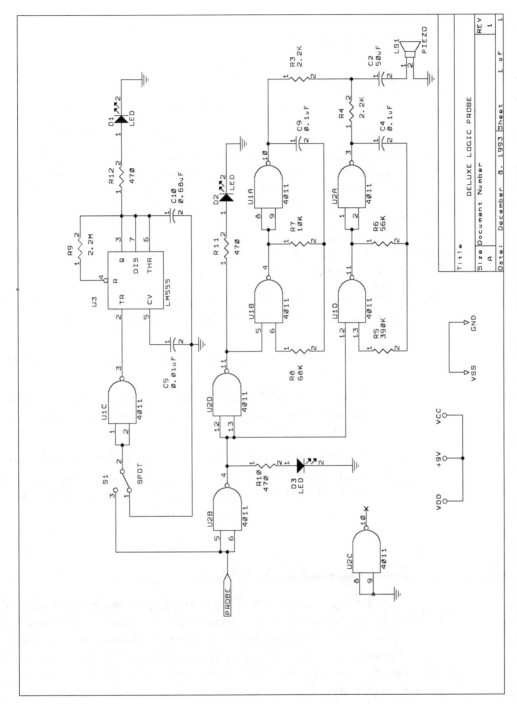

Figure 7-8b. Schematic after Back Annotation (*Source*: Delton T. Horn, *50 CMOS IC Projects*, 1988, p.64, used with permission from TAB Books Inc.)

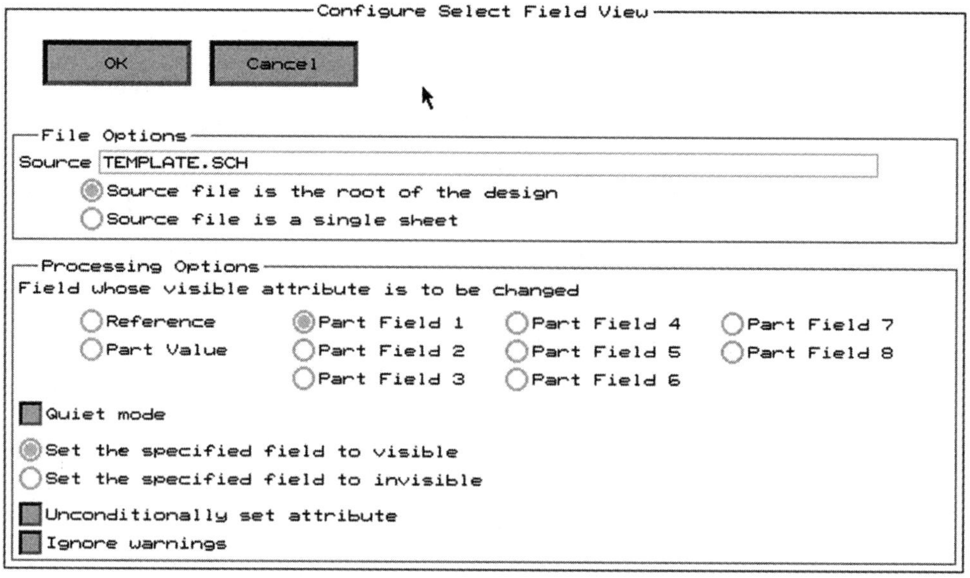

Figure 7-9. Select Field View Local Configuration Screen (Courtesy of OrCad Inc.)

Also, only one type of field at a time can be modified for all the parts in a schematic.

5. If the contents of the specified field are desired to be visible, click on the bubble beside the field. If they are desired to be invisible, click on the next bubble.//
6. Exit from the Local Configuration menu by clicking on the OK icon.
7. Invoke the Transit Menu of the tool again, select **Execute,** and click on it. The schematic will be processed by the tool. Invoke the schematic to verify the desired operation.

7-7. CLEANUP SCHEMATIC

The **Cleanup Schematic** tool removes the more obvious problems of your schematic. Some of the obvious problems it repairs are duplicated or overlapping wires, buses, and junctions. After the clean-up process, the original schematic is renamed with a BAK extension and the cleaned-up version of the file is saved under the original name with an SCH extension. Use the following procedure:

1. Invoke the Transit Menu of the **Cleanup Schematic** tool, located under **Processors** in the SDT_MENU screen.
2. Select **Local Configuration** and click on it. The Local Configuration of the tool shown in Figure 7–10 will appear.
3. Under **File Options,** provide the name of the schematic file you want to clean up. You do not have to provide the destination filename.

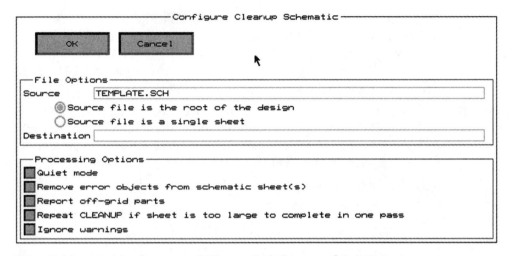

Figure 7-10. Local Configuration of Cleanup Tool (Courtesy of OrCAD Inc.)

However, if you do provide it, the cleaned-up version of the schematic will be saved under this name.

7-8. M2EDIT TEXT EDITOR

The M2EDIT editor is a general purpose full-screen text editor. During the schematic and artwork design process, a text editor will be required to create the following text files. You may use another text editor, but having the M2EDIT editor within the environment is very convenient when you are performing the following operations:

- Creating Was/Is files for Back Annotate
- Creating Stuff (.stf) files for Update Field Contents
- Creating library source files
- Viewing and editing Netlist files (.net)
- Viewing ERC files generated by Check Electrical Rules

7-8.1 Editing a Text File

The M2EDIT text editor can be invoked from the main menu screen of all tools. Here we are going to access the editor from the SDT_MENU screen. Using the following steps, a text file can be created:

1. Use the Design Management Tools to set the appropriate design environment for the text file.
2. Invoke ESP_MENU and click on the **Schematic Design Tools** icon. The SDT_MENU screen will appear on the screen.

3. Left-click on the **Edit File** icon. The Edit File window (EDIT_FL) will appear. The list box may already have many files. If you need to create a new text file that does not exist, do not click on any filename in the list box. Just click on the OK icon, and the M2EDIT screen will appear on the screen.

4. If you want to edit a file that already exists, scroll up and down the list box to find the file, and click on it. The filename will be highlighted. Click on the OK icon. The M2EDIT screen with the text file will appear on the screen.

5. Common keys and prompts for the M2EDIT editor:

    a) The text cursor is a flashing underline, _.
    b) The mouse prompt is a solid block, ■.
    c) **Shift** ↑ is used for capitalization.
    d) **Delete** is used for deleting characters.
    e) **Return** is used for end of line.
    f) ← **Backspace** is used for delete and backspacing.
    g) ↓↑ →← keys can be used to move around in the file.
    h) →← ↓↑ these keys, located in the edit screen border, can be used to move around the text.

6. Any text can be created on the M2EDIT screen by entering the characters available on the keyboard.

7-9. UPDATE FIELD CONTENTS

Each part in a schematic diagram has ten part fields. The first field is the part's Reference Designator. The second field is the part's Value. The eight other fields contain user-defined information, such as type of packaging, power/voltage/current rating, and any other pertinent value. A very important piece of information needs to be placed in one of these part fields: i.e., packaging type—as an example, for a 14 pin DIP, it must be 14DIP300. Using the procedure described earlier, each part can be edited to enter the packaging type in Part Field 1. But this is a very tedious process, especially for a large schematic. The **Update Field Contents** tool can enter this data in one operation. OrCAD provides a list of various packaging available for parts. This information can also be generated from the PCB package library. If you generate a netlist file for a schematic without this information, the file cannot be used by the OrCAD/PCB software.

The **Update Field Contents** tool constructs a string using the configuration of the Key Field designators. Suppose that the dialogue box of Combine for Field 1, located under Key Fields of the SDT Configuration screen, shows **R V**. If you execute the **Update Field Contents** tool for a schematic that has this configuration, the software will scan each part in the schematic

```
─Key Fields──────────────────────────────────────────
Annotate Schematic
      Part Value Combine   [                    ]
Update Field Contents
      Combine for Value    [                    ]
      Combine for Field 1  [R                   ]
      Combine for Field 2  [                    ]
      Combine for Field 3  [                    ]
      Combine for Field 4  [                    ]
      Combine for Field 5  [                    ]
      Combine for Field 6  [                    ]
      Combine for Field 7  [                    ]
      Combine for Field 8  [                    ]
Create Netlist
      Part Value Combine   [V                   ]
      Module Value Combine [1                   ]
Create Bill of Materials
      Part Value Combine   [                    ]
      Include File Combine [                    ]
Extract PLD
      PLD Part Combine     [                    ]
      PLD Type Combine     [                    ]
```

Figure 7-11. Key Field Designators for Update Field Contents (Courtesy of OrCAD Inc.)

and will create a string using the content of the Reference Designator (R), a space, and the content of the Part Value (V) of parts. Figure 7–11 shows the Key Field designators for **Update Field Contents.** The tool will then try to find a match with the string listed in the update file (a file with an STF extension). The update file contains a list of Reference Designators and Part Values separated by a space within a single quote of all the parts of the schematic on the left and the matching string, such as 14DIP300, whose destination is Part Field 1 on the right. Remember that a space is considered as a character in ASCII. Therefore, if you have a space between **R** and **V** in the Key Fields, you must have a space between Reference Designator and Part Value in the Stuff file. The following is an example of an Update file.

⌐ Space between reference designator and part value is important.

'R1 10K'   'RC07'
'U1 74LS04'   '14DIP300'
'C1 50uf'   'RC02'

In this file, information in the left column is the string created for every part of the schematic by the Update tool, based on the configuration of the Key Fields. Information in the right column is the parts' packaging type, whose destination is Part Field 1 in the schematic.

The following procedure can be used to update the content of Part Field 1 of all parts of a schematic in one shot:

1. Invoke the Configure Schematic Design Tools screen and scroll down to Key Fields. Under Update Field Contents enter the letter **R**

in the dialogue box of Combine for Field 1. Exit the screen by using the OK icon.

2. Create an ASCII file similar to the one shown below, using the M2EDIT editor. If you use other word processing software, make sure the file is free from any formatting characters. This way, you put the value of all the parts of your schematic on the left between single quotes and the associated packaging type number on the right, also between single quotes. Obtain the packaging number from the list provided by OrCAD and also provided in Appendix B and E of this text. Figure 7–12 is the Update file for the schematic shown in Figure 7–7b.

'10K'	'RC01'
'100K'	'RC04'
'741'	'14DIP300'
'74LS08'	'14DIP300'

3. The ASCII file in Figure 7–12 has no set format as long as you put the part value and its associated packaging number in a single line and between single quotes. Give the file the same name as the schematic filename, with an STF extension.

4. Invoke SDT_MENU, and click on the **Update Field Contents** icon. Select **Local Configuration** from the Transit Menu. The Local Configuration screen of the **Update Field Contents** tool shown in Figure 7–13a will appear. Provide the schematic filename in the

Figure 7–12. Update File for Schematic of Figure 7–7b

LOGIC PROBE STUFF FILE

'C1'	'CK06'
'C2'	'CK06'
'C3'	'CK06'
'C4'	'CK06'
'C5'	'CK06'
'D1'	'DO7'
'D2'	'DO7'
'D3'	'DO7'
'R1'	'RC07'
'R2'	'RC07'
'R3'	'RC07'
'R4'	'RC07'
'R5'	'RC07'
'R6'	'RC07'
'R7'	'RC07'
'R8'	'RC07'
'R9'	'RC07'
'R10'	'RC07'
'U1'	'14DIP300'
'U2'	'8DIP300'
'U3'	'14DIP300'
'LS1'	'PIEZO'
'S1'	'RCKRSPDT'

7-9. Update Field Contents

Source dialogue box and the name of the file you just created with STF extension in the Update file dialogue box.

5. Under processing options, check to see if the Part Field 1 bubble is green. If not, click on it to activate.

6. Click on Create an Update Report and provide a filename with a TXT extension in the dialogue box.

7. Click on Unconditionally Update Field option square, and click on Set the Specified Field to Visible bubble. Exit the local configuration by using the OK icon.

8. Click on the **Update Field Contents** icon and select **Execute** from the Transit Menu. The software will scan the Part Value field of all parts of the schematic, find a matching packaging number from the STF file, and place it in 1st Part Field of the schematic. Figure 7–13b shows the schematic after processing through the **Update Field Contents** tool.

9. If the process fails due to error, check the STF file carefully for any undue space. Remember, space is also an ASCII character. The matching string in the STF file and in the Key Field dialogue box must be exactly the same. Also check errors between the letter "O" and the number "0" because the first two digits of many package numbers are letters. For example, it is TO92, not T092.

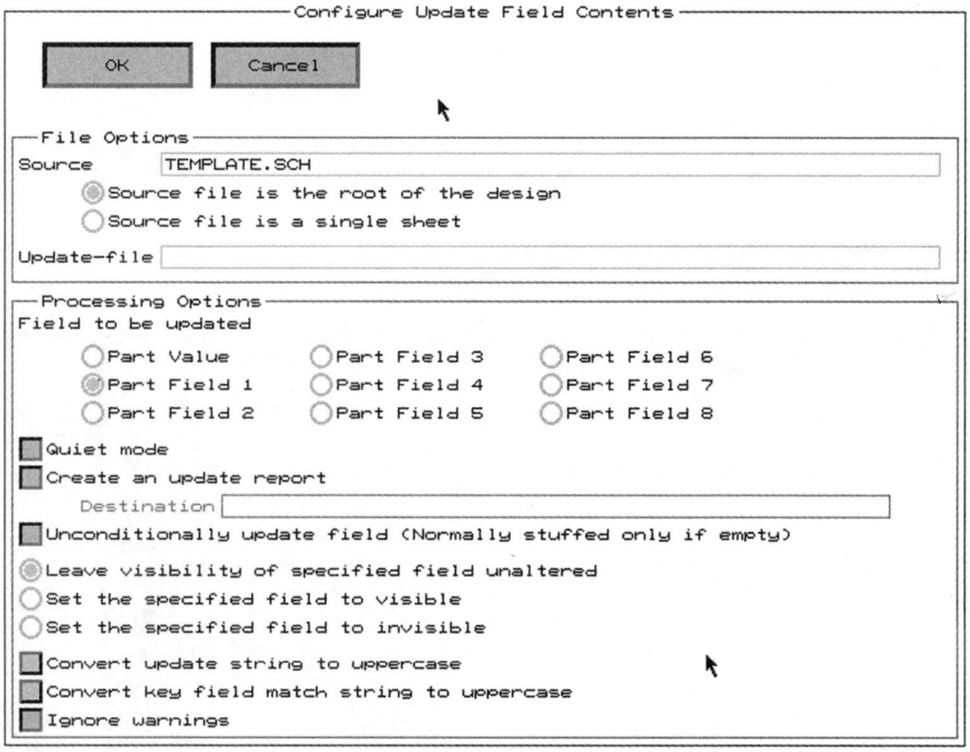

Figure 7–13a. Local Configuration of Update Field Contents Tool (Courtesy of OrCAD Inc.)

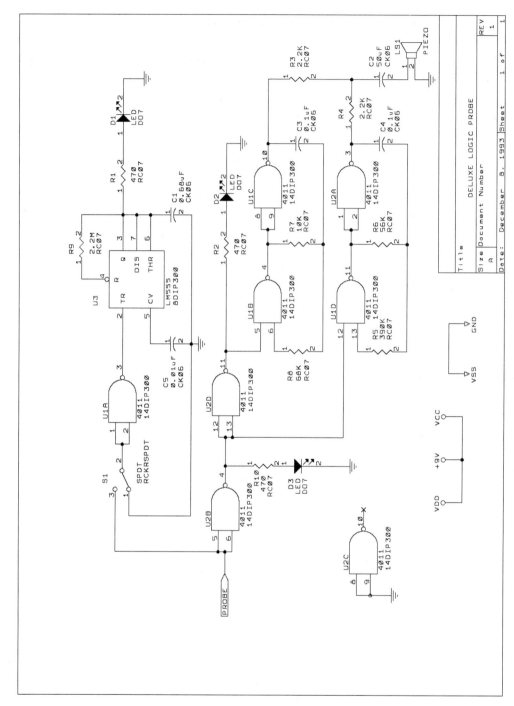

Figure 7-13b. Schematic after Processing by Update Field Contents (Source: Delton T. Horn, *50 CMOS IC Projects*, 1988, p.64, used with permission from TAB Books Inc.)

If the schematic has only a few parts, using **Update Field Contents** may be more time consuming than editing part fields one component at a time.

7-10. ARCHIVE SCHEMATIC PARTS

In OrCAD/SDT, schematic parts and symbols are grouped in libraries. These libraries are named after the name of their manufacturer, with Moto standing for Motorola, etc. If a product has many manufacturers, its library name is given after the name of its type, such as TTL or Analog. When a schematic is drawn by fetching parts from these libraries, the schematic file keeps only the name of the parts. This means that if a schematic is drawn using parts from ten different libraries, all those libraries must be available in the Configured Library list box to display that schematic. Also, if this schematic is to be displayed under another OrCAD/SDT environment, all those libraries must be available in the Configured Library list box in that environment. To avoid these sorts of limitations, library parts of a schematic are archived into a single library source file. This library source file can be transported with the schematic to another OrCAD/SDT environment, and the schematic can be displayed without configuring any of the library. However, if the archived library for the schematic is not configured automatically by the software, you may have to insert the file in the Configured Library list box manually by using the **Insert** icon. This is done in the SDT configuration screen shown in Figures 6–19a through 6–19g.

Archiving parts of a single-sheet schematic or a design into a single library is accomplished in two steps. First, the tool builds a library source file of all parts in the design or in a single-sheet schematic. Second, it compiles the library source file into a form usable by the schematic **Editor, Processors, Reporters,** and **Transfer** tools. Each of these two processes can be performed separately by turning them **ON** or **OFF.** The following procedure will archive parts of a schematic into a single library file.

1. Invoke the SDT_MENU and click on the **Archive Parts in Schematic** icon. A Transit Menu for the Library Archive tool will appear. Click on **Local Configuration.** The Transit Menu of the **Configure Library Archive** tool shown in Figure 7–14a will appear at the upper left corner of the screen.

2. Select **Configure LIBARCH** and click on it. The LIBARCH configuration screen shown in Figure 7–14b will appear. Under **File Options** for Source file, provide the name of the single-sheet schematic or the design name file with the extension SCH. Click on the appropriate bubble, such as *Source file is the name of the root design* or *Source file is a single sheet*. For destination, provide the same filename with SRC extension. Exit LIBARCH by clicking on the OK icon.

3. Go to the Transit Menu again, and this time click on **Configure Composer.** The Composer configuration screen shown in Figure

Chaper 7 / Computer-Aided Schematic Capture

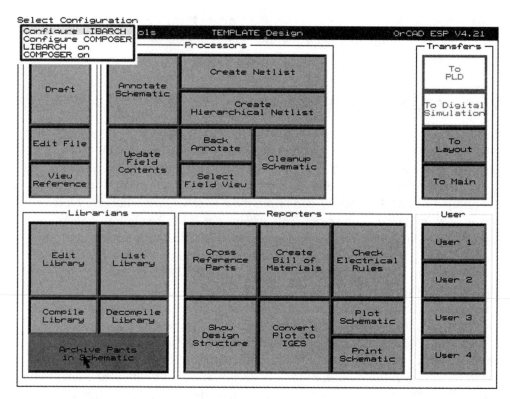

Figure 7–14a. Transit Menu of Archive Parts (Courtesy of OrCAD Inc.)

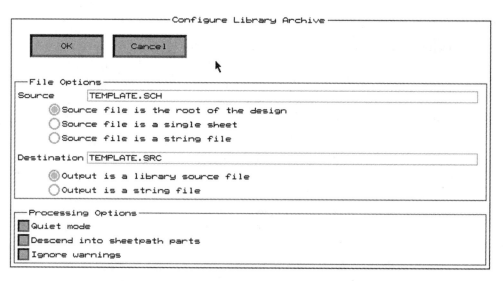

Figure 7–14b. LIBARCH Configuration Screen (Courtesy of OrCAD Inc.)

7–10. Archive Schematic Parts

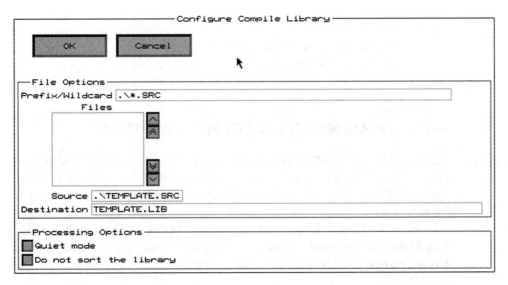

Figure 7-14c. Local Configuration of Compiling Library Tool (Courtesy of OrCAD Inc.)

7–14c will appear. Under **File Options,** select the file you want to compile from the list box. Provide the name of the destination file with the same name of your design or single sheet with LIB extension. Exit the screen by clicking on the OK icon.

4. Go to the Transit Menu again and click on **LIBARCH.** Choose **ON.** Similarly, choose **ON** for **COMPOSER** also.

5. Go to the **Archive Parts in Schematic** Transit Menu again, and select and click on **Execute.** The tool will create a library file for the design or single-sheet schematic in whatever manner you defined during configuration.

6. To test that the process has been done properly, you need to go to the **Library Options** portion of the Configure Schematic Tools screen. Check to see that a library file exists in the Configured Libraries list box whose name is the same as the name you provided as the destination library file under the Configure Compile Library screen.

7. Click on **Remove** and remove all the libraries from the Configured Libraries list box except the library file for your design. Exit the Configuration menu by clicking on the OK icon.

8. Go to **Draft** and try to display the schematic. It should appear on the screen. If you encounter any problems, you need to go through the archive process again.

You may view the parts of the archive library or any library by using the LIBRARY command from the SDT_MAIN menu. Use the following procedure.

1. Click on the **LIBRARY** command from the SDT_MAIN menu.
2. Click on **Browse** from the Transit Menu.

3. Click on **All Parts** from the options.
4. Click on the library name you want to view. You can view the parts by clicking on **Forward** or **Backward.** To exit, click on **Quit.**

7-11. CREATE NETLIST FOR PCB SOFTWARE TOOLS

Creating a netlist for PCB tools is one of the major steps toward the printed circuit board artwork design process. The netlist file is the link between OrCAD/SDT and OrCAD/PCB software tools. It contains all the connectivity information, packaging information, reference, and value of parts of a schematic or design. To generate the correct netlist format for PCB tools, Key Fields (located under Configure Schematic Design Tools) dedicated to **Create Netlist** must be configured appropriately.

7-11.1 Configure Key Field before Generating Netlist File for PCB

1. Invoke the Configure Schematic Tools screen by clicking on the option for the **Draft** Transit Menu.
2. Scroll down to Key Fields dedicated to **Create Netlist** and configure it as follows:

Create Netlist

| Part Value Combine | V |
| Module Value Combine | 1 |

There are two netlist formats:

- Flat netlist, which is produced by the **Create Netlist** tool
- Hierarchical netlist, which is produced by the **Create Hierarchical Netlist** tool

7-11.2 Generate Netlist

The **Create Netlist** tool first converts all parts, ports, and signals into a flat file, and then the flat netlist file is generated.

Hierarchical netlist. The **Create Hierarchical Netlist** tool creates a netlist file in hierarchical format. Here, sheets and subsheets remain in their hierarchical form and are used to reference nets and subnets. This text will describe only the flat netlist format and the procedure to create it.

Flat netlist file. A flat netlist file is generated in three steps:

1. INET
2. ILINK
3. IFORM

INET produces the incremental connectivity database for the design. Next, ILINK converts the incremental database into a single database where all parts have unique references and all nodes have unique names. Finally, IFORM produces the final flat netlist file for the entire design. The following procedure will create a flat netlist file for a design or a single-sheet schematic:

1. Invoke SDT_MENU on the screen. Click on the **Create Netlist** icon located under **Processors.** The Transit Menu will appear. Click on **Local Configuration** to display the Transit Menu of the **Create Netlist** tool, shown in Figure 7–15a. Since this is a three-step process, configuration of tools in each step is required.

2. Click on **Configure INET,** and the Local Configuration screen of **INET** shown in Figure 7–15b will appear. Under **File Options,** provide the source as the name of your schematic file or design, with an

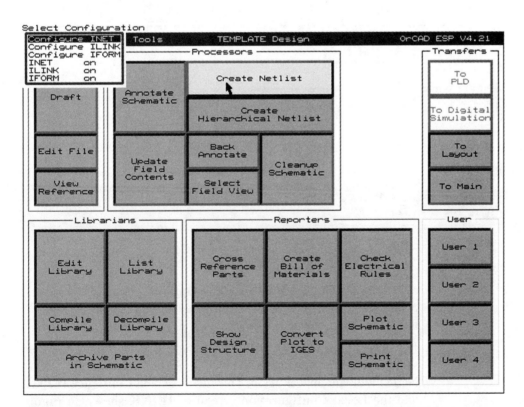

Figure 7–15a. Configuration Transit Menu of Create Netlist (Courtesy of OrCAD Inc.)

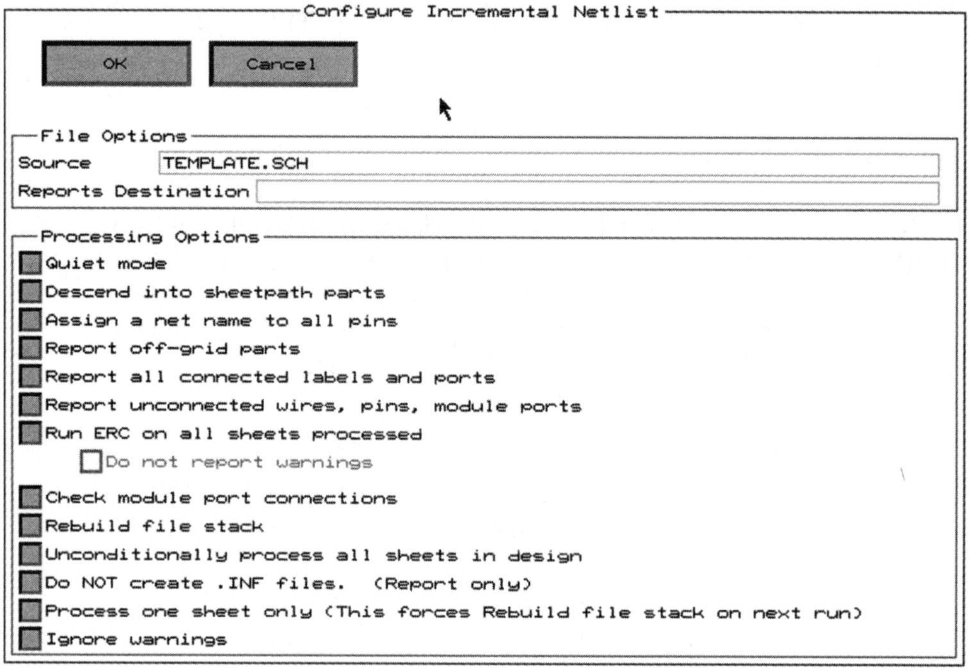

Figure 7-15b. Local Configuration of Netlist Tool (Courtesy of OrCAD Inc.)

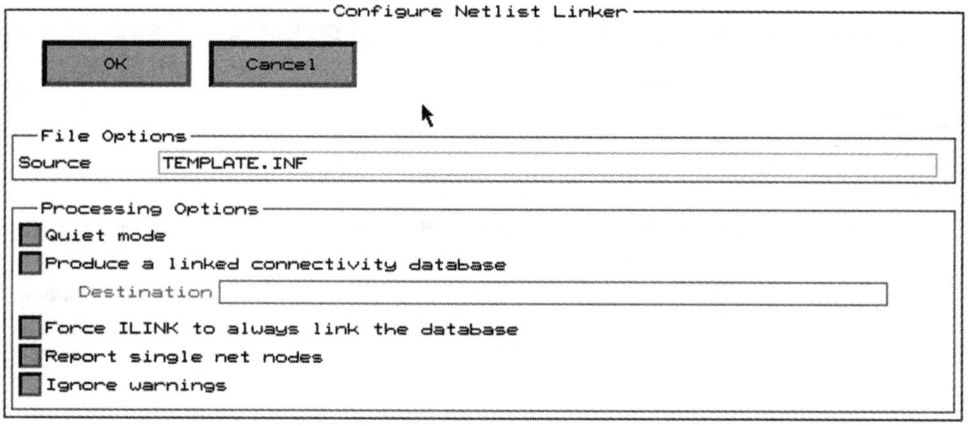

Figure 7-15c. Local Configuration of Netlist Linker (Courtesy of OrCAD Inc.)

extension of SCH. Under **Processing Options,** you may choose any combination of different options provided by clicking on the individual buttons. Exit the screen by clicking on the OK icon.

3. If you are back to SDT_MENU, click on Create Netlist and then select **Local Configuration.** The Configuration menu of the **Create Netlist** tool will appear again. This time, click on **Configure ILINK,** and the Local Configuration screen of **ILINK** shown in Figure 7-15c will appear. Under **File Options,** provide the name of the source as the name of your schematic file or design, with an exten-

7-11. Create Netlist for PCB Software Tools   215

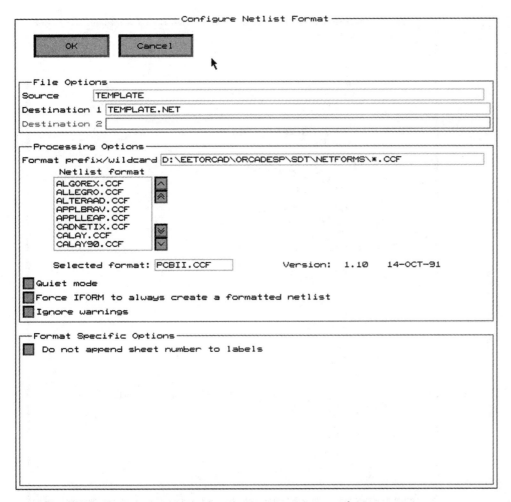

Figure 7–15d. Local Configuration of Netlist Format (Courtesy of OrCAD Inc.)

sion of INF. Again, under the processing option you may choose any combination. Exit from the screen by clicking on the OK icon.

4. If you are back to SDT_MENU, go back to the Configuration menu of the **Create Netlist** tool. This time, click on **Configure IFORM.** The Local Configuration screen of **IFORM** shown in Figure 7–15d will appear. Under **File Options,** provide the source as the name of your design or schematic filename (in case of a single-sheet design), without any extension. Note that the source file name must be without extension. Provide under **Destination 1** the name of your file or design, with an extension NET. Under **Processing Options,** choose the netlist format for OrCAD/PCB tools for the list box, which is PCB II.CCF. Scroll through the list box and click on **PCB II.CCF.** You may choose another processing option if you want to generate a netlist in another format. A netlist for PCB tools must be in PCB II.CCF format. Exit from the screen by clicking on the OK icon.

Figure 7-16. Netlist File of a Schematic

```
( ( OrCAD/PCB II Netlist Format        Revised:        July  9, 1992
                                       Revision:

        Time Stamp-   )
      ( AF917152 RC07 (R1) 10K
        ( 1 N00001 )
        ( 2 N00004 )                   ── Part Reference
      )
      ( AF917157 CK12 (C1) (1MF)
        ( 1 N00001 )
        ( 2 N00006 )                   ── Part Value
      )
      ( AF917154 RC07 R3 (2.2K)
        ( 1 N00008 )
        ( 2 N00006 )
      )
      ( AF917153 RC07 R2 2.7K
        ((1) N00002 )
        ( 2 N00005 )
      )
Pin Number
      ( B093B98E TO220 Q1 2N2102
        ( 1 N00001 )
        ( 2 N00003 )
        ((3) N00002 )
      )
      ( AF917155 RC01 VR1 5K
        ( 1 N00008 )
        ( 2 N00007 )
        ( 3 N00005 )
      )
      ( AF917156 DO7 D1 6.8V
        ( 1 N00006 )
        ( 2 (N00004) )                 ── Netname or node
      )                                   name of each
      ( AF917151 8DIP300 U1 741          connecting point.
        ( 1 ?1 )
        ( 2 (N00004) )
        ( 3 N00007 )
        ( 4 N00001 )
        ( 5 ?2 )
        ( 6 N00003 )
        ( 7 N00006 )
      )
  )
```

5. Invoke the Transit Menu of **Create Netlist** by clicking on the icon from the SDT_MENU. This time, instead of **Local Configuration,** click on **Execute.** If everything is in order, the software will create a netlist file for the entire design or, if the design is a single sheet, then for the single-sheet schematic. The processing may take a couple of seconds. Since the netlist file is in ASCII, it can be inspected by using the M2EDIT text editor. Figure 7–16 is an example of a netlist file formatted for OrCAD/PCB.

7–11.3 Create Netlist for a Hierarchical Design

A different procedure needs to be used to create a netlist for a hierarchical design. If you have a design whose worksheets have been arranged in complex hierarchical structure, the file must be converted to simplified form before creating the netlist. In other words, designs that have sheet symbols must be converted from complex hierarchical to simple format before using the **Create Netlist** tool. To do this, OrCAD provides a tool called **Complex to Simple,** located under Design Management Tools.

Converting hierarchical design to simple design

1. Invoke the Design Management Tools set menu (DM_MENU). Select the Design name from the left list box by left-clicking the mouse on it.

2. Click on the **Complex to Simple** icon.

3. Select the design name from the new list box, and notice that the selected name will appear in the Source Design dialogue box.

4. Click inside the Destination Design dialogue box and enter your design name with the character **S** in front of it. The software will simplify the design and put it under this name. You can use any filename as your destination filename. However, keeping the name like this will help you recognize it easily.

5. Click on the OK icon for the software to create the simplified design and its corresponding files.

6. Select the **Cancel** icon to go to the DM_MENU.

7. Scroll through the left list box of the DM_MENU to find the new design name. Select the new design as your current design. The parts in the new design files will have unique reference designators. You must use **Draft** to examine the files of the design before generating the netlist. You will later use this simplified design and its files to create the netlist.

Explanation of the simplified file. Let's suppose that a netlist file needs to be generated for the hierarchical design shown in **Figure 5–8a**. This design has two half adder sheets, both referring to the same half adder schematic shown in Figure 5–8b. As far as the schematic is concerned, it does not make any difference. However, if this circuit is to be used for printed circuit board layout design, each part of the design must have the unique reference designator. This means that there must be only one U1A, U2C, etc., in the entire design. After executing the **Complex to Simple** command for the full adder, each of the half adder sheets will have uniquely labeled parts. However, if you have simplified the design after annotation, the simplified file will not have the unique part reference numbers. To solve this problem, you must unannotate the simplified file and then annotate it again. Figures 5–8c through 5–8e show the full adder and the uniquely labeled half adder schematics.

7–12. CREATE BILL OF MATERIALS

The **Create Bill of Materials** tool scans an entire design or a single-sheet schematic and creates a list of all parts, called a Bill of Materials (BOM), used in the design. It creates a text file arranged by grouping the same part

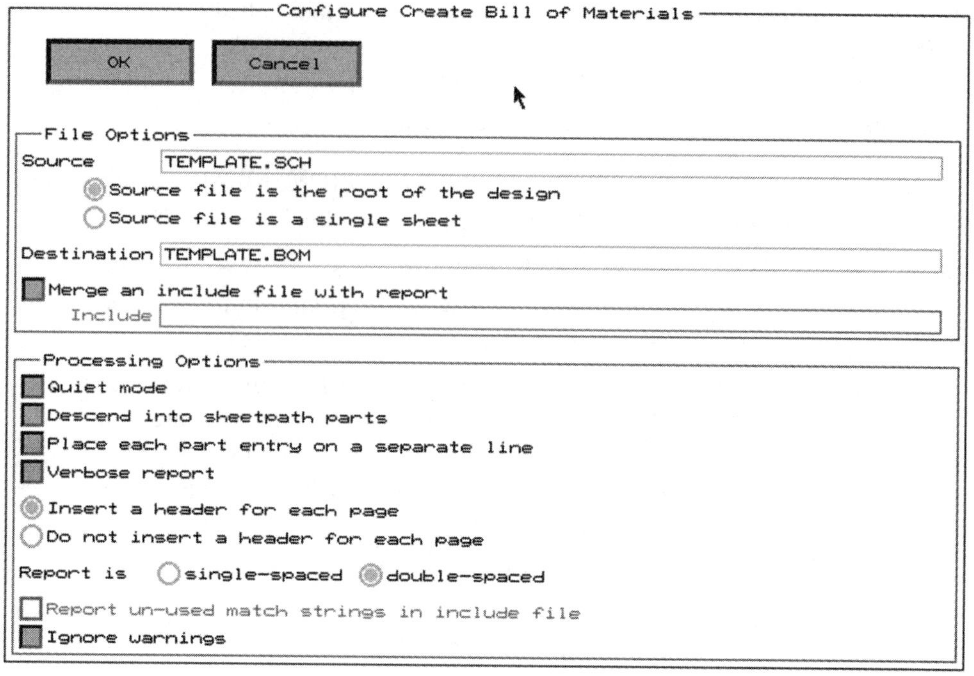

Figure 7-17. Local Configuration of Bill of Materials (Courtesy of OrCAD Inc.)

values together. Any other text information and special format for the file can be added by using the M2EDIT text editor. Special information such as order code, vendor number, location, and any other pertinent information can also be added to the Bill of Materials file.

Key Fields in the SDT configuration and the Local Configuration shown in Figure 7-17 together determine the format of the Bill of Materials file. **Create Bill of Materials** has two Key Fields located in the SDT configuration screen.

- Part Value Combine
- Include File Combine

Part value combine. If the **Part Value Combine, Key Field** contains V (for Part Value) followed by any number 1 through 8 (for Part Fields 1 through 8), the **Create Bill of Materials** tool will create the part summary that will contain Part Value and the content of the Part Field corresponding to the number. For example, if the Key Field contains V 1, then the Bill of Materials will combine the Part Value and the content of Part Field 1 of each part of the schematic and report them in the BOM summary. However, if this field is empty, the tool will always report the Part Value in the summary.

Include file combine. **Include File Combine** is used to build a lookup string to match in the Include File. However, if this key field is not specified, the Part Value will be the matching string. The Include File is a text

7-12. Create Bill of Materials

Figure 7-18. Include File

' '	Description	Source
'0.68uF'	Capacitor Ceramic Disk	Radio Shack
'0.01uF'	Capacitor Ceramic Disk	Radio Shack
'0.1uF'	Capacitor Ceramic Disk	Radio Shack
'50uF'	Capacitor Electrolytic	Radio Shack
'470'	Resistor 1/4 Watts, 5%	Radio Shack
'2.2M'	Resistor 1/4 Watts, 5%	Radio Shack
'68K'	Resistor 1/4 Watts, 5%	Radio Shack
'10K'	Resistor 1/4 Watts, 5%	Radio Shack
'390K'	Resistor 1/4 Watts, 5%	Radio Shack
'56K'	Resistor 1/4 Watts, 5%	Radio Shack
'2.2K'	Resistor 1/4 Watts, 5%	Radio Shack
'4011'	CMOS Quad NAND Gate	JDR Electronics
'LM555'	LM555 Monostable Timer	JDR Electronics
'PIEZO'	Piezo Buzzer	Digikey
'LED'	RED, GREEN, YELLOW	JDR Electronics
SPDT'	PC Board Type	Digikey

file that has the additional information to be placed in the Bill of Material file. The string described by the Key Fields is within single quotes on the left side of the Include File and on the right contains the part of the line that will be inserted in the BOM file. Figure 7–18 shows an Include File for the Logic Probe project. For this to work properly, the **Local Configuration** of **Create Bill of Materials** must be formatted properly.

The following procedure will generate a Bill of Materials file for an entire design or for a single-sheet schematic:

1. Edit the Include File shown in Figure 7–18 if you desire to add special information into the BOM file.

2. Configure the Key Fields located in the SDT configuration. If you do not specify any Key Fields character, the BOM file will be generated by grouping the parts with the same Part Value. Figure 7–19 shows the simplest form of a BOM file, a BOM file that is created without any Key Fields configuration characters and without any Include File.

3. Invoke the Create Bill of Materials Local Configuration screen by clicking on the icon located under **Reporters** in SDT_MENU and selecting **Local Configuration** from the Transit Menu. Figure 7–17 shows the Local Configuration screen of **Create Bill of Materials.**

4. Under **File Options,** for source file provide the name of the source file as the name of your design or single-sheet schematic file name.

5. Select the appropriate button, depending on single-sheet schematic or multisheet design.

6. Under **File Options,** for destination file provide the name of your design file with a BOM extension. This is the Bill of Materials file, where all the parts' information will be written.

7. Also under **File Options,** activate the *Merge an Include File with Report* option button by clicking on it.

Figure 7-19. Simplest Form of a Bill of Material File

```
DELUXE LOGIC PROBE                              Revised: January 26, 1993
EET-SP-91                                       Revision: 1
Bill Of Materials                               June 15, 1994   16:50:53   Page 1
  Item      Quantity        Reference       Part              Package
```

Item	Quantity	Reference	Part	Package
1	1	C1	0.68uF	CK06
2	1	C2	0.01uF	CK06
3	2	C3,C4	0.1uF	CK06
4	1	C5	50uF	CK06
5	3	R1,R2,R4	470	RC07
6	1	R3	2.2M	RC07
7	1	R5	68K	RC07
8	1	R6	10K	RC07
9	1	R7	390K	RC07
10	1	R8	56K	RC07
11	2	R10,R9	2.2K	RC07
12	2	U1,U3	4011	14DIP300
13	1	U2	LM555	8DIP300
14	1	LS1	PIEZO	
15	3	D1,D2,D3	LED	DO7
16	1	S1	SPDT	RCKRSPDT

Figure 7-20. Bill of Material Created by Using Include File

```
DELUXE LOGIC PROBE                              Revised:  January 26, 1993
EET-SP-91                                       Revision: 1
Bill Of Materials                               June 15, 1994   16:50:53   Page 1
```

Item	Quantity	Reference	Part	Package	Description	Source
1	1	C1	0.68uF	CK06	Capacitor Ceramic Disk	Radio Shack
2	1	C2	0.01uF	CK06	Capacitor Ceramic Disk	Radio Shack
3	2	C3,C4	0.1uF	CK06	Capacitor Ceramic Disk	Radio Shack
4	1	C5	50uF	CK06	Capacitor Electrolytic	Radio Shack
5	3	R1,R2,R4	470	RC07	Resistor 1/4 Watts, 5%	Radio Shack
6	1	R3	2.2M	RC07	Resistor 1/4 Watts, 5%	Radio Shack
7	1	R5	68K	RC07	Resistor 1/4 Watts, 5%	Radio Shack
8	1	R6	10K	RC07	Resistor 1/4 Watts, 5%	Radio Shack
9	1	R7	390K	RC07	Resistor 1/4 Watts, 5%	Radio Shack
10	1	R8	56K	RC07	Resistor 1/4 Watts, 5%	Radio Shack
11	2	R10,R9	2.2K	RC07	Resistor 1/4 Watts, 5%	Radio Shack
12	2	U1,U3	4011	14DIP300	CMOS Quad NAND Gate	JDR Electronics
13	1	U2	LM555	8DIP300	LM555 Monostable Timer	JDR Electronics
14	1	LS1	PIEZO		Piezo Buzzer	Digikey
15	3	D1,D2,D3	LED	DO7	RED, GREEN, YELLOW	JDR Electronics
16	1	S1	SPDT	RCKRSPDT	PC Board Type	Digikey

8. Under **Processor Options,** choose any combination of processing options by clicking and activating the appropriate buttons.

9. Go to the Transit Menu of **Create Bill of Materials** again and click on **Execute.**

10. The software tool will create the BOM file. Figure 7–20 shows a BOM file generated by using **Include File.** The BOM is a text file; therefore, it can be viewed by the M2EDIT text editor. Any information in the file can also be edited by using the editor.

7-13. SCHEMATIC HARD COPY

There are two types of output devices that are used to produce schematic hard copy of a design, plotter and printer. A plotter accepts vector commands. A vector is a series of points with a specific function attached to each. The plotting device needs to know what the vector information is, but does not need every point along the vector. On the other hand a printer accepts raster commands. A raster is an array of dots. For example, to draw a line, a raster device (such as a printer) would need all the points along the line. In OrCAD, a plotter will create higher resolution output than a printer. However, plotting a schematic will take longer than printing. Therefore, only the final schematics of a design should be plotted. A printer can be used during the design process.

7-13.1 Plotter

The **Plot Schematic** tool can used for both printing and plotting schematics. This tool has more options for reproduction of a schematic than the **Print Schematic** tool. The options are as follows:

- Adjustable scale factor
- Various paper sizes
- X,Y offsets for a specified sheet

The following procedure will plot the schematics of a design:

1. Invoke the Configure SDT screen under **Driver Options,** verifying that the appropriate plotter and printer drivers are loaded. If appropriate drivers are not loaded, scroll through the printer and plotter list and click on the pertinent driver name. The plotter and printer driver will appear in the configured printer driver box. If the printer you are using does not have a driver, contact OrCAD to find out which driver will be compatible for your printer. If it is a name-brand printer, OrCAD should already have a driver for it. For some printers, you may copy the printer driver from OrCAD's bulletin board.

2. Under Printer/Plotter Output Option, specify the appropriate port for printer and plotter. If any one of the ports happens to be serial, configure the data rate and other parameters necessary to match the configuration of the output device. Plotters are generally connected to a serial port.

3. Invoke the SDT_MENU and click on the **Plot Schematic** icon. The Transit Menu of the tool will appear.

4. Select **Local Configuration,** and the Configure Plot Schematic screen shown in Figure 7-21 will appear. Under **File Options,** provide the name of your design or single-sheet schematic. Although

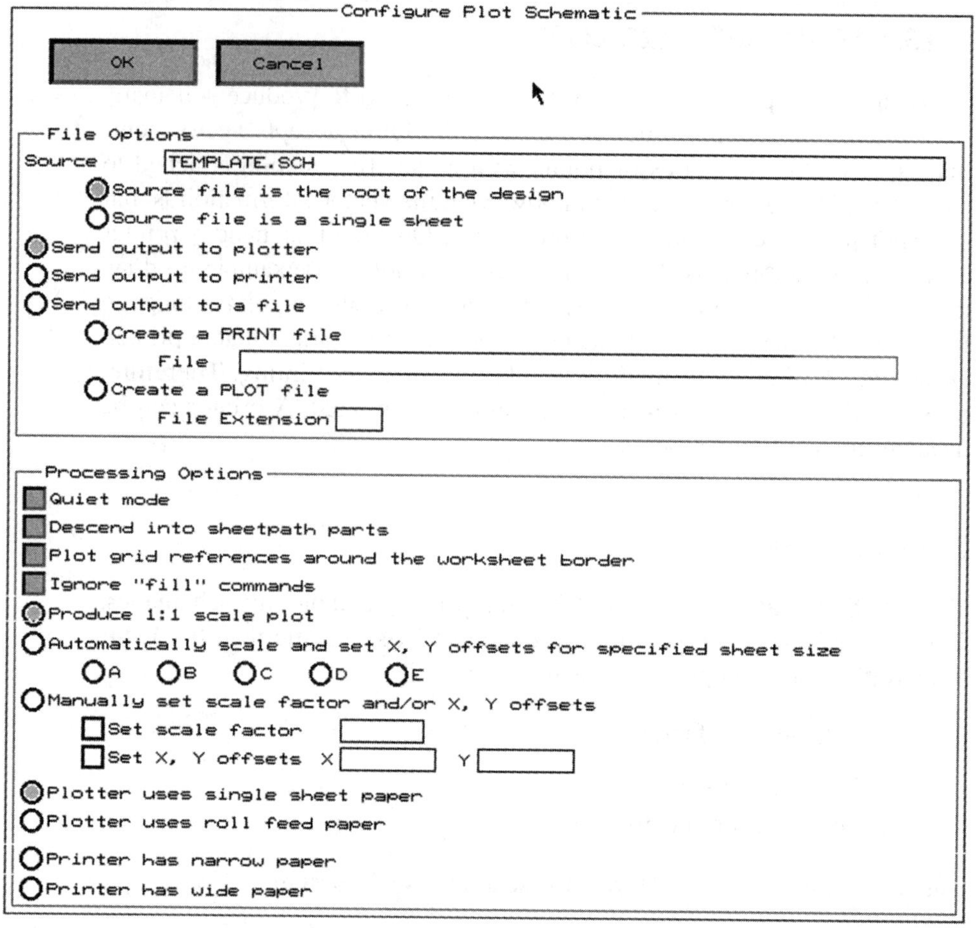

Figure 7–21. Local Configuration Screen of Plot Schematic Tool (Courtesy of OrCAD Inc.)

you are under plot utility, you can send the output to a printer. If you want to create a soft copy of the design, you need to send the output to a file, whose name you need to provide under this option.

5. Under **Processing Options,** select a desired sheet size and a scale factor. Exit from the Local Configuration screen by clicking on the OK icon.

6. Click on the **Plot Schematic** icon, and click on **Execute** from the Transit Menu. The software will send the output to the appropriate printer or plotter.

7–13.2 Printer

Schematics can be printed by using the **Print Schematic** tool. This tool is less flexible and does not have as many options as the **Plot Schematic** tool. However, the use of it is not much different from **Plot Schematic.** Since **Plot**

Schematic can be used for both plotting and printing, **Print Schematic** is hardly used. The following procedure will print the hard copy of your schematic:

1. Left-click the mouse on the **Print Schematic** icon from the SDT_MENU, and select **Configure Schematic Tools.**

2. Under **Driver Options,** select the correct printer driver for your printer from the Available Printer Driver list box. The driver should appear in the Configured Printer Driver list box. If the driver for your printer is not available in the list box, you must contact OrCAD for the driver. However, it is possible that you did not load the driver during the software installation.

3. Click on the OK icon to go back to the SDT_MENU.

4. Click on the **Print Schematic** icon and select **Local Configuration.** The Local Configuration screen for the tool will appear. Provide the source filename in the dialogue box. Select *Send Output to Printer* by clicking on it.

5. If you want to send the output to a file for making a soft copy for later use, select *Send Output to a File*. Click on the OK icon to go back to the SDT_MENU.

6. Click on the **Print Schematic** icon again, and this time select **Execute.** The printer should start printing your schematic.

7-13.3 Schematic Soft Copy

If a soft copy of the schematic is desired, it can be generated in many different industry-accepted printer formats, such as DXF, HPLASER, and Postscript. Once a soft copy of the schematic file is generated in another format, it can be printed using the respective software. This situation arises when the computer where OrCAD is loaded does not have the desired printer or plotter. In another situation, if you want to transport a schematic file to another location where OrCAD is unavailable, you can generate a soft copy of the file in another format. Later the schematic can be printed by using the appropriate software. The following procedure will create a soft copy in another printer format:

1. Invoke the SDT configuration screen, and under **Driver Options** select the correct plotter driver for which you are generating the soft copy from the Available Plotter Driver list box. If the desired plotter driver is not available in the list box, contact OrCAD. Exit the configuration by clicking on the OK icon.

2. Click on the **Plot Schematic** icon from the SDT_MENU. Select **Local Configuration** from the Transit Menu. Local Configuration for **Plot Schematic** will appear. Under **Processing Options** of the Configuration menu, click on the bubble for *Send output to a file*. Also, enter in the dialogue box the output filename with appropriate extension and path. Exit the configuration by clicking the OK icon.

3. Invoke the Local Configuration menu again, and this time click on **Execute.** The software will generate the file.

For example, if you format the schematic file in DXF format, it can be printed by using AutoCAD software.

7-14. AN EXAMPLE USING THE LOGIC PROBE PROJECT

At this point, you should have enough information about the Logic Probe project to begin capturing the schematic. Chapter 2 discusses a few important factors that you need to consider for planning. This section will discuss file organization and schematic capture by doing them for the Logic Probe project.

7-14.1 File Organization

Create design named LOG_PRO

1. Invoke the ESP_MENU by typing **orcad** at the C:\MYORCAD> DOS prompt.
2. Left-click the mouse inside the **Design Management Tools** icon. The Design Management Tools menu, DM_MENU, will appear.
3. Left-click the mouse on the design named **TEMPLATE** for the Design list box. A design named TEMPLATE will be highlighted.
4. Left-click the mouse inside the **Create Design** icon. The Create Design window, DM_MENU2, will appear.
5. Left-click inside the New Design Name dialogue box. Provide the name of the design **LOG_PRO** and left-click the mouse again.
6. Left-click the mouse on the OK icon. The design name LOG_PRO will appear in the Design list box. All the files from the TEMPLATE design subdirectory will be copied in the LOG_PRO design subdirectory.
7. Left-click on the new design name LOG_PRO, and the design name will be highlighted. Click on the OK icon. The design environment for Logic Probe is now set and the design name LOG_PRO will appear at the top of the screen.

If the design name has already been created, you need only to set the environment for it.

7-14.2 Configure Schematic Design Tools

At this point most of the configuration for the SDT tool is complete. However, you need to configure Library Options, Worksheet Options, and Key Fields. After reviewing the hand-drawn version of the logic probe schematic, it was found that the library that may have the symbols for parts

7-14. An Example Using the Logic Probe Project

are: CMOS, PCB DEVICE, ANALOG. If the appropriate symbols for the parts are not found in these libraries, it is always possible to configure the desired library later. Sometimes it may be necessary for you to browse through the symbols of several libraries to find the desired symbol. Use the following procedure:

1. Invoke the Schematic Design Tools menu by clicking the mouse on the icon. The SDT_MENU will appear.
2. Click on the **Draft** icon once, and the Transit Menu will appear.
3. Click on **Configure Schematic Tools,** and the Configuration Screen will appear. Scroll down to **Library Options.**
4. Click on the library name **CMOS** in the Available Libraries list box and then click on the **Insert** icon. CMOS will appear in the Configured Libraries list box. In this way configure all the libraries required for the Logic Probe schematic.
5. Scroll to **Worksheet Options.** Click on the dialogue box for Sheet Size and provide the size **A.**
6. Click and provide the following information inside the appropriate dialogue boxes:
   a. Document number: 01-1995
   b. Revision: 01
   c. Title: Deluxe Logic Probe
   d. Organization name: ABC Technology Inc.
   e. Organization address: 1475 Glenwood Dr.
      Hammond, IN 46323
7. Scroll down to Key Fields and configure them as follows.
   a. Under **Update Field Contents,** enter **V** in the dialogue box for *Combine for Field 1*.
   b. Under **Create Netlist,** enter **V** in the dialogue box for *Part Value Combine* and **1** in the dialogue box for *Module Value Combine*.
   c. Under **Create Bill of Materials,** enter **V** in the dialogue box for *Part Value Combine*.

At this point, do not make any changes in the SDT tool set configuration unless you are absolutely sure about them.

8. Scroll to the top of the screen either by using the mouse or pressing the Home key. Exit the configuration by clicking on the OK icon.

7-14.3 Schematic Capture for the Logic Probe

1. Invoke the SDT_MENU.
2. Click on the **Draft** icon located under **Editor.** The SDT worksheet, SDT_MAIN, will appear. At this point the tool is ready for fetching

parts from the schematic library and connecting them to create the circuit schematic for the design.

3. Look at the final hand-drawn schematic for the Logic Probe shown in Figure 2–6 and fetch the following schematic symbols from the appropriate library onto the worksheet, using the **GET** command from the SDT_MAIN menu.

Part Value	Quantity	Library	
CD 4011	7	CMOS	
LM555	1	Analog 1	
LED	3	PCB Device	
CAPACITOR	4	PCB Device	Ceramic
CAPACITOR	1	PCB Device	Electrolytic
RESISTOR	10	PCB Device	
SPDT Switch	1	PCB Device	
PIEZO BUZZER	1	PCB Device	

You should save the schematic onto the hard drive under the Logic Probe design subdirectory. To do this, click on **QUIT** and choose **Write to File;** the software will prompt you for the filename to write into. Provide the name as LOG_PRO for the schematic, with an SCH extension. From now on, you only have to update the file as you add symbols or make changes to the worksheet.

If you have not placed the components at the appropriate place on the worksheet, this is the best time to move them. Use **BLOCK-Move** or **-Drag** to do so. You may have to move the symbols again as you connect them with wires and buses. Try to position them in such a way on the worksheet that they are not crowded at one place yet also not placed too far apart. Place parts as much as possible from left to right and from top to bottom. Also place all the external inputs on the left and outputs on the right. The key is to avoid placing far apart parts that have the most number of connections among themselves.

Fetch wires and junction boxes by using **PLACE** from the SDT_MAIN menu and choosing *Wire* or *Junction* as required from the submenu. Following the procedure described earlier, place wires and junction boxes to create the circuit according to the hand-drawn circuit. Remember to **Quit-Update File** every 7 to 10 minutes. The complete schematic after all parts are connected, according to the hand-drawn schematic, is shown in Figure 7–22. You may also clean up the schematic by using the **Cleanup Schematic** tool located under **Processors** in SDT_MENU.

7-14.4 Annotate Schematic

The schematic you just have drawn needs annotation. When you fetch parts from the symbol library, they do not have unique designators and in some cases do not have appropriate part values. Each schematic symbol must be provided with a unique reference designator. This can be done in two ways:

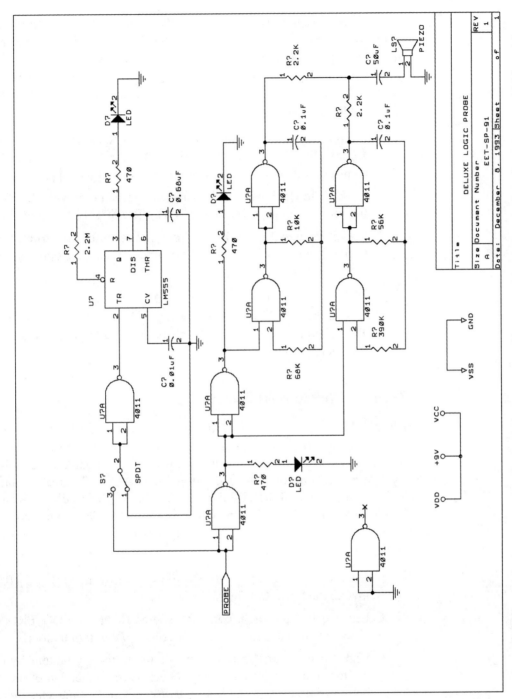

Figure 7-22. Schematic of Deluxe Logic Probe before Annotation (*Source: Delton T. Horn 50 CMOS IC Projects, 1988, p.64, used with permission from TAB Books Inc.*)

(1) by editing each part, using the **EDIT** command from the SDT_MAIN menu; or (2) by using the **Annotate** tool located under **Processors** in SDT_MENU. For a large circuit with many parts, using **Annotate** is much faster than editing each part. However, a combination of both (**Annotate** and **EDIT**) can also be done. After annotating the entire circuit, parts can still be edited as required.

1. From SDT_MENU, click on the **Annotate Schematic** icon. Click on **Local Configuration** from the SDT Transit Menu. The Local Configuration of the **Annotate Schematic** tool will appear.
2. Under **File Options** in the Configuration menu, provide the name of the source file; in this case it is LOG_PRO.SCH.
3. Under **Processing Options,** click on **Incremental Annotation.** If needed, you may click on a few other processing options, except **Unannotate.**
4. Exit the screen by clicking on the OK icon.
5. Invoke the SDT Transit Menu again, and this time click on **Execute.** The schematic will be annotated. Figure 7–23 shows the annotated schematic of the Logic Probe.

7-14.5 Update Field Content

Part Field 1 of each part of the schematic must be updated with the packaging information. This packaging information will later be used by the PCB software tool to download modules from the PCB module library. To use **Update Field Content,** you need a stuff file in ASCII. To create a stuff file you need to use the M2EDIT editor. Invoke the M2EDIT editor screen, create the following ASCII file, and call it LOG_PRO.STF. The left column of the stuff file shown in Figure 7–24a depends on the Key Field configuration.

1. Click on the **Update Field Contents** icon from SDT_MENU and invoke the Local Configuration screen.
2. Under **File Options,** provide the source filename, LOG_PRO.SCH. Also activate the *Source file is the root of the design* button.
3. Under **Processing Options,** activate the following buttons by clicking the mouse pointer on them. Green is the indication of an active button: *Part Field 1, Unconditionally update, Set the specified field to visible,* and *Convert stuff to uppercase Field.*
4. Exit the Configuration screen by clicking on the OK icon.
5. Invoke the **Update Field Contents** Transit Menu, and this time click on **Execute.** The software will update the Part Field 1 of each symbol in the schematic with the packaging information provided in the stuff file. Figure 7–24a is the stuff file for the Logic Probe schematic.

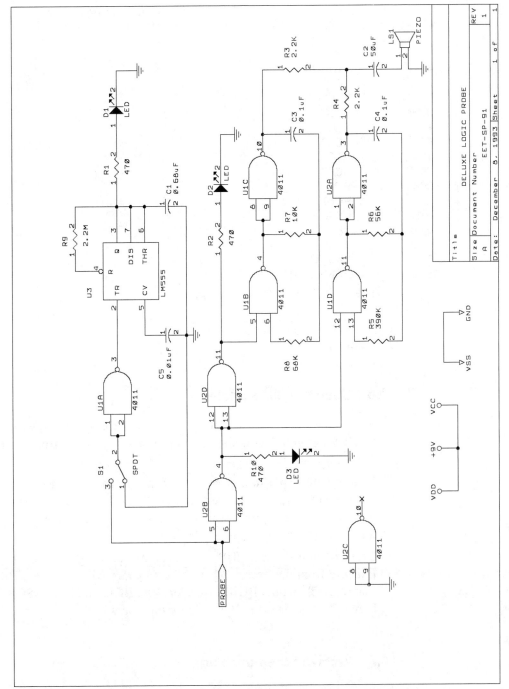

Figure 7-23. Schematic of Deluxe Logic Probe after Annotation (*Source: Delton T. Horn 50 CMOS IC Projects, 1988, p.64*, used with permission from TAB Books Inc.)

```
Content of      Packaging          ←This information must be excluded from
Value Field     Identification     ←the actual Stuff file and is included for
---------------------------        ←explanation only.

'4011'          '14DIP300'
'LED'           'DO7'
'LM555'         '8DIP300'
'2.2M'          'RC07'
'470'           'RC07'
'68K'           'RC07'
'10K'           'RC07'
'390K'          'RC07'
'56K'           'RC07'
'2.2K'          'RC07'
'0.01uF'        'CK06'
'0.68uF'        'CK06'
'0.1uF'         'CK06'
'50uF'          'CK06'
'SPDT'          'RCKRSPDT'
```

Figure 7-24a. Stuff File for the Logic Probe

Figure 7–24b shows the schematic after processing by the **Update Field Contents** tool.

7-14.6 Generate Bill of Materials

1. Invoke the Local Configuration screen of **Create Bill of Materials** by clicking on the icon and selecting **Local Configuration** from the Transit Menu.

2. Under **File Options,** provide the name of the source file as LOG_PRO.SCH and the destination file as LOG_PRO.BOM. You may choose a few processing options as needed. Exit the configuration screen by clicking on the OK icon.

3. Click on the **Create Bill of Materials** icon again and choose **Execute** from the Transit Menu. The software will produce a Bill of Materials file with a BOM extension. The BOM file is an ASCII file. Figure 7–25 shows the BOM file for the Logic Probe.

7-14.7 Archive Schematic Parts

Archiving schematic parts will make your schematic portable from one OrCAD system to another. Also, this is convenient, because you do not need to have all the libraries for your schematic. You need only the archive library for the schematic to display it.

1. Invoke the Local Configuration screen for the **Archive Parts in Schematic** tool by clicking on the icon once and selecting **Local Configuration** from the Transit Menu.

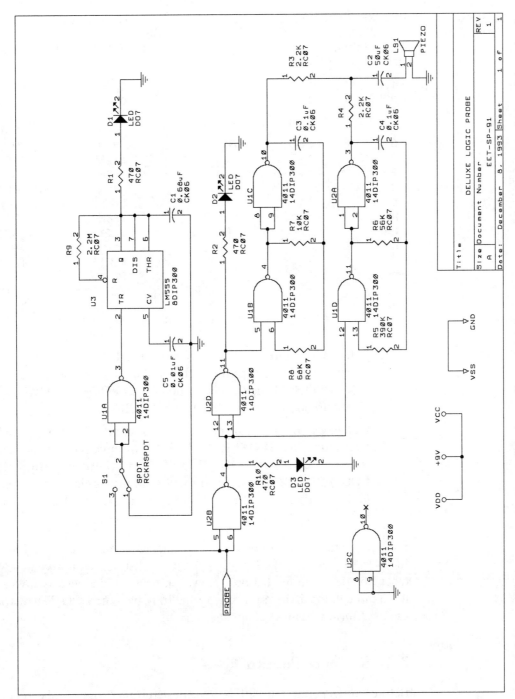

Figure 7-24b. Schematic of Deluxe Logic Probe after Stuff (*Source*: Delton T. Horn, *50 CMOS IC Projects*, 1988, p.64, used with permission from TAB Books Inc.)

Figure 7-25. The Simplest Bill of Materials for the Deluxe Logic Probe

```
DELUXE LOGIC PROBE              Revised:  January 26, 1993
EET-SP-91                       Revision: 1
Bill Of Materials               June 15, 1994   16:50:53   Page  1
```

Item	Quantity	Reference	Part	Package
1	1	C1	0.68uF	CK06
2	1	C2	0.01uF	CK06
3	2	C3,C4	0.1uF	CK06
4	1	C5	50uF	CK06
5	3	R1,R2,R4	470	RC07
6	1	R3	2.2M	RC07
7	1	R5	68K	RC07
8	1	R6	10K	RC07
9	1	R7	390K	RC07
10	1	R8	56K	RC07
11	2	R10,R9	2.2K	RC07
12	2	U1,U3	4011	14DIP300
13	1	U2	LM555	8DIP300
14	1	LS1	PIEZO	
15	3	D1,D2,D3	LED	DO7
16	1	S1	SPDT	RCKRSPDT

2. Select **Configure LIBARCH** from the Library Archive Transit Menu. Under **File Options,** provide the name of the source file as LOG_PRO.SCH and the destination file as LOG_PRO.SRC. Exit the Configuration screen by clicking on the OK icon.

3. Get back to the Library Archive Transit Menu and click on **Configure COMPOSER.** Under **File Options,** provide the name of the source file as LOG_PRO.SRC and the destination file as LOG_PRO.LIB. The software will scan the Logic Probe schematic and archive all the symbols in a single library called LOG_PRO.LIB.

From now on, for you to work on the schematic, only the LOG_PRO.LIB file is to be placed in the Configured Library list box. You may remove all other libraries from the list box. However, if you add a new part, you need to go through the archive process again. To review the LOG_PRO.LIB file, use the **Library Browse** command.

7-14.8 Check Electrical Rules

Before you generate the netlist, you must run the Logic Probe schematic through the electrical rules check.

1. Invoke the **Check Electrical Rules** Local Configuration screen by clicking on the icon and selecting **Local Configuration** from the Transit Menu.

2. Under **File Options,** provide the name of the source file as LOG_PRO.SCH and the destination file as LOG_PRO.ERC. You

7-14. An Example Using the Logic Probe Project

may choose other options, such as *check module port connections*. Exit the configuration screen by clicking on the OK icon.

3. Click on the **Check Electrical Rules** icon again, and click on **Execute** from the Transit Menu. The software will scan the circuit and report any violation of electrical rules in the circuit.

7-14.9 Generate Netlist for OrCAD/PCB Software

The netlist file for the schematic must be generated. This file is the link between the SDT and PCB software. Generation of a netlist is done in the following steps:

1. From SDT_MENU, click on the **Create Netlist** icon. Click on **Configure INET** from the **Create Netlist** Transit Menu. In the Local Configuration screen for **INET,** provide the name of the source file as LOG_PRO.SCH. You may choose to activate a few of the processing buttons. Exit the configuration screen by clicking on the OK icon.

2. Click on **Configure ILINK** from the **Create Netlist** Transit Menu. In the Local Configuration screen for **ILINK,** provide the name of the source file as LOG_PRO.INF. You may choose to activate a few of the processing buttons. Exit the configuration screen by clicking on the OK icon.

3. Click on **Configure IFORM** from the **Create Netlist** Transit Menu. In the Local Configuration screen for **ILINK,** provide the name of the source file as LOG_PRO and the destination file as LOG_PRO.NET. Under **Processing Options,** click on PCB II.CCF from the Netlist Format list box in order to determine the format of the netlist. You may choose to activate a few more of the processing buttons. Exit the configuration screen by clicking on the OK icon.

4. Make sure that in the **Create Netlist** Transit Menu, **INET, ILINK,** and **IFORM** all are in **ON** condition.

5. Get back to the SDT Transit Menu by clicking on the **Create Netlist** icon once. This time, choose **Execute.** The software will generate a netlist file of the schematic.

6. A netlist is an ASCII file, so you may be able to inspect its proper format by using the M2EDIT editor.

Figure 7–26 shows the netlist file for the LOG_PRO. After generating the netlist file, the function of the SDT tool effectively ends. This netlist file will be called by the PCB tools for the artwork design. At this time you should make a final printout of the Logic Probe schematic by invoking the **Plot Schematic** tool. The final printout of the schematic for the logic probe will look like the schematic shown in Figure 7–24b.

Figure 7-26. Netlist File for Deluxe Logic Probe

DELUXE LOGIC PROBE
EET-SP-91

Revised: January 26, 1993
Revision: 1

Time Stamp - }
(C563B41F CK06 C2 0.01UF
 (1 N00006)
 (2 VSS)
)
(C563B41C CK06 C3 0.1UF
 (1 N00011)
 (2 N00014)
)
(C563B41D CK06 C4 0.1UF
 (1 N00017)
 (2 N00019)
)
(C563B41E CK06 C1 0.68UF
 (1 N00002)
 (2 VSS)
)
(C563B423 RC07 R6 10K
 (1 N00010)
 (2 N00014)
)
(C563B427 RC07 R9 2.2K
 (1 N00017)
 (2 N00015)
)
(C563B428 RC07 R10 2.2K
 (1 N00011)
 (2 N00015)
)
(C563B425 RC07 R3 2.2M
 (1 N00001)
 (2 N00002)
)
(C563B421 RC07 R7 390K
 (1 N00018)
 (2 N00019)
)
(C5639E11 14DIP300 U1 4011
 (1 N00007)
 (2 N00007)
 (3 N00008)
 (4 N00007)
 (5 PROBE)
 (6 PROBE)
 (7 VSS)
 (8 N00003)
 (9 N00003)
 (10 N00004)
 (11 ?1)
 (12 VSS)
 (13 VSS)
 (14 +9V)
)

7-14. An Example Using the Logic Probe Project

Figure 7-26. Continued

```
( C563B437 14DIP300 U3 4011
 ( 1 N00010 )
 ( 2 N00010 )
 ( 3 N00011 )
 ( 4 N00010 )
 ( 5 N00008 )
 ( 6 N00012 )
 ( 7 VSS )
 ( 8 N00007 )
 ( 9 N00018 )
 ( 10 N00016 )
 ( 11 N00017 )
 ( 12 N00016 )
 ( 13 N00016 )
 ( 14 +9V )
)
( C563B420 RC07 R1 470
 ( 1 N00007 )
 ( 2 N00013 )
)
( C563B426 RC07 R2 470
 ( 1 N00008 )
 ( 2 N00009 )
)
( C563B42C RC07 R4 470
 ( 1 N00002 )
 ( 2 N00005 )
)
( C563B42D CK06 C5 50UF
 ( 1 N00015 )
 ( 2 N00020 )
)
( C563B422 RC07 R8 56K
 ( 1 N00016 )
 ( 2 N00019 )
)
( C563B424 RC07 R5 68K
 ( 1 N00012 )
 ( 2 N00014 )
)
( C563B429 DO7 D1 LED
 ( 1 N00013 )
 ( 2 VSS )
)
( C563B43C DO7 D2 LED
 ( 1 N00009 )
 ( 2 VSS )
)
( C563B42B DO7 D3 LED
 ( 1 N00005 )
 ( 2 VSS )
)
( C5639E10 8DIP300 U2 LM555
 ( 1 VSS )
 ( 2 N00004 )
```

Figure 7-26. Continued

```
    ( 3 N00002 )
    ( 4 N00001 )
    ( 5 N00006 )
    ( 6 N00002 )
    ( 7 N00002 )
    ( 8 +9V )
)
( C563B42F PIEZO LS1 PIEZO
    ( 1 N00020 )
    ( 2 VSS )
)
( C^563B42E RCKRSPDT S1 SPDT
    ( 1 VSS )
    ( 2 N00003 )
    ( 3 PROBE )
)
)
```

7-15. THE OTHER PROJECTS

Describing the other three projects step by step would be repetitive. Therefore, the rest of this chapter provides schematics and other pertinent information for you to follow during the projects.

7-15.1 The Electronic Cricket

File organization. Set the design environment for the Cricket. If the design name has not been created, you need to create it now, using the **Create Design** tool located under Design Management Tools.

Configure schematic design tools. Configuration of SDT tools for this project has been provided in Section 7-3.4.

Schematic capture for the cricket. Look through the final hand-drawn schematic for the Electronic Cricket shown in Figure 2-10, and get from the library the parts mentioned below:

Part Value	Quantity	Library
4069	1	CMOS
741	2	ANALOG
Diode Silicon	5	PCB DEVICE
Resistor	19	PCB DEVICE
Resistor VAR 2	1	PCB DEVICE
Varistor	1	PCB DEVICE
Capacitor Non Pol	5	PCB DEVICE
Capacitor Pol	3	PCB DEVICE
Buzzer	1	PCB DEVICE
Microphone	1	PCB DEVICE
Battery	1	PCB DEVICE

7-15. The Other Projects

Once you have all the symbols on the worksheet, save the schematic file by using **QUIT,** then choose **Write to File** or **QUIT,** and then **Update File.** At this point you may use **BLOCK Move** or **BLOCK Drag** to move the symbols on the worksheet and arrange them almost like your hand-drawn circuit. Connect them by using **PLACE Wire** and **PLACE Junction.** The schematic shown in Figure 7–27 is the schematic after all the symbols have been connected.

Annotate schematic. The schematic must be annotated before proceeding further. Use the **Annotate** tool and follow the procedure to annotate the schematic for the Cricket. Figure 7–28 shows the schematic after annotation.

Update field contents. Part Field 1 of all the schematic symbols must be updated with packaging identification before generating the netlist file. Use the **Update Field Contents** tool and the following update file (CRICKET.STF) to perform the operation.

You may need to manually update the content of Part Field 1 of the 1K variable resistor by using **EDIT.** Choose the packaging identification provided by OrCAD. Note that when you create this file, do not provide any characters other than the content of the part value and the corresponding packaging identification within single quotes. Figure 7–29a is the stuff file for the Electronic Cricket, and Figure 7–29b shows the schematic after processing by **Update Field Contents.**

Generate bill of materials. Process the schematic by using the **Bill of Materials** tool. Figure 7–30 shows the BOM file after processing.

Archive schematic parts. It is a good idea to archive all the schematic parts in one single library at this time. Use the **Archive Parts in Schematic** tool to perform the operation. You may view the archive part library by using the **Library Browse** command.

Check electrical rules. At this point it is necessary to see if any of the electrical rules have been violated while drawing the circuit. You need to process the circuit, using this tool.

Generate netlist for OrCAD/PCB. A netlist file must be generated before the schematic can be used by the OrCAD/PCB tool to design artwork. Process the schematic by using the **Create Netlist** tool. Figure 7–31 shows the netlist file for the Cricket.

Print the schematic. Print the hard copy of the schematic for your reference and documentation by using the **Plot Schematic** tool located under **Reporters.** If you have not made any changes since processing by **Update Field Contents,** it will look like the schematic shown in Figure 7–29b.

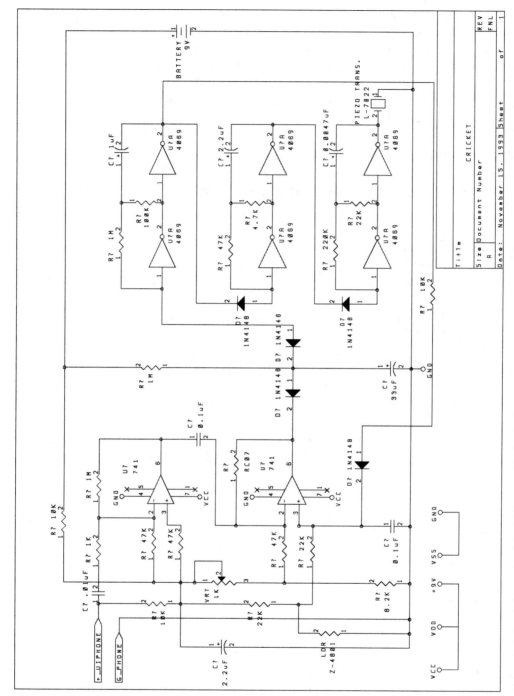

Figure 7-27. Schematic of the Electronic Cricket before Annotation (Source: Adapted from Dick Smith, SAMS *Fun Way into Electronics*, Howard W. Sams & Co., 1986, p.47)

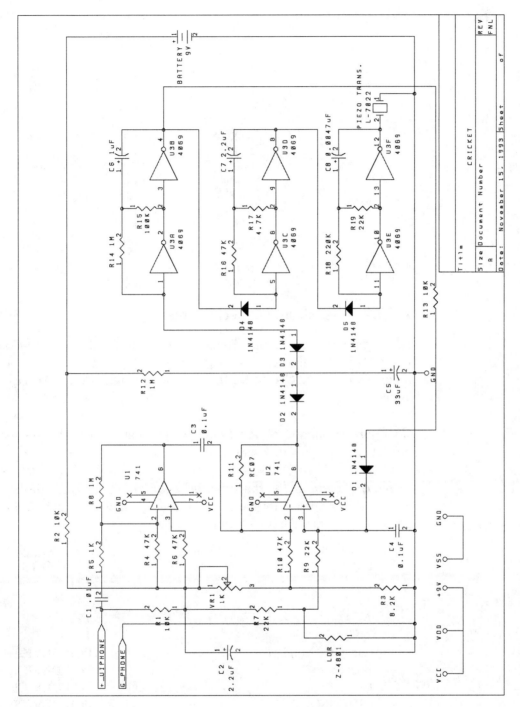

Figure 7-28. Schematic after Annotation (*Source:* Adapted from Dick Smith, *SAMS Fun Way into Electronics*, Howard W. Sams & Co., 1986, p.47)

Figure 7–29a. Stuff File for Cricket

```
'741'        '8DIP300'
'4069'       '14DIP300'
'1K'         'RC07'
'4.7K'       'RC07'
'8.2K'       'RC07'
'10K'        'RC07'
'22K'        'RC07'
'47K'        'RC07'
'100K'       'RC07'
'220K'       'RC07'
'1M'         'RC07'
'.0047uF'    'CK05'
'.01uF'      'CK05'
'1uF'        'CK05'
'2.2uF'      'CK05'
'33uF'       'CK05'
'9V'         'RC05'
'1N4148'     'D07'
'L-7022'     'RC07'
```

7–15.2 The Infrared Object Counter

File organization. Set the design environment for the Infrared Counter (**IR_COUNT**). If the design name has not been created, you need to create it now by using the **Create Design** tool located under Design Management Tools.

Configure schematic design tools. Configuration of SDT tools for this project has been provided in Section 7–3.4.

Schematic capture for the infrared counter. Look through the final hand-drawn schematic for the Infrared Counter shown in Figure 2–14, and get from the library the parts mentioned below:

Part Value	Quantity	Library
LM555	1	ANALOG 3
7420	2	TTL
7447	2	TTL
7474	1	TTL
7490	2	TTL
Capacitor Non Pol	3	PCB DEVICE
Buzzer	1	PCB DEVICE
Diode	1	PCB DEVICE
OPTO ISO	1	PCB DEVICE
Resistor	10	PCB DEVICE
8 PIN HEADER	2	PCB DEVICE
Transistor NPN	1	PCB DEVICE

Once you have all the symbols on the worksheet, save the schematic file by using **QUIT,** then choose the **Write to File** or **Quit**, and then **Update File.** At this point you may use **BLOCK Move** or **BLOCK Drag** to move the

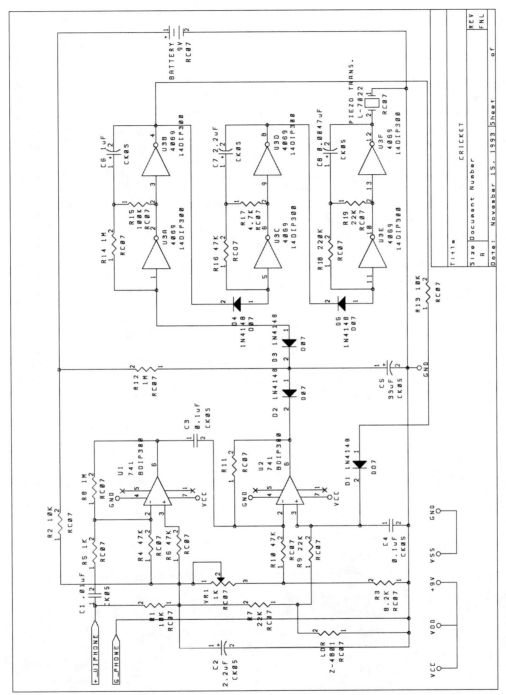

Figure 7-29b. Schematic after Processing by Update Field Contents (*Source:* Adapted from Dick Smith, SAMS *Fun Way into Electronics,* Howard W. Sams & Co., 1986, p.47)

Figure 7-30. Bill of Materials for the Cricket

CRICKET Revised: November 15, 1993
         Revision: FNL
Bill Of Materials   June 29, 1994 12:26:29 Page 1

Item	Quantity	Reference	Part	Packaging ID
1	1	BATTERY	9V	RC07
2	1	C1	01uF	CK05
3	2	C2,C7	2.2uF	CK05
4	2	C3,C4	0.1uF	CK05
5	1	C5	33uF	CK05
6	1	C6	1uF	CK05
7	1	C8	0.0047uF	CK05
8	1	D1	1N4148	DO7
9	4	D2,D3,D4,D5	1N4148	D07
10	1	LDR	Z-4801	RC07
11	1	PIEZO TRANS.	L-7022	RC07
12	3	R1,R2,R13	10K	RC07
13	1	R3	8.2K	RC07
14	3	R4,R6,R10	47K	RC07
15	2	R5,VR1	1K	RC07
16	3	R7,R9,R19	22K	RC07
17	4	R8,R12,R14, R11	1M	RC07
18	1	R15	100K	RC07
19	1	R16	47K	RC07
20	1	R18	220K	RC07
21	1	R17	4.7K	RC07
22	2	U1,U2	741	8DIP300
23	1	U3	4069	14DIP300

Figure 7-31. Netlist File for the Cricket

CRICKET   Revised: July 8, 1993
          Revision:

```
Time Stamp - }
( C2519C1F RC05 BATTERY +9V
 ( 1 N00002 )
 ( 2 -_UPHONE )
)
( C2519BF8 CK05 C8 0.0047UF
 ( 1 N00022 )
 ( 2 N00023 )
)
( C2519BF9 CK05 C1 0.01UF
 ( 1 +_UPHONE )
 ( 2 N00003 )
)
( C2519BF6 CK05 C3 0.1UF
 ( 1 N00005 )
 ( 2 N00012 )
)
( C2519BF7 CK05 C4 0.1UF
 ( 1 N00020 )
 ( 2 -_UPHONE )
)
( C2519C15 RC07 R15 100K
```

7-15. The Other Projects

Figure 7-31. Continued

```
  ( 1 N00007 )
  ( 2 N00009 )
)
( C2519C09 RC07 R1 10K
  ( 1 N00001 )
  ( 2 +_UPHONE )
)
( C2519C0D RC07 R2 10K
  ( 1 N00001 )
  ( 2 N00002 )
)
( C2519C1B RC07 R13 10K
  ( 1 N00025 )
  ( 2 N00008 )
)
( C2519C03 RC07 R5 1K
  ( 1 N00003 )
  ( 2 N00004 )
)
( C2519C0C RJ26X VR1 10K
  ( 1 N00001 )
  ( 2 N00001 )
  ( 3 N00016 )
)
( C2519C01 RC07 R8 1M
  ( 1 N00004 )
  ( 2 N00005 )
)
( C2519C02 RC07 R11 1M
  ( 1 N00012 )
  ( 2 N00017 )
)
( C2519C0F RC07 R12 1M
  ( 1 N00011 )
  ( 2 N00002 )
)
( C2519C1A RC07 R14 1M
  ( 1 N00006 )
  ( 2 N00007 )
)
( C2519BFE CK05 C6 1UF
  ( 1 N00007 )
  ( 2 N00008 )
)
( C2519C00 CK05 C2 2.2UF
  ( 1 N00001 )
  ( 2 -_UPHONE )
)
( C2519BFF CK05 C7 2.2UF
  ( 1 N00014 )
  ( 2 N00015 )
)
( C2519C18 RC07 R18 220K
  ( 1 N00021 )
  ( 2 N00022 )
)
```

Figure 7-31. Continued

```
( C2519C0A RC07 R7 22K
  ( 1 N00019 )
  ( 2 N00001 )
)
( C2519C07 RC07 R9 22K
  ( 1 N00019 )
  ( 2 N00020 )
)
( C2519C17 RC07 R19 22K
  ( 1 N00022 )
  ( 2 N00024 )
)
( C2519BFD CK05 C5 33UF
  ( 1 N00011 )
  ( 2 -_UPHONE )
)
( C2519C16 RC07 R17 4.7K
  ( 1 N00014 )
  ( 2 N00018 )
)
( C2519BF0 14DIP300 U3 4069
  ( 1 N00006 )
  ( 2 N00009 )
  ( 3 N00009 )
  ( 4 N00008 )
  ( 5 N00013 )
  ( 6 N00018 )
  ( 7 -_UPHONE )
  ( 8 N00015 )
  ( 9 N00018 )
  ( 10 N00024 )
  ( 11 N00021 )
  ( 12 N00023 )
  ( 13 N00024 )
  ( 14 +9V )
)

( C2519C0 RC07 R4 47K
  ( 1 N00001 )
  ( 2 N00004 )
)
( C2519C05 RC07 R6 47K
  ( 1 N00001 )
  ( 2 N00010 )
)
( C2519C06 RC07 R10 47K
  ( 1 N00016 )
  ( 2 N00012 )
)
( C2519C19 RC07 R16 47K
  ( 1 N00013 )
  ( 2 N00014 )
)
```

7-15. The Other Projects

Figure 7-31. Continued

```
( C2519BFB 8DIP300 U1 741
  ( 1 ?1 )
  ( 2 N00004 )
  ( 3 N00010 )
  ( 4 -_UPHONE )
  ( 5 ?2 )
  ( 6 N00005 )
  ( 7 +9V )
)
( C2519BFC 8DIP300 U2 741
  ( 1 ?3 )
  ( 2 N00012 )
  ( 3 N00020 )
  ( 4 -_UPHONE )
  ( 5 ?4 )
  ( 6 N00017 )
  ( 7 +9V )
)
( C2519C08 RC07 R3 8.3K
  ( 1 -_UPHONE )
  ( 2 N00016 )
)
( C2519C12 DO7 D1 IN4148
  ( 1 N00025 )
  ( 2 N00020 )
)
( C2519C10 DO7 D2 IN4148
  ( 1 N00011 )
  ( 2 N00017 )
)
( C2519C11 DO7 D3 IN4148
  ( 1 N00006 )
  ( 2 N00011 )
)
( C2519C14 DO7 D4 IN4148
  ( 1 N00013 )
  ( 2 N00008 )
)
( C2519C13 DO7 D5 IN4148
  ( 1 N00021 )
  ( 2 N00015 )
)
( C2519C1D RC05 PIEZO L7022
  ( 1 -_UPHONE )
  ( 2 N00023 )
)
( C2519C08 RC05 LDR PHOTOCEL
  ( 1 -_UPHONE )
  ( 2 N00019 )
)
)
```

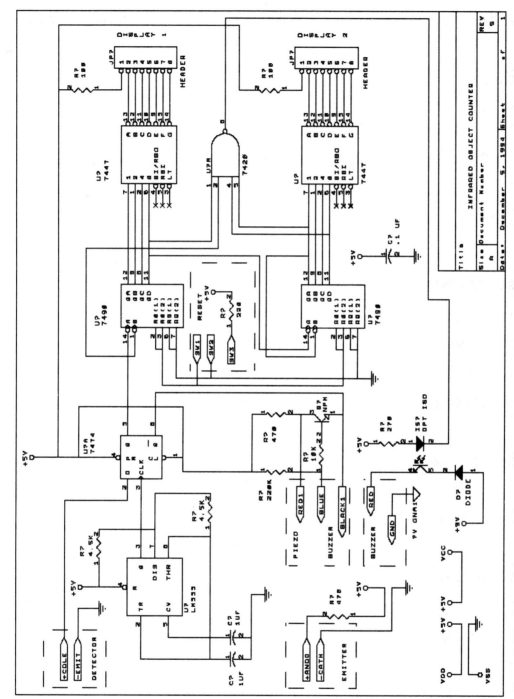

Figure 7-32. Infrared Counter Schematic before Annotation (*Source:* Adapted from Edward Cherbak, *EET250*, Student Project, Fall 1991.)

symbols on the worksheet and arrange them almost like your hand-drawn circuit. Connect them by using **PLACE Wire** and **PLACE Junction**. The schematic shown in Figure 7–32 is the schematic in which all the symbols have been connected.

Annotate schematic. The schematic must be annotated before proceeding further. Use the **Annotate** tool and follow the procedure to annotate the schematic for the Infrared Counter. Figure 7–33 shows the schematic after annotation.

Update field contents. Part Field 1 of all the schematic symbols must be updated with packaging identification before generating the netlist file. Use the **Update Field Contents** tool and the following update file (IR_COUNT.STF) to perform the operation.

Note that when you create this file, do not provide any characters other than the contents of the part value and the corresponding packaging identification within single quotes. Figure 7–34a is the stuff file for the Infrared Object Counter schematic. Figure 7–34b shows the schematic after processing by the **Update Field Contents** tool.

Generate bill of materials. Process the schematic by using the **Bill of Materials** tool. Figure 7–35 shows the BOM file after processing.

Archive schematic parts. It is a good idea to archive all the schematic parts in one single library at this time. Use the **Archive Parts in Schematic** tool to perform the operation. You may view the archive part library by using the **Library Browse** command.

Check electrical rules. At this point it is necessary to see if any of the electrical rules have been violated while drawing the circuit. You need to process the circuit by using this tool.

Generate netlist for OrCAD/PCB. A netlist file must be generated before the schematic can be used by the OrCAD/PCB tool to design artwork. Process the schematic by using the **Create Netlist** tool. Figure 7–36 shows the netlist file for the Infrared Counter.

Print the schematic. Print the hard copy of your schematic for your reference and documentation by using the **Plot Schematic** tool located under **Reporters**. If you have not made any changes since processing by **Update Field Contents,** it will look like the schematic shown in Figure 7–34b.

7–15.1 The Mini Stereo Amplifier

File organization. Set the design environment for the **Mini Stereo Amplifier (AMPLIFIE).** If the design name has not been created, you need

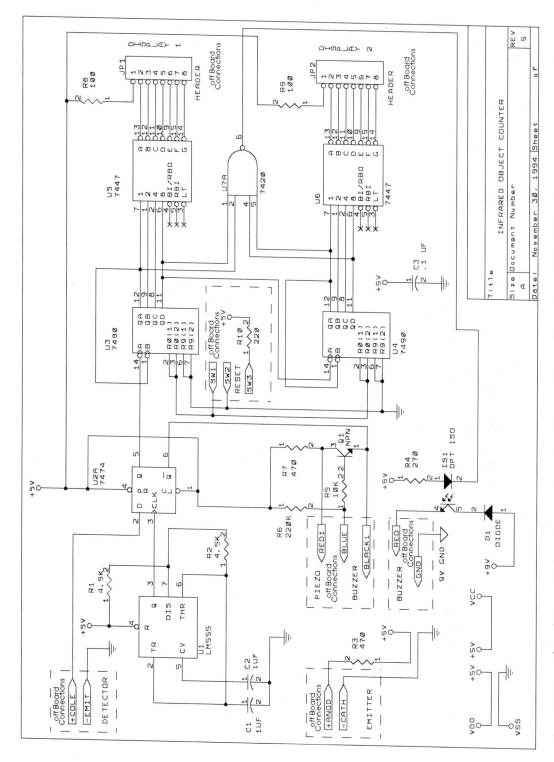

Figure 7-33. Infrared Counter Schematic after Annotation (*Source:* Adapted from Edward Cherbak, *EET 250*, Student Project, Fall 1991)

'7420'	'14DIP300'
'7447'	'16DIP300'
'7474'	'14DIP300'
'7490'	'14DIP300'
'LM555'	'8DIP300'
'HEADER'	'8SIP100'
'OPT ISO'	'8DIP300'
'NPN'	'TO92'
'.1uF'	'CK05'
'1uF'	'CK05'
'100'	'RC07'
'10K'	'RC07'
'220'	'RC07'
'220K'	'RC07'
'270'	'RC07'
'4.5K'	'RC06'
'470'	'RC07'
'DIODE'	'DO7'

Figure 34a. Stuff File for IR Counter

to create it now, using the **Create Design** tool located under Design Management Tools.

Configure schematic design tools. Configuration of SDT tools for this project has been provided in Section 7–3.4.

Schematic capture for the mini stereo amplifier. Look through the hand-drawn schematic for the Mini Stereo Amplifier shown in Figure 2–18 and get from the library the parts mentioned below:

Part Value	Quantity	Library
LM384	2	ANALOG 4
LM741	2	ANALOG
Capacitor Non Pol	2	PCB DEVICE
Capacitor Pol	11	PCB DEVICE
Resistor	13	PCB DEVICE
Resistor (POT)	2	PCB DEVICE
Battery	1	PCB DEVICE

Once you have all the symbols on the worksheet, save the schematic file by using **QUIT,** then choose **Write to File** or **QUIT,** and then **Update File.** At this point you may use **BLOCK Move** or **BLOCK Drag** to move the symbols on the worksheet and arrange them almost like your hand-drawn circuit. Connect them by using **PLACE Wire** and **PLACE Junction.** The schematic shown in Figure 7–37 is the schematic after all the symbols have been connected.

Annotate schematic. The schematic must be annotated before proceeding further. Use the **Annotate** tool and follow the procedure to annotate the

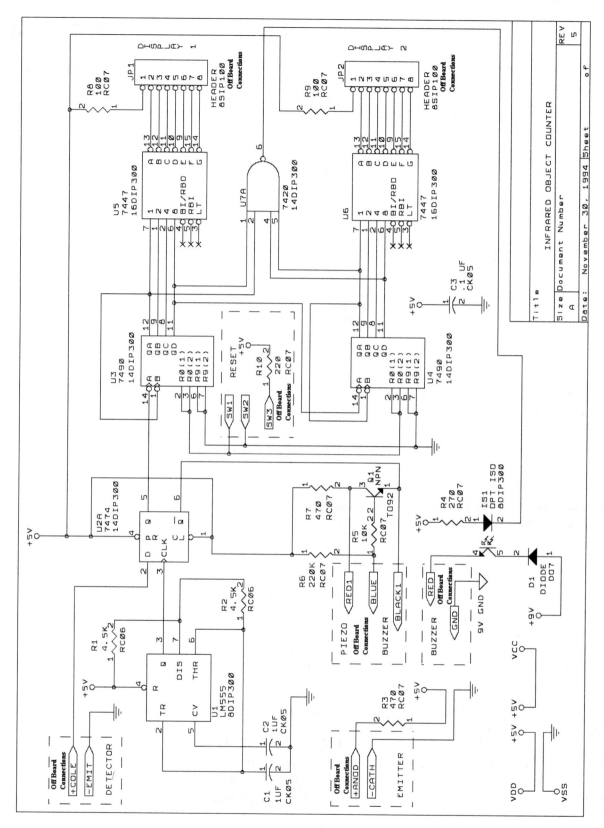

Figure 7-34b. Schematic after Processing by Update Field Contents (*Source:* Adapted from Edward Cherbak, *EET 250*, Student Project, Fall 1991)

7-15. The Other Projects

Revised: November 30, 1994
Bill Of Materials

Revision: 5
November 30, 1994 1:50:09 Page 1

Item	Quantity	Reference	Part Value	Packaging ID
1	2	C1,C2	1UF	CK05
2	1	C3	0.1 UF	CK05
3	1	D1	DIODE	DO7
4	1	IS1	OPT ISO	8DIP300
5	2	JP2,JP1	HEADER	8SIP100
6	1	Q1	NPN	TO92
7	2	R1,R2	4.5K	RC06
8	2	R3,R7	470	RC07
9	1	R4	270	RC07
10	1	R5	10K	RC07
11	1	R6	220K	RC07
12	2	R8,R9	100	RC07
13	1	R10	220	RC07
14	1	U1	LM555	8DIP300
15	1	U2	7474	14DIP300
16	2	U4,U3	7490	14DIP300
17	2	U5,U6	7447	16DIP300
18	1	U7	7420	14DIP300

Figure 7-35. Bill of Materials for the Infrared Object Counter

```
({OrCad/PCB II Netlist Format
INFRARED OBJECT COUNTER          Revised: November 30, 1994
                                 Revision: 5
Time Stamp - }
( 9756DBF  CK05 C3 .1 UF
 ( 1 VDD )
 ( 2 GND )
)
( 96B80E1B RC07 R8 100
 ( 1 N00002 )
 ( 2 VDD )
)
( 96B80E1C RC07 R9 100
 ( 1 N00021 )
 ( 2 VDD )
)
( 96B80E13 RC07 R5 10K
 ( 1 BLUE )
 ( 2 N00025 )
)
( 96B80E22 CK05 C1 1UF
 ( 1 N00008 )
 ( 2 GND )
)
( 96B80E23 CK05 C2 1UF
 ( 1 N00016 )
 ( 2 GND )
)
( 96D3EBCB RC07 R10 220
 ( 1 SW3 )
 ( 2 VDD )
)
```

Figure 7-36. Netlist File for the Infrared Object Counter

Figure 7-36. Continued

```
( 96B80E12 RC07 R6 220K
 ( 1 VDD )
 ( 2 BLUE )
)
( 96B80E1D RC07 R4 270
 ( 1 VDD )
 ( 2 N00032 )
)
( 96B80E18 RC06 R1 4.5K
 ( 1 VDD )
 ( 2 N00001 )
)
( 96B80E19 4RC06 R2 4.5K
 ( 1 N00008 )
 ( 2 N00001 )
)
( 96B80E1A RC07 R3 470
 ( 1 VDD )
 ( 2 +ANOD )
)
( 96B80E11 RC07 R7 470
 ( 1 VDD )
 ( 2 RED1 )
)
( 96BF933C 14DIP300 U7 7420
 ( 1 N00012 )
 ( 2 N00004 )
 ( 3 ?1 )
 ( 4 N00019 )
 ( 5 N00020 )
 ( 6 N00018 )
 ( 7 GND )
 ( 8 ?2 )
 ( 9 ?3 )
 ( 10 ?4 )
 ( 11 ?5 )
 ( 12 ?6 )
 ( 13 ?7 )
 ( 14 VDD )
)
( 96B80E0B 16DIP300 U5 7447
 ( 1 N00006 )
 ( 2 N00010 )
 ( 3 ?8 )
 ( 4 ?9 )
 ( 5 ?10 )
 ( 6 N00012 )
 ( 7 N00004 )
 ( 8 GND )
 ( 9 N00014 )
 ( 10 N00013 )
 ( 11 N00011 )
 ( 12 N00007 )
 ( 13 N00005 )
 ( 14 N00017 )
 ( 15 N00015 )
```

Figure 7-36. Continued

```
  ( 16 VDD )
)
( 96B80E0C 16DIP300 U6 7447
  ( 1 N00023 )
  ( 2 N00026 )
  ( 3 ?11 )
  ( 4 ?12 )
  ( 5 ?13 )
  ( 6 N00019 )
  ( 7 N00020 )
  ( 8 GND )
  ( 9 N00029 )
  ( 10 N00028 )
  ( 11 N00027 )
  ( 12 N00024 )
  ( 13 N00022 )
  ( 14 N00031 )
  ( 15 N00030 )
  ( 16 VDD )
)
( 96B80E0F 14DIP300 U2 7474
  ( 1 VDD )
  ( 2 +COLE )
  ( 3 N00009 )
  ( 4 VDD )
  ( 5 N00003 )
  ( 6 BLACK1 )
  ( 7 GND )
  ( 8 ?14 )
  ( 9 ?15 )
  ( 10 ?16 )
  ( 11 ?17 )
  ( 12 ?18 )
  ( 13 ?19 )
  ( 14 VDD )
)
( 96B80E09 14DIP300 U3 7490
  ( 1 N00004 )
  ( 2 SW1 )
  ( 3 SW1 )
  ( 4 ?20 )
  ( 5 VDD )
  ( 6 GND )
  ( 7 GND )
  ( 8 N00010 )
  ( 9 N00006 )
  ( 10 GND )
  ( 11 N00012 )
  ( 12 N00004 )
  ( 13 ?21 )
  ( 14 N00003 )
)
( 96B80E0A 14DIP300 U4 7490
  ( 1 N00020 )
  ( 2 SW1 )
  ( 3 SW1 )
  ( 4 ?22 )
  ( 5 VDD )
```

Figure 7-36. Continued

```
( 6 GND )
( 7 GND )
( 8 N00026 )
( 9 N00023 )
( 10 GND )
( 11 N00019 )
( 12 N00020 )
( 13 ?23 )
( 14 N00012 )
)
( E3E2D3D9 DO7 D1 DIODE
 ( 1 +9V )
 ( 2 N00033 )
)
( E3E2D3DB 8SIP100 JP1 HEADER
 ( 1 N00002 )
 ( 2 N00005 )
 ( 3 N00007 )
 ( 4 N00011 )
 ( 5 N00013 )
 ( 6 N00014 )
 ( 7 N00015 )
 ( 8 N00017 )
)
( E3E2D3DC 8SIP100 JP2 HEADER
 ( 1 N00021 )
 ( 2 N00022 )
 ( 3 N00024 )
 ( 4 N00027 )
 ( 5 N00028 )
 ( 6 N00029 )
 ( 7 N00030 )
 ( 8 N00031 )
)
( E3E2D3D8 8DIP300 U1 LM555
 ( 1 GND )
 ( 2 N00008 )
 ( 3 N00009 )
 ( 4 VDD )
 ( 5 N00016 )
 ( 6 N00008 )
 ( 7 N00001 )
 ( 8 VDD )
)
( E3E2D3DA TO92 Q1 NPN
 ( 1 BLACK1 )
 ( 2 N00025 )
 ( 3 RED1 )
)
( 96B80E15 8DIP300 IS1 OPT ISO
 ( 1 N00032 )
 ( 2 N00018 )
 ( 3 ?24 )
 ( 4 RED )
 ( 5 N00033 )
)
)
```

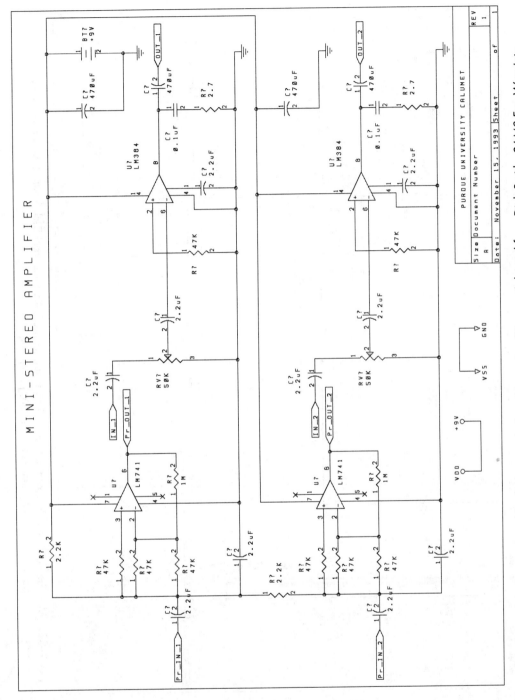

Figure 7-37. Mini Stereo Amplifier Schematic before Annotation (*Source:* Adapted from Dick Smith, *SAMS Fun Way into Electronics*, Indianapolis: Howard W. Sams & Co., 1986, p.51)

255

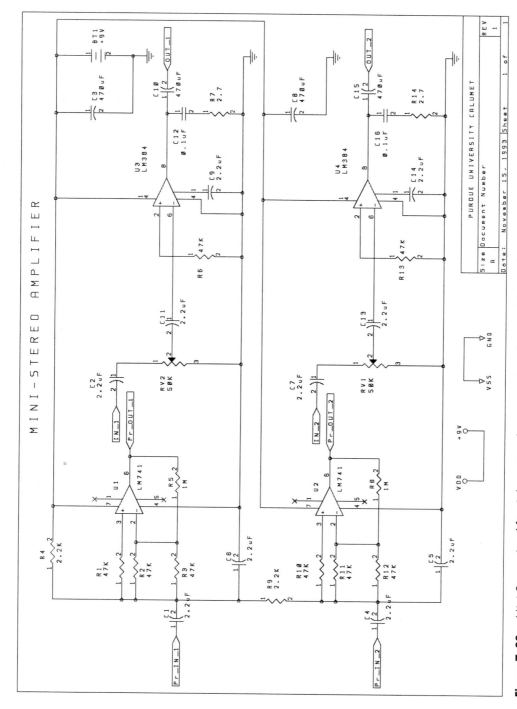

Figure 7-38. Mini Stereo Amplifier Schematic after Annotation (Source: Adapted from Dick Smith, *SAMS Fun Way into Electronics*, Indianapolis: Howard W. Sams & Co., 1986, p.51)

schematic for the Mini Amplifier. Figure 7–38 shows the schematic after annotation.

Update field contents. Part Field 1 of all the schematic symbols must be updated with packaging identification number before generating netlist file. Use the **Update Field Contents** tool and the following update file (AMPLI-FIE.STF) to perform the operation.

Note that when you create this file, do not provide any characters other than the contents of the part value and the corresponding packaging identification within single quotes. Figure 7–39a is the stuff file for the schematic, and Figure 7–39b shows the schematic after processing by the **Update Field Contents** tool.

Generate bill of materials. Process the schematic using the **Bill of Materials** tool. Figure 7–40 shows the BOM file after processing.

Archive schematic parts. It is a good idea to archive all the schematic parts in one single library at this time. Use the **Archive Parts in Schematic** tool to perform the operation. You may view the archive part library by using the **Library Browse** command.

Check electrical rules. At this point it is necessary to see that none of the electrical rules has been violated while drawing the circuit. You need to process the circuit by using this tool.

Generate netlist for OrCAD/PCB. A netlist file must be generated before the schematic can be used by the OrCAD/PCB tool to design artwork. Process the schematic by using the **Create Netlist** tool. Figure 7–41 shows the netlist file for the Mini Stereo Amplifier.

Print the schematic. Print the hard copy of the schematic for your reference and documentation by using the **Plot Schematic** tool located under **Reporters.** If you have not made any changes since processing by **Update Field Contents,** it will look like the schematic shown in Figure 7–39b.

7-16. MODIFY LIBRARY PARTS

During the schematic design process, you may need to create a new symbol because it is not available in any of the OrCAD-provided libraries. This situation is not common, yet having this ability is a lifesaver in some situations. OrCAD/SDT has provided four tools—**Edit Library, List Library, Compile Library,** and **Decompile Library**—located under **Librarians** in SDT_MENU. Using these tools, you may edit a graphic library symbol or describe a graphic part by using symbolic language.

Figure 7-39a. Stuff File for the Mini Stereo Amplifier

'LM784'	'14DIP300'
'LM741'	'8DIP300'
'50K'	'RJ24X'
'47K'	'RC07'
'2.7'	'RC07'
'2.2K'	'RC07'
'1M'	'RC07'
'.1uF'	'CK06'
'2.2uF'	'CK06'
'470uF'	'CK06'
'+9V'	'RC07'

There are two ways you can create a symbol in OrCAD/SDT: (1) using symbolic description language and (2) using the graphical part editor.

7-16.1 Symbolic Description Language

There are key words and special syntax by which you can define a part and create a library source file in ASCII. This ASCII library file is compiled to convert it into a graphic symbol by using the **Compile Library** tool. This graphic part can then be stored in a custom library. After storing the part, you can call this library part symbol onto your schematic design screen and design schematics by using this custom graphic symbol.

The **Decompile Library** tool can be used to convert a graphic symbol into symbolic language. This means that the custom part that you have created or any existing library part symbol can be converted into symbolic description language by using this tool. This tool is a big help in the custom module design process. However, to create a part using this symbolic description language is beyond the scope of this book. Therefore, to do this you need to refer to the OrCAD/SDT manual.

7-16.2 Using the Graphical Part Editor

Often, you just want to modify an existing library symbol rather than create it from scratch. The **Edit Library** tool will let you do that. By using this tool, you can create a new part by modifying an existing library part in graphic editor mode. The graphic part symbol can be saved under a custom library. Using this tool, you may call any existing graphic symbol and modify it according to your needs and save it under a different name in the custom library. This is a quick way to satisfy your need for a special part. The **Edit Library** tool is not difficult to use. However, detailed description of this tool is not within the scope of this book. If you need to use this tool, refer to the OrCAD/SDT reference guide.

There is another tool called **List Library** that is used to create a list of the parts of a schematic. This tool scans a library and generates a list of

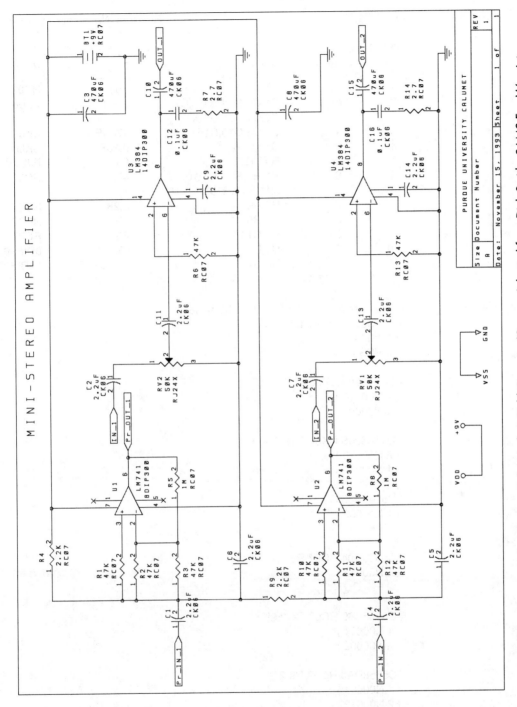

Figure 39b. Schematic after Processing by Update Field Contents (*Source:* Adapted from Dick Smith, *SAMS Fun Way into Electronics*, Howard W. Sams & Co., 1986, p.51)

Figure 7-40. Bill of Materials for Amplifier

AMPLIFIER   Revised: November 15, 1993
                                                       Revision: 1
Bill Of Materials   June 16, 1994   15:27:37   Page 1

Item	Quantity	Reference	Part	Packaging ID
1	1	BT1	+9V	RC07
2	10	C1,C2,C4,C5,C6,C7, C9,C11,C13,C14	2.2uF	CK06
3	4	C3,C8,C10,C15	470uF	CK06
4	2	C12,C16	0.1uF	CK06
5	2	RV1,RV2	50K	RJ24X
6	8	R1,R2,R3,R6,R10, R11,R12,R13	47K	RC07
7	2	R4,R9	2.2K	RC07
8	2	R5,R8	1M	RC07
9	2	R7,R14	2.7	RC07
10	2	U1,U2	LM741	8DIP300
11	2	U3,U4	LM384	14DIP300

Figure 7-41. Netlist for Amplifier

```
({ Time Stamp -  15-NOV-1993   16:32:38 }
( C5658D88 RC07 BT1 +9V
 ( 1 N00002 )
 ( 2 VSS )
)
( E44A6209 CK06 C12 0.1UF
 ( 2 N00010 )
 ( 1 N00007 )
)
( E44A620F CK06 C16 0.1UF
 ( 2 N00020 )
 ( 1 N00017 )
)
( C564E426 RC07 R5 1M
 ( 1 N00005 )
 ( 2 PR_OUT_1 )
)
( C5658D94 RC07 R8 1M
 ( 1 N00015 )
 ( 2 PR_OUT_2 )
)
( C5658D89 RC07 R4 2.2K
 ( 1 N00001 )
 ( 2 N00002 )
)
( C5658DA0 RC07 R9 2.2K
 ( 1 N00001 )
 ( 2 N00012 )
)
( C564E42E CK06 C1 2.2UF
 ( 2 N00001 )
 ( 1 PR_IN_1 )
)
( C564E42C CK06 C2 2.2UF
 ( 2 IN_1 )
 ( 1 N00003 )
```

Figure 7-41. Continued

```
)
( C5658D8C CK06 C4 2.2UF
 ( 2 N00012 )
 ( 1 PR_IN_2 )
)
( C5658D8D CK06 C5 2.2UF
 ( 2 VSS )
 ( 1 N00012 )
)
( C564E42F CK06 C6 2.2UF
 ( 2 VSS )
 ( 1 N00001 )
)
( C5658D96 CK06 C7 2.2UF
 ( 2 IN_2 )
 ( 1 N00013 )
)
( E44A6208 CK06 C9 2.2UF
 ( 2 VSS )
 ( 1 N00011 )
)
( C564E42B CK06 C11 2.2UF
 ( 2 N00008 )
 ( 1 N00009 )
)
( E44A620E CK06 C13 2.2UF
 ( 2 N00018 )
 ( 1 N00019 )
)
( E44A620C CK06 C14 2.2UF
 ( 2 VSS )
 ( 1 N00021 )
)
( C564E432 RC07 R7 2.7
 ( 1 N00010 )
 ( 2 VSS )
)
( E44A620B RC07 R14 2.7
 ( 1 N00020 )
 ( 2 VSS )
)
( C564E42A CK06 C3 470UF
 ( 2 VSS )
 ( 1 N00002 )
)
( C5658D9B CK06 C8 470UF
 ( 2 VSS )
 ( 1 N00002 )
)
( C564E42D CK06 C10 470UF
 ( 2 OUT_1 )
 ( 1 N00007 )
)
( E44A6210 CK06 C15 470UF
 ( 2 OUT_2 )
 ( 1 N00017 )
```

Figure. 7–41 Continued

```
)
( C564E424 RC07 R1 47K
 ( 1 N00001 )
 ( 2 N00004 )
)
( C564E425 RC07 R2 47K
 ( 1 N00001 )
 ( 2 N00005 )
)
( C564E423 RC07 R3 47K
 ( 1 N00001 )
 ( 2 N00005 )
)
( C564E422 RC07 R6 47K
 ( 1 N00006 )
 ( 2 VSS )
)
( C5658D91 RC07 R10 47K
 ( 1 N00012 )
 ( 2 N00014 )
)
( C5658D92 RC07 R11 47K
 ( 1 N00012 )
 ( 2 N00015 )
)
( C5658D93 RC07 R12 47K
 ( 1 N00012 )
 ( 2 N00015 )
)
( E44A620A RC07 R13 47K
 ( 1 N00016 )
 ( 2 VSS )
)
( C5658D99 RJ24X RV1 50K
 ( 1 N00013 )
 ( 2 N00018 )
 ( 3 VSS )
)
( C564E431 RJ24X RV2 50K
 ( 1 N00003 )
 ( 2 N00008 )
 ( 3 VSS )
)
( E44A6206 14DIP300 U3 LM384
 ( 1 N00011 )
 ( 2 N00006 )
 ( 3 ?00001 )
 ( 4 VSS )
 ( 5 VSS )
 ( 6 N00009 )
 ( 7 VSS )
 ( 8 N00007 )
 ( 9 ?00002 )
 ( 10 VSS )
 ( 11 VSS )
 ( 12 VSS )
```

7-16. Modify Library Parts

Figure 7-41. Continued

```
  ( 13 ?00003 )
  ( 14 N00002 )
)
( E44A620D 14DIP300 U4 LM384
  ( 1 N00021 )
  ( 2 N00016 )
  ( 3 ?00004 )
  ( 4 VSS )
  ( 5 VSS )
  ( 6 N00019 )
  ( 7 VSS )
  ( 8 N00017 )
  ( 9 ?00005 )
  ( 10 VSS )
  ( 11 VSS )
  ( 12 VSS )
  ( 13 ?00006 )
  ( 14 N00002 )
)
( C564E41E 8DIP300 U1 LM741
  ( 1 ?00007 )
  ( 2 N00005 )
  ( 3 N00004 )
  ( 4 VSS )
  ( 5 ?00008 )
  ( 6 PR_OUT_1 )
  ( 7 N00002 )
  ( 8 ?00009 )
)
( C5658D95 8DIP300 U2 LM741
  ( 1 ?00010 )
  ( 2 N00015 )
  ( 3 N00014 )
  ( 4 VSS )
  ( 5 ?00011 )
  ( 6 PR_OUT_2 )
  ( 7 N00002 )
  ( 8 ?00012 )
)
( E44B2D0B mhole2 M1 INPUT1
  ( XXXX PR_IN_1 )
)
( E44B2E4A mhole2 M2 INPUT2
  ( XXXX PR_IN_2 )
)
( E44B2F12 mhole2 M3 PR_OUT1
  ( XXXX PR_OUT_1 )
)
( E44B2F9C mhole2 M4 PR_OUT2
  ( XXXX PR_OUT_2 )
)
( E44B3BD2 mhole2 M7 AMP_OUT1
  ( XXXX OUT_1 )
)
( E44B3BD3 mhole2 M8 AMP_OUT2
  ( XXXX OUT_2 )
```

Figure 7-41. Continued

```
        )
      ( E44B3CF6 mhole2 M9 +9V
       ( XXXX N00002 )
      )
      ( E44B3CF7 mhole2 M10 GND
       ( XXXX VSS )
      )
      ( E44FD8E5 mhole2 M5 AMP_IN1
       ( XXXX IN_1 )
      )
      ( E44FD958 mhole2 M6 AMP_IN2
       ( XXXX IN_2 )
      )
    )
```

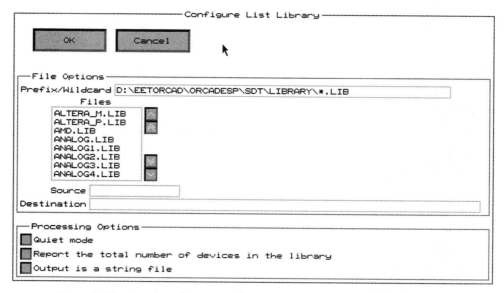

Figure 7-42. Local Configuration of List Library (Courtesy of OrCAD Inc.)

parts in ASCII format. This tool is convenient for creating a stuff file for a schematic. Use the following procedure:

1. Left-click on the **List Library** icon located in SDT_MENU, and click on **Local Configuration** from the Transit Menu. The Local Configuration screen of the **List Library** tool shown in Figure 7-42 will appear.

2. Under **File Options,** locate the library for which you want to generate the list, and click on it. Provide a name of a destination file in the Destination dialogue box.

3. Under **Processing Options,** activate the *Report the total number of devices in library* and *Output is a string file* buttons. Exit the configuration screen by clicking on the OK icon.

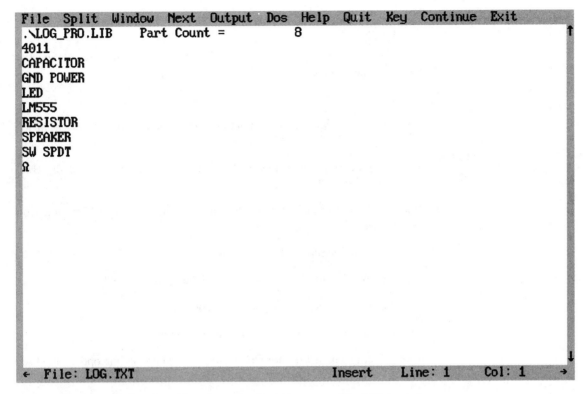

Figure 7-43. Parts List Produced by Library List Tool (Courtesy of OrCAD Inc.)

4. Click on the **List Library** icon again from SDT_MENU, and this time select **Execute** from the Transit Menu. The tool will scan the schematic library (for a schematic it is the Archive library) and generate a list of parts. The list will be written in the destination file.

Figure 7-43 is a list of parts created by this tool from the archive library file for the Logic Probe project. By using this tool, you can create a hard-copy list of parts for quick reference.

QUESTIONS

1. What are two basic criteria that you will use to decide on the number of drawings and their sheet size for a project?

2. How would you decide which part should be mounted off-board?

3. If you know that your schematic will ultimately lead to a PC board design, from which library should you fetch your resistors, capacitors, transistors, diodes, etc.?

4. If your 'Device.Lib' is also configured, where should it be in the Configured List Box?

5. If two wires meet end-to-end, will OrCAD consider this as a connection without a junction box?

6. If two wires intersect each other, will OrCAD consider this as a connection without a junction box?

7. What is the difference between a bus and a wire connection?

8. Normally, if you connect two outputs together, the electrical rules check will give you errors. What could you do to stop that?

9. What is the difference between the Unannotate command and the Back Annotate command?

10. What tool should you use to make the content of the Part Field 1 invisible from the schematic?

11. Can the M2EDIT editor edit a schematic?

12. Why and how should you use the Update Field Contents tool?

13. Why is it necessary to archive parts of a schematic?

14. What are the three different steps in creating a netlist?

15. When creating a netlist file for OrCAD/PCB, what processing option should be used?

16. What are the three important pieces of information that a netlist file provides to the PCB tool?

17. Is it possible to create a netlist file and edit its information without executing the netlist tool?

18. How would you find the packaging number for a module?

19. How can you customize a bill of materials for a project?

20. If a good printer is not available with the computer station you are using, and a computer station that has a good printer does not have OrCAD, what should you do to print your schematic on the good printer?

OrCAD/PCB Tool Set

8

PRINTED CIRCUIT BOARD LAYOUT TOOL SET

8-1. WHAT ARE THE COMPONENTS OF A PRINTED CIRCUIT BOARD?

A printed circuit board (PCB) refers to any electrical circuit in which individual wire lead connections have been replaced by two-dimensional metallic conductive patterns bonded to one or more dielectric substrates. The metallic conductive patterns on the base material are formed by using both subtractive methods (etching) and additive methods (plating). Later components mounted on the board are soldered in place. There are, in general, three types of printed circuit boards: single-sided, double-sided, and multilayer. Only single- and double-sided boards are within the scope of this book.

8-1.1 Single-Sided Boards

A printed circuit board with conductive patterns only on one side of the board is normally used for simple electronic circuits and does not require copper-plated through-holes. Single-sided boards are easy to fabricate but difficult to design.

8-1.2 Double-Sided Boards

A printed circuit board with a conductive pattern on both sides involves a comparatively more complicated electronic circuit and provides greater circuit density than single-sided boards do. Side-to-side connections are made possible by holes that are plated through. Double-sided boards are most desirable for contemporary circuits. However, single-sided boards are less expensive.

Objectives

After completing this chapter, you should be able to

1. Describe single-sided and double-sided printed circuit boards
2. Describe the functions of the PC Board Layout tool set
3. Configure the PC Board Layout tool set
4. Describe various commands and functions of PC Board Layout Tools

8-2. DESCRIPTION OF THE PC BOARD LAYOUT TOOL SET

The PC Board Layout tool set is used to design the master artwork layout for printed circuit boards. It is a tool that transforms logical connections among components of a schematic into physical tracks of desired widths. These tracks are the segments of conducting paths from one electronic component to another. These tracks cannot overlap each other on a board surface. The process is not straightforward. The Layout Tools reads the netlist file, which contains all connectivity information and packaging size for each component. The logical connection among components, packaging size, value, and reference for components in a schematic are expressed in the form of a netlist file by a tool called **Create Netlist** that is located within the Schematic Design Tools. As the Layout Tools downloads netlist information, it retrieves the appropriate module from the PCB library, depending on the packaging information available for each component in the netlist file. Modules are downloaded onto a board of specified size. Among many other things, the Layout Tools downloads the logical connection scheme for each component. Modules on the board require rearrangement so that each logical connection can be converted into physical tracks by the **Route Board** tool. The tool set window can be invoked by using the following steps:

1. Invoke ESP_MENU by typing **orcad** under the MYORCAD subdirectory.
2. Click on the **PC Board Layout Tools** icon, and the PC Board Layout Tools menu (PCB_MENU) will appear on the screen.

The PC Board Layout Tools is arranged into five categories: Editors, Processors, Reporters, Transfers, and User. Each of these subdivisions has several tools. To invoke any of these tools, use the following procedure:

1. Place the mouse pointer on the desired tool icon and left-click the mouse once. The Transit Menu will appear at the upper left corner of the screen.
2. Select **Execute** from the options and click on it once. This action will either execute the respective action of the tool or display another menu screen to choose from.

The behavior of the tool that has Local Configuration can be altered in various different ways by configuring the various options in the Local Configuration screen. More details about each of these tools and how they are used for PC board master artwork design will be discussed later in this chapter. The function of each tool in the PCB_MENU screen will be discussed next to give the designer a clear idea about which tool to use under what circumstances. The tools are divided into various categories depending on their functions.

8-2.1 Editors

Editors has three tools: Route Board, Edit File, and View Reference.

Route board. Route Board is the tool set that is used to do most of the artwork design. This set of tools loads the board, places the library module, configures part of the design environment, edits the PC board's worksheet file, and does the auto and manual routing. To invoke this set of tools, perform the following steps:

1. Using the Design Management Tools, load the design whose board file you want to access or create.
2. In ESP_MENU, click on the **PC Board Layout Tools** icon, and the PCB_MENU will be displayed.
3. Click on the **Route Board** located under Editors, and the PC Board Design Worksheet (PCB_MAIN) will appear. More details about this tool set will be described later during the artwork design procedure.

Edit file. Edit File is a general purpose, full screen text editor. OrCAD calls this M2EDIT. It is the same tool that is also available in the SDT tool set.

View reference. View Reference will allow you to view reference material regarding the tool set. Click on the icon and follow the directions on the screen to review the reference material.

8-2.2 Processors

Processors has three tools: Modify Modules, Create NC Drill File, and Reannotate Board File.

Modify modules. Modify Modules will modify pad shape, size, type, drill size, and the sides of the board for modules either in a layout or in a module library. However, the capability of modifying the attributes of pads is more useful when you are in a board file. The same functions can be performed by using **EDIT** under the Route Board tools.

Create NC drill files. Create NC Drill Files will generate files containing drill information for a board file. This information includes size and location of drill in ASCII or Excellon format.

Reannotate board file. Reannotate Board File is used to reannotate or renumber the modules of a board file. In the local configuration of the tool, you can specify to scan the board by column or by row for renumbering the modules. Unconditional or conditional renumbering of the board can be achieved by activating the proper button.

8-2.3 Reporters

Reporters has four tools: Print PCB, Compare Netlists, Convert Plot to IGES, and Module Report.

Print PCB. Print PCB is used to print artwork, silkscreen, drill files, etc. The local configuration of the tool can be configured for several different types of output. Hard-copy or soft-copy output can be produced by sending the output to a printer or to a file, respectively. Scaled, inverted output, and output with drill holes can be produced by choosing the appropriate processing options. The inverted image of the artwork is used as a negative mask for certain imaging processes in PC board fabrication. This is only a print tool. The file must be generated earlier in **Route Board** for this tool to be used for printing or plotting.

Compare netlist. Two netlist files can be compared using Compare Netlist. The differences between two netlists can be produced as a report.

Convert plot to IGES. By using Convert Plot, the plot of a board file can be translated into Initial Graphics Exchange Specification (IGES) full ASCII format. A source file with PLT extension is converted to an IGES extension.

Module report. Module Report produces a list of modules used in the design and writes into a file with the extension LOC.

8-2.4 Transfers

The following are four tools that will transfer design information from one tool to the other. These tools will perform the necessary work to transfer the information. Before the transfer, **Local Configuration** will provide processing options for the transfer.

- **To Schematic:** to transfer necessary information from the PC Board Layout Tools to the Schematic Design Tools
- **To PLD:** to transfer necessary information from the PC Board Layout Tools to the Programmable Logic Design Tools
- **To Digital Simulation:** to transfer necessary information from the PC Board Layout Tools to the Digital Simulation Tools
- **To Main:** to transfer you to the main ESP menu from where you may branch out to other tools or exit the software.

8-2.5 User

The four **User** buttons can be used to run any system command, allowing other programming environments to be invoked from OrCAD tools. Each of these buttons can be programmed to invoke certain design tools that are not part of the OrCAD software environment.

8-3. CONFIGURATION OF PC BOARD LAYOUT TOOL SET

The configuration screen of the PC Board Layout Tool set can be accessed from the Transit Menu of **Route Board.** This configuration process allows the designer to set many global configuration parameters of the tool set. The configuration screen is quite long and cannot be seen all at once on the screen. Generally, the mouse is more convenient to use to scroll through the configuration screen. Options in the configuration screen are grouped into several chunks: **Driver, Board File, Netlist File, Module, Strategy, Memory Allocation, Serial, Design, Conditions, Net Conditions, Photo-Wheel (D-Code) Configuration, Pen Carrousel,** and **Color Table.** Figures 8–1a through 8–1f show the configuration screens for the PC Board Layout Design Tools.

8-3.1 Driver Options

Driver Prefix is the directory path or disk drive where the PC Board Layout Tools find and load the display, printer, and plotter driver programs. Driver prefixes are set automatically during the installation process; therefore, if they are running, do not disturb them needlessly. You may search through the list box of each driver type to locate the desired type of driver. The list box will contain only those drivers that you have loaded during software installation. When you find the appropriate one, click the mouse on it to configure it as the respective driver for display, plotter, and printer. Be very careful about the display driver. If you select one that is incompatible with your display monitor, PCB_MENU will not display on the screen. Even if you reboot the computer, you will not be able to display PCB_MENU. If such a situation arises, you need to edit the PCB.CFG file located under the design subdirectory. One such file will be under every design subdirectory. However, you need to edit only the one that is incompatible.

Perform the following operation:

1. From the ESP_MENU, click on the **Design Management Tools** icon.

2. Click on the design name whose .CFG file is incompatible.

3. Click on the **File View** and select the file called PCB.CFG.

4. Click on **Edit File.** The PCB.CFG file will be displayed on the M2EDIT screen.

5. Scroll down to DD (display driver) and replace the driver name with the appropriate one suitable for your monitor. Do not change anything else if you are not sure about it.

6. Click on OK. Save the modified PCB.CFG file. Return to ESP_MENU and invoke PCB_MENU for the design again.

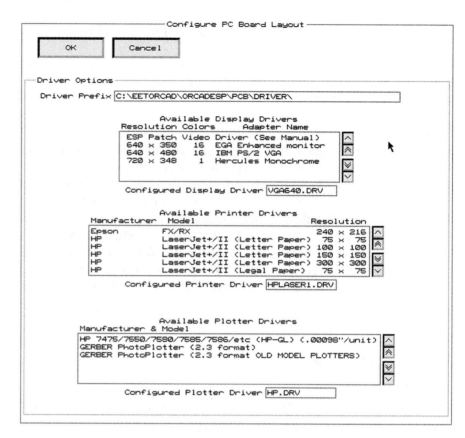

Figure 8–1a. Configuration Screen of the PC Board Layout Tool (Courtesy of OrCAD Inc.)

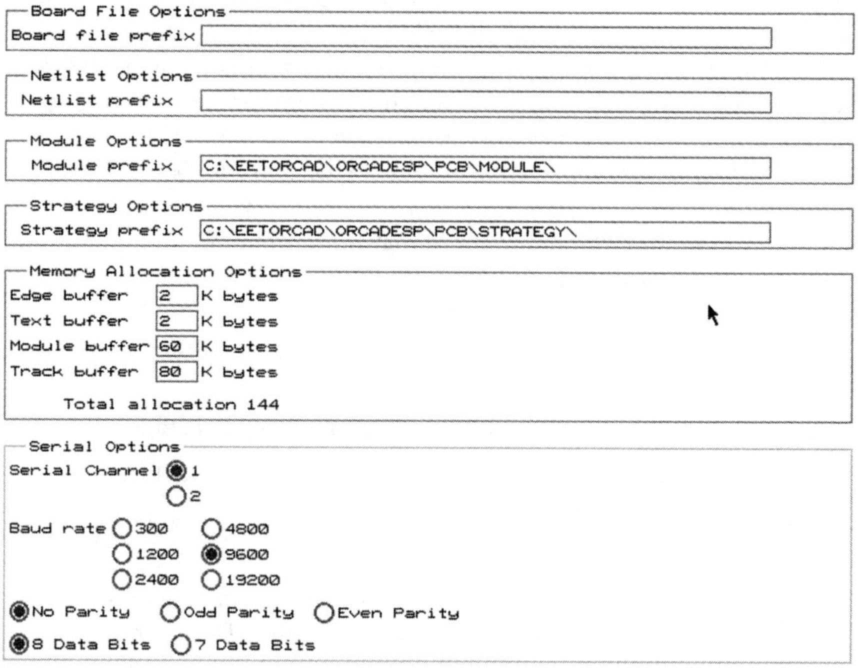

Figure 8–1b. (Courtesy of OrCAD Inc.)

```
┌─Design Conditions─────────────────────────────────────────┐
│ Track width      .015                                     │
│ Pad diameter     .055      Pad drill    .035              │
│ Via diameter     .045      Via drill    .025              │
│ Solder mask guard .020                                    │
│ Text dimensions                                           │
│       Horizontal .050      Vertical     .050              │
│ Isolations                                                │
│       Track-Track .015     Via-Via      .020              │
│ Routing grid     .050                                     │
│ Number of layers 2                                        │
│ Working layer  A 1              layer B 2                 │
│ Pass 1 strategy         Pass 2 strategy                   │
│ ● Normal                ○ Normal                          │
│ ○ Flexible              ● Flexible                        │
│ ○ Extensive             ○ Extensive                       │
│ ○ 90 Degree             ○ 90 Degree                       │
│ ○ No Via                ○ No Via                          │
│ ○ Power                 ○ Power                           │
│ Net pattern                                               │
│       ○ Chain net pattern                                 │
│       ● Tree net pattern                                  │
│       ○ Comb net pattern                                  │
│ ● Short cross cursor                                      │
│ ○ Long cross cursor                                       │
│ ○ Disable automatic backups                               │
│ ● Enable automatic backups                                │
│       ○ 1   ● 2   ○ 4   ○ 8                               │
│ Plot X Offset .000                                        │
│ Plot Y Offset .000                                        │
└───────────────────────────────────────────────────────────┘
```

Figure 8-1c. (Courtesy of OrCAD Inc.)

8-3.2 Board File Options

You may define a path or disk drive in the dialogue box into which the PC Board Layout Tools can load the board file. If you are using the standard OrCAD directory structure, leave this option blank.

8-3.3 Netlist File Options

This option defines the drive path for the PC Board Layout Tools to find and load the netlist file. If you are using the standard OrCAD design directory structure, leave this dialogue box empty.

8-3.4 Module Options

This option defines the path where PC Board Layout Tools find and load the module files. If you are using the standard OrCAD design directory structure, leave this dialogue box as follows: C:\MYORCAD\ORCADESP\PCB\MODULE\.

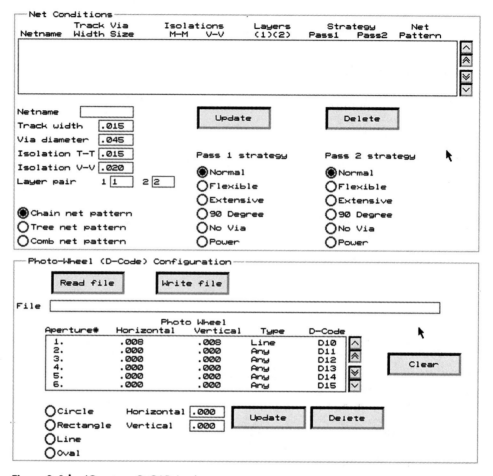

Figure 8-1d. (Courtesy OrCAD Inc.)

8-3.5 Strategy Options

The Strategy Option prefix defines the path where PC Board Layout Tools finds and loads the auto routing strategy files. If you need to change the prefix, left-click on the dialogue box and edit the pathname by using the standard editing techniques.

8-3.6 Memory Allocation

This option defines the size of the allocated memory buffers. Keep the default value of the memory allocation. During board design if the *not enough memory* problem appears, you may need to change the memory allocation. For example, if the *not enough memory* signal appears before autorouting, you may need to allocate more memory for the Track Buffer than for other buffers. Default memory allocation is provided as follows:

    Edge Buffer: 4 Kbytes
    Text Buffer: 4 Kbytes
    Module Buffers: 80 Kbytes
    Track Buffers: 100 Kbytes

8-3. Configuration of PC Board Layout Tool Set

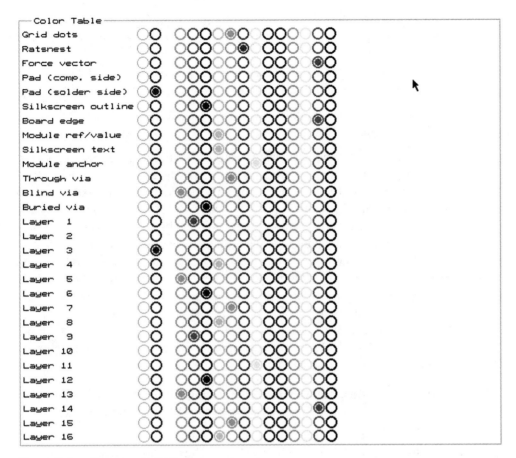

Figure 8-1e. (Courtesy OrCAD Inc.)

Figure 8-1f. (Courtesy OrCAD Inc.)

The total RAM available depends on the operating system. For the current version of MS-DOS, the operating system recognizes only 640 Kbytes of memory, no matter how many Mbytes of memory you have in your computer. To find out the available amount of memory, type CHKDSK at the C:> prompt. A typical result can be as follows:

655360 bytes total memory
570320 bytes free

To increase the free memory, change the AUTOEXEC.BAT file such that it does not load any memory resident programs when you boot up your computer. The mouse driver is also a memory resident program, but you definitely need that. If you are using MS Windows 3.1 or higher, you may use the **Memmaker** command to free up your system memory. The result may provide you more available memory for autorouting. If you are unable to increase your system memory, you may reallocate the available system memory by using OrCAD tools. Remember that you may have two or three megabytes of RAM available in your computer, but DOS recognizes only 640 Kbytes. This is the reason that most systems, when you invoke the **CHKDSK** (check disk) command from the DOS prompt, respond by showing 640 Kbytes of RAM. The following information will help you reallocate the system memory for OrCAD:

- **Edge Buffer:** Memory allocated for this buffer is reserved to develop board edge. The allowed range is 1 to 64 K. The default is 2 K.
- **Text Buffer:** Memory allocated for this buffer is reserved to create text. The allowed range is 1 to 64 K. The default is 2 K.
- **Module Buffer:** Memory allocated for this buffer is reserved for modules. The allowed range is 1 to 128 K. The default is 60 K.
- **Track Buffer:** Memory allocated for this buffer is reserved for layout tracks, including vias. The allowed range is 1 to 512 K. The default is 80 K.

For other operations, such as **Design Rule Check, Zone Placement,** and **Auto Route,** the PC Board Layout Tools require RAM memory. Since more memory allocation reduces the free system memory, you need to be conservative about allocation. Altogether, do not allocate more than 240 K for the Edge, Text, Module, and Track buffers. Autorouter will not work without at least 65 K of system memory.

8-3.7 Serial Options

Leave them at their default conditions at this time.

8-3.8 Design Conditions

Design conditions define dimensions, strategy, net pattern, cursor type, etc. Dimensions such as track width and pad diameter have allowed range and default values. Leave them at their default values; you may change them during the PC board design session. To change any one of these values, left-click the mouse in the dialogue box and edit the desired value. If you change these values now, you can alter them again during layout design.

8-3.9 Net Conditions

Net Conditions define special conditions assigned for all the nets of a design. Leave these also at their default values. You should alter them during layout design. Whatever is specified under the Net Conditions list box has the highest priority. This means that a certain net that has been specified with conditions will always be routed in that manner, disregarding the condition specified during the design and under Design Conditions.

8-3.10 Photo-Wheel (D-Code) Configuration

This part of the configuration specifies the dimensions and type of each aperture used by a photoplotter. Leave these dimensions at their default value. When you are ready to plot, you can alter them to their appropriate values.

8-3.11 Pen Carrousel

Pen Carrousel specifies pen width, velocity, and acceleration for plotter pens. Use the default values.

8-3.12 Color Table

The Color Table selects the screen display colors for the different layers and items in the board file. Use the default values.

8-4. COMMON FUNCTIONS OF THE PCB TOOL SET

8-4.1 Again

AGAIN repeats the last main menu command executed. For example, if you execute a command in **EDIT,** the command may be repeated any number of times by selecting **AGAIN.** Since it is the first item in the menu, using it saves time. When a command is required to be repeated several times, instead of scrolling down through the main menu, **AGAIN** will repeat the command and provide the appropriate option for the next step. **AGAIN** repeats only the commands from the main menu and not from any submenu. The command does not have a submenu.

8-4.2 Block

BLOCK, coupled with its submenu, is used to manipulate specific areas of a layout. By selecting a specified part of an artwork layout, it can be moved around the worksheet, duplicated, imported, or exported. The specified part of the board can be moved, copied, and saved. By selecting the appropriate command, you can move, save, or copy everything in the area or save only the routes, module, or text. Figures 8–2a through 8–2d show the **BLOCK** submenus.

The **BLOCK Move** command will cut out the selected objects from the specified area and move them to the desired location. To perform the **BLOCK Move** operation, follow the steps below:

1. Move the mouse pointer to the upper or lower corner of the area you are planning to move.

2. Click on **BLOCK** from the PCB_MAIN menu, and then select **Move.** A bar menu, shown below, will appear. Type **B,** for **Begin,** or

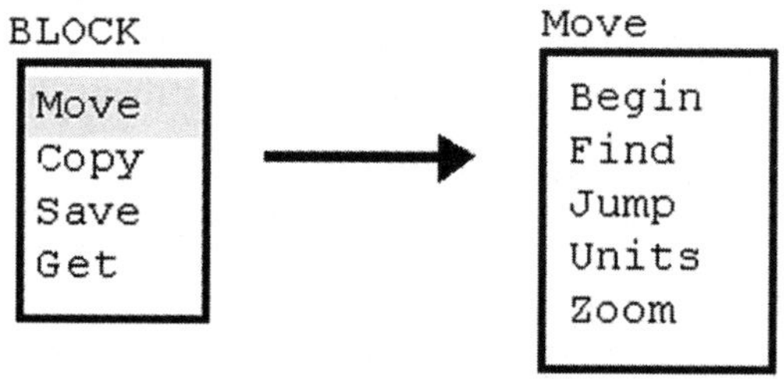

Figure 8–2a. BLOCK Move Submenu

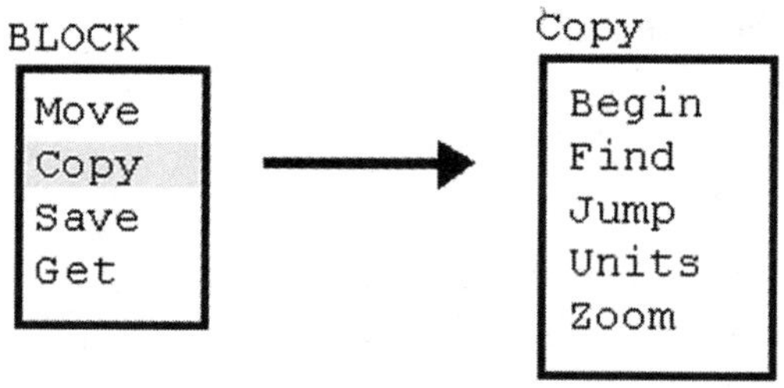

```
BLOCK            Move
Move            Begin
Copy     →      Find
Save            Jump
Get             Units
                Zoom
```

Figure 8–2b. BLOCK Copy Submenu

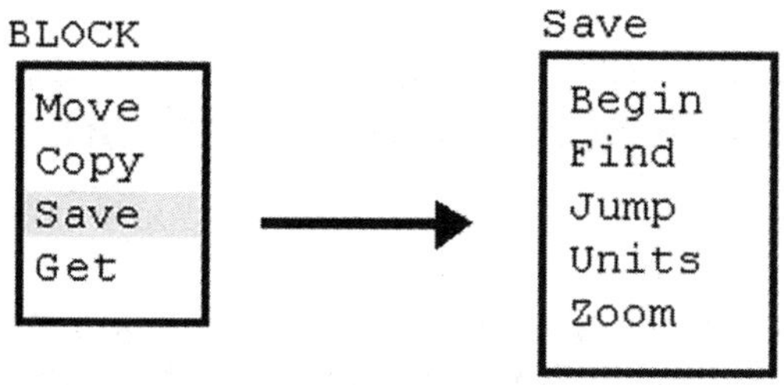

```
BLOCK            Copy
Move            Begin
Copy     →      Find
Save            Jump
Get             Units
                Zoom
```

Figure 8–3c. BLOCK Save Submenu

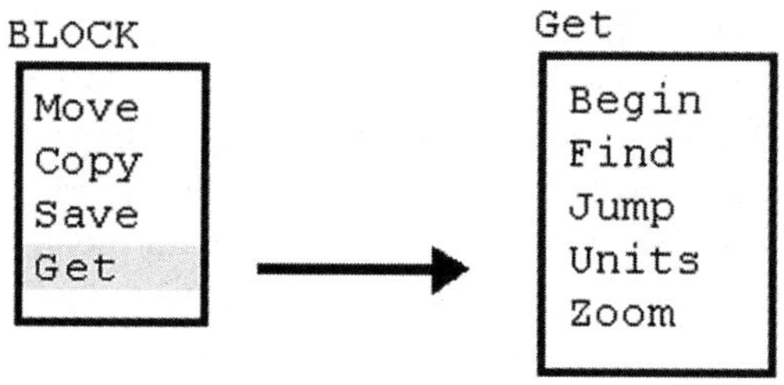

```
BLOCK            Save
Move            Begin
Copy     →      Find
Save            Jump
Get             Units
                Zoom
```

Figure 8–2d. BLOCK Get Submenu

```
BLOCK            Get
Move            Begin
Copy     →      Find
Save            Jump
Get             Units
                Zoom
```

click again to use the mouse to select **Begin.** Draw the box around the desired area by using the mouse.

| **Begin** | **Find** | **Jump** | **Units** | **Zoom** | **Escape** |

3. Left-click the mouse and select **End** from the bar or pull-down menu. Select from the options ([1] All; [2] Edge & All) the items you want to move. Next, move your mouse to the location where this part of the board is to be placed.
4. Click the mouse, and a bar menu as shown below will appear. Select **Place** from the options, and the software will place the block in the new position.

| **Place** | **Find** | **Jump** | **Units** | **Zoom** | **Escape** |

The **BLOCK Copy** command copies a selected group of objects from a specified area of a PCB layout and places it in another part of the board. Unlike **Move, Copy** will not cut part of the board; it will create a copy only. To perform the operation, follow the steps below:

1. Move the mouse pointer to one of the corners of the area you are planning to copy and then move to another part of the board.
2. Click on **BLOCK** and then on **Copy,** and the bar menu shown below will appear. Type **B,** for **Begin,** or click again to use the mouse to select **Begin.** Draw the box around the desired area by moving the mouse diagonally to the opposite corner of the area.

| **Begin** | **Find** | **Jump** | **Units** | **Zoom** | **Escape** |

3. Left-click the mouse and select **End** from the bar or pull-down menu. Select from the following options the items you want to copy:

   Module
   Routes
   Text
   Zone
   All

4. Now the mouse pointer will have everything that was selected. Move the mouse pointer to the location where this part of the board is to be placed.
5. Click the mouse, and the bar menu shown below will appear. Select **Place** from the options shown below, and the software will place the copy of the block in the new position.

| **Place** | **Find** | **Jump** | **Units** | **Zoom** | **Escape** |

The **BLOCK Save** command stores a group of objects from a specified area of a PCB layout in memory and lets you place them in another part of the same board or on another board. **Save** is coupled with **Get.** This means that to retrieve the saved objects, you need to use the **Get** command. Like **Copy, Save** will not cut the selected part of the board, but it will create a copy. To perform the operation, follow the steps below:

1. Move the mouse pointer to one of the corners of the area you are planning to save and then move to another part of the board.

2. Click on **BLOCK,** and then on **Save,** and the bar menu shown below will appear. Type **B,** for **Begin,** or click again to use the mouse to select **Begin.** See Figure 8–2c. Draw the box around the desired area by moving the mouse diagonally to the opposite corner of the area.

Begin	Find	Jump	Units	Zoom	Escape

3. Left click the mouse and select **End** for the bar or pull-down menu. After saving the objects, you may return to the main menu. To retrieve the saved information, you need to use the **BLOCK Get** command.

BLOCK Get is a command that will retrieve the objects saved by the **BLOCK Save** command. The saved block can be placed in as many locations as you wish. The objects remain in the memory until they are replaced by another block of objects. Use the following steps to get the objects:

1. Click on **BLOCK Get.** The following menu will appear. Select from the options the items you want to retrieve. See Figure 8–2d. When you use the **Save** Command, everything in that specified area is saved, but by using the **Get** command you may selectively retrieve the objects.

    Module
    Routes
    Text
    Zone
    All

2. When you have made the selection from the menu, the mouse pointer will have the object. Move the mouse pointer to the location where the part of the board is to be placed.

3. Click the mouse, and the bar menu shown below will appear. Select **Place** from the options shown below, and the software will place the selected objects of the block in the new position.

Place	Find	Jump	Units	Zoom	Escape

8-4.3 Conditions

CONDITIONS shows the current memory allocation and grid size. It reports the allocation but does not let you change it. To make changes in the allocation, you need to get out of the worksheet and go to the PC Board Layout configuration screen. During autorouting, reallocation of buffer size may be necessary. The following is a typical memory allocation screen:

```
Free Edge Buffer     2048
Free Module Buffer   102400
Free Text Buffer     2048
Free Track Buffer    102400
Free System Memory        90528
Grid                 12.70 mm/0.050"
```

8-4.4 Delete

DELETE will erase an object or a block of objects from the PC board layout. The DELETE deletes only objects or a block object from the current layer.

DELETE Object will erase the selected object from the current layer of a PC board layout. To delete an object, perform the following steps:

1. Position the arrowhead of the mouse pointer over the object you want to delete, select **DELETE** from the PCB_MAIN menu, and then select **Object** from the **DELETE** submenu shown in Figure 8–3a. The following bar menu will appear:

Delete	Find	Jump	Units	Zoom	Escape

2. If you wish, you may reposition the mouse arrowhead on the object again. Select **Delete** from the bar menu. If the layer is right and the mouse arrowhead points to the right place, the object that you selected will be erased from the layout.

3. If the displayed message says *nothing to delete on this layer,* you are not in the correct layer for the selected object, the placement of the mouse pointer is not right, or you selected a module. To switch between layers, select **LAYER** from the PCB_MAIN menu and then select 1 or 2 for a double-layer board. A module cannot be deleted by **DELETE Object**; it requires **DELETE Block.**

4. If there is more than one object at the location indicated by the mouse pointer, the following menu will appear, asking you to select among the objects at that location. Select the object you desire to delete, and the object will be erased.

   Segment
   Track
   Net
   Zone

DELETE Block is a command that will delete the objects within a certain area of the PC board layout. To delete a block of objects, follow the steps below:

1. Position the mouse pointer at one corner of the area containing the objects you wish to delete. Click on **DELETE** for the PCB_MAIN menu; then click on **Block** for the **DELETE** submenu shown in Figure 8–3b. The following bar menu will appear:

   | **Begin** | **Find** | **Jump** | **Units** | **Zoom** | **Escape** |

2. Use **Begin** and **End** to select the block or area from which you wish to delete the objects.

3. Select the desired objects to be deleted from the submenu shown in Figure 8–3b. The following list will help you select the objects:

   - **Modules** will erase all modules and tracks connected to those modules on the current layer within the block.
   - **Only Modules** will erase all modules within the block. Tracks are not erased.
   - **Routes** will erase all tracks that have terminals within the block from the current layer.
   - **Text** will erase all text within the block from the current layer. The text that is written with copper can be erased with this command. The text that is written as silkscreen text cannot be erased with this command.
   - **Zone** will erase all zones within the block from the layout.
   - **All** will erase all the objects within the block from both layers of the layout.

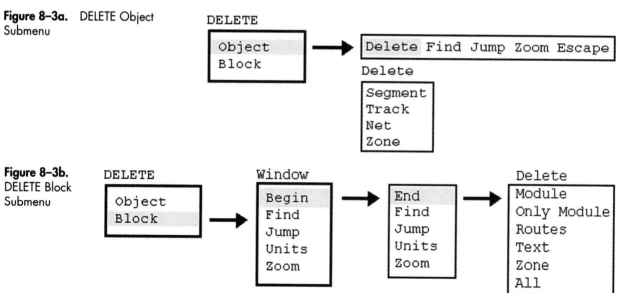

Figure 8–3a. DELETE Object Submenu

Figure 8–3b. DELETE Block Submenu

8–4.5 Edit

EDIT can be used to change the attributes of certain items on the layout. To use the **EDIT** command, follow this procedure: Select **EDIT** from PCB_MAIN menu. The bar menu shown below will appear.

| Edit | Find | Jump | Layer | Units | Zoom | Escape |

Edit will let you go further with the edit process. **Find** will find a reference designator or a part value on the board. **Jump** will move the mouse pointer to a selected tag location. The submenu for **EDIT Jump** is shown in Figure 8–4a. **Layer** will let you move from the current layer to the other. The submenu for **EDIT Layer** is shown in Figure 8–4b. OrCAD/PCB supports up to 16-layer boards. You can route only two layers at one time. **Unit** will let you change the units of measure. The submenu for **EDIT Units** is shown in Figure 8–4c. Sometimes you need to zoom in or out of the worksheet while editing. The **Zoom** submenu can be accessed from **EDIT** also and is shown in Figure 8–4d.

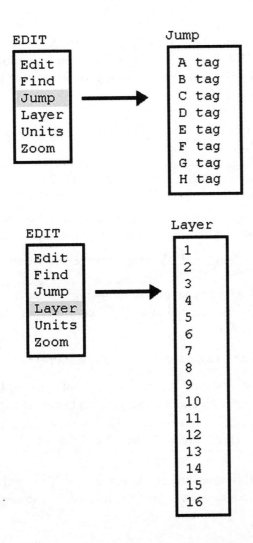

Figure 8–4a. EDIT Jump Submenu

Figure 8–4b. EDIT Layer Submenu

Figure 8–4c. EDIT Units Submenu

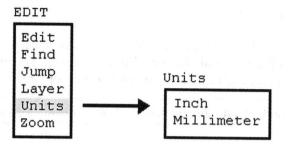

Figure 8–4d. EDIT Zoom Submenu

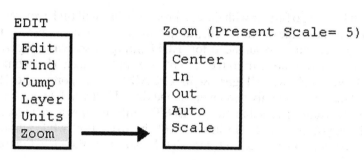

Position the mouse pointer on the object you desire to edit and click on **Edit** from the **EDIT** submenu. If the pointer is at a location where there is more than one object, the software will display the following submenu:

Pad
Text
Route
Name

- **Pad** will let you change the following attributes of the pad: Netname, Type, Vertical, Horizontal, Orientation, Sides, Drill Size.
- **Text** will let you change the following attributes of the silkscreen text: Place, Rotate, Vertical, Horizontal, Origin, Unit, Zoom.
- **Route** will let you change the following attributes of the route: Segment, Track, Net, Width.
- **Name** will let you change the following attributes of the reference designator of the selected module: Place, Rotate, Vertical, Horizontal, Origin, Unit, Zoom.

For example, to edit a route width, follow the steps below:

1. Place the mouse pointer on a segment of a route. Click on **EDIT** from PCB_MAIN. The bar menu shown below will appear:

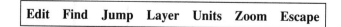

2. Click on **Edit** from the above menu, and if more than one item is under the mouse pointer, then the *Select Item* submenu shown below

will appear. If, however, only a track is under the mouse pointer, then the *Route Width* submenu will appear.

> *Select Item*
> **Pad**
> **Text**
> **Route**
> **Name**

3. Click on **Route**, and the following *Route Width* submenu will appear:

> *Route Width*
> **Segment**
> **Track**
> **Net**
> **Width**

4. To change the width of a track or a segment, select **Width.** When the software displays another submenu, select **Track**. (from **Track/Via**). The current width of the track will appear at the bottom of the layout screen, and the field will be highlighted. This highlighting means you may change the value. Move your mouse or the arrow key to adjust it to a desired value. The mouse may be too sensitive to adjust the number to a new value, and in that case you should use the arrow key.

5. Right-click the mouse (or escape) when the desired value is set. The software will return to **Route Width.** Select **Track** if you want to give the new track dimension to the entire track. Select **Segment** if you want to give the dimension to only a segment of a track. Track is the entire route from one net to another, whereas a segment is only the straight portion of the track. Move the mouse pointer to another track or segment for which you want to give the same dimension, and select **Track** or **Segment** from the **Route Width** menu to repeat the procedure.

8-4.6 Find

FIND locates a particular item anywhere in the PCB layout file. You can search for an item by its reference designator or part value. The search is not case-sensitive. To find an item, click on **FIND** and provide the search string at the question mark.

8-4.7 Jump

JUMP moves the mouse pointer to a specified location defined earlier by a Tag. There are eight Tag locations that can be selected on the PCB layout. Jumping from one part of the board to another can save time during editing.

8-4.8 Layer

LAYER is used to select the working layer. The display screen, **X-Y-L**, displays the number of the current layer. The total number of layers the board will have is selected by the **SET** command. You may go from one layer to another simply by clicking on **LAYER** and selecting the number from the menu. There are other ways to go from one layer to another, such as by left-clicking the mouse on **OTHER**, located in the PCB_MAIN menu.

8-4.9 Place

PLACE will let you define the edges of a PC board, place a module, create special zones, and place text on the layout. A left click on the mouse will bring up the **PLACE** submenu shown below.

Module
Text
Zone
Edge

PLACE Module is used to place the module on the layout. Select **PLACE** from the PCB_MAIN menu, then select **Module** from the **PLACE** submenu, and the **PLACE Module** submenu shown in Figure 8–5a will appear.

    PLACE Module Move is used to reposition a module on the PCB layout. To reposition a module, select it by positioning the mouse pointer on the module and click on **PLACE→ Module→ Move.**

    PLACE Module Load is used to load a module from the library onto the PCB layout. If you do not know the name of the module, you may browse through the module in the library, select it from the list, and finally place it onto the board.

    PLACE Module Browse is used to browse through the module in the library.

    PLACE Module Get will bring a module from another position on the board. This command can also be used to get a module by using the reference designator.

Figure 8–5a. PLACE Module Submenu

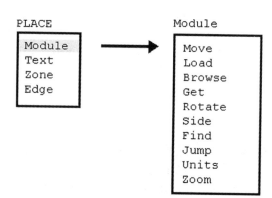

PLACE Module Rotate will rotate the module by 90°.

PLACE Module Side is used to reposition a module on the opposite side of a PCB layout.

PLACE Text places text on the PCB layout. The text can be rotated, mirrored, sized, etc. Also, the text for silkscreen and for any of the copper layer can be placed on the layout. The **PLACE Module Text** submenu is shown in Figure 8–5b.

PLACE Zone is used to place special areas on the layout. Three types of zones are placed on the board: **Copper, Forbidden,** and **No Via.** Defined **Copper zones** are filled with copper, and **Forbidden zones** contain no tracks or segments. **No Via** zones contain no vias, but tracks are permitted to pass through. Figure 8–5c shows the **PLACE Zone Width** submenu.

PLACE Zone Type is used to specify one of the three types of zones. Figure 8–5d shows the submenu.

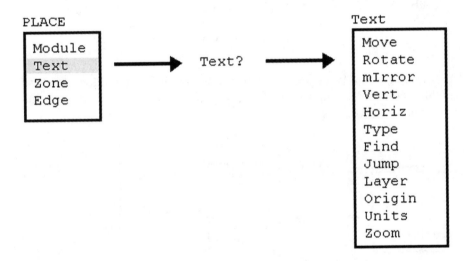

Figure 8–5b. PLACE Module Text Submenu

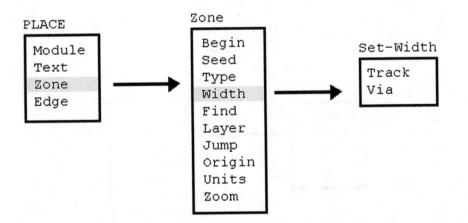

Figure 8–5c. PLACE Zone Width Submenu

290 Chapter 8 / Printed Circuit Board Layout Tool Set

PLACE Zone Seed is used to fill the specified zone area with a Copper, Forbidden, or No Via symbol.

PLACE Edge is used to place the edge of the board. The board area should be generally specified before downloading the module from the PCB library. Figure 8–5e shows the submenu for the **PLACE Edge** command.

8–4.10 Quit

The **QUIT** command from PCB_MAIN menu will let you load, update, or write to files. The command also allows the user to create new or modify existing library modules, create various plot files, produce reports, suspend to DOS, and exit from the layout worksheet without saving the changes. Figures 8–6a through 8–6e show some of the submenus for the options of **QUIT**.

Abandon Program is used to exit the layout worksheet without saving changes.

QUIT Initialize Use Netlist is used to load modules and associated net information onto the board from a netlist file. To do that, click on **QUIT→ Initialize→ Use Netlist,** and the following bar menu will appear:

| Edit | Find | Jump | Layer | Units | Zoom | Escape |

Use **Begin** and **End** to specify an area within the board edge to load the modules from the netlist file. At this point the software will prompt for the net filename as follows:

Figure 8–5d. PLACE Zone Type Submenu

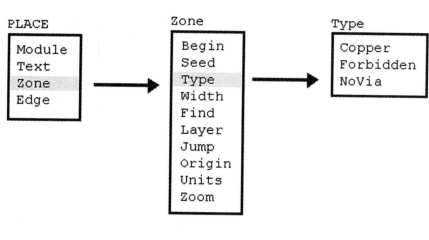

Figure 8–5e. PLACE Edge Submenu

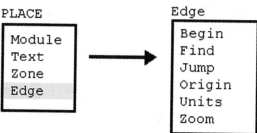

8-4. Common Functions of the PCB Tool Set 291

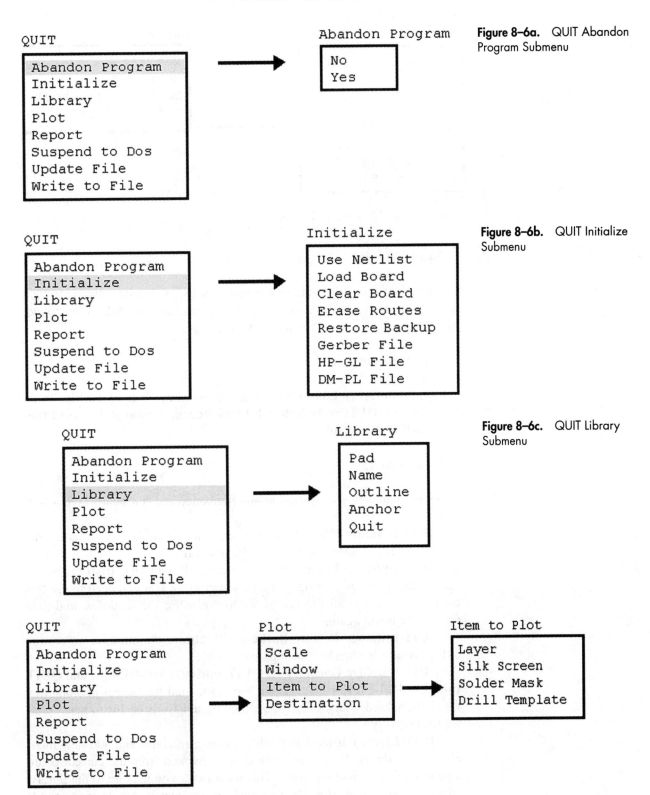

Figure 8-6a. QUIT Abandon Program Submenu

Figure 8-6b. QUIT Initialize Submenu

Figure 8-6c. QUIT Library Submenu

Figure 8-6d. QUIT Item to Plot Submenu

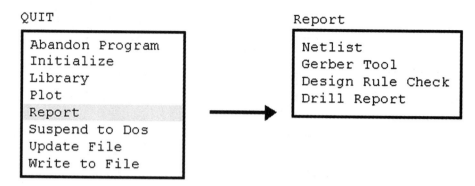

Figure 8-6e. QUIT Report Submenu

Read Net File?

Enter the net filename, and the software will use netlist information to load modules from the PCB module library. If a matching module is not found, the software will display the following message:

XX Module not found

QUIT Initialize Load Board loads the existing PCB layout file. To do this, click on **QUIT→ Initialize→ Load Board.** A prompt for board filename will appear as follows:

Load Board?

Enter the name of the desired board filename, and the board layout will appear.

 QUIT Initialize Clear Board will erase the entire layout, including the modules. **Clear Board** will not save the board file.

 QUIT Initialize Erase Routes will clear all the routes from the board, but not the modules. During routing and rerouting you will need to use this command to erase all the routes without erasing the modules, and then reroute the board again.

 QUIT Initialize Restore Backup will retrieve the most recent version of the layout file from the backup disk.

 QUIT Initialize Garber File, QUIT Initialize HP-GL File, and **QUIT Initialize DM-PL File** display plot files formatted for Garber photoplotter, Hewlett-Packard Graphic Language plotter, and Houston Instruments plotter, respectively.

 QUIT Library is used to modify existing modules and add new modules to the library. When this command is invoked from the current board layout, the layout will not erase. This means that you may go to the library, create a module part, store the part, and restore your layout design again. If you create or modify a module part, put that in a custom library by using the command **QUIT→Library→Update** or **QUIT→Library→Write** and

8-4. Common Functions of the PCB Tool Set

providing the library name. If you put the new module in an OrCAD-provided module library, the part will be destroyed when the regular library update is loaded from OrCAD.

QUIT→Library→Pad will let you modify the attributes of a module pad.

QUIT→Library→Name will allow you to assign a reference designator and part value to a module.

QUIT→Library→Outline will let you draw or edit the outline of a module.

QUIT→Library→Anchor will define the point around which a module is rotated by the **Rotate** command.

QUIT→Library→Quit will allow you to browse and load existing library modules. It will also let you update the present version of a module and write the newly created module in the desired area of the library.

QUIT Plot will allow you to generate different kinds of files for plotting or saving the soft copy of the file in a specific format.

QUIT Plot Scale is used to scale the item 2, 4, 8, and 16 times larger than its actual size.

QUIT Plot Window will let you select the portion of the layout to be plotted. It will also let you rotate and size the plot.

QUIT Plot Item to Plot is used to select the item to be plotted. Layers, silkscreen, soldering mask, and drill template are the items associated with a board that can be plotted.

QUIT Plot Destination is used to specify the destination for the plotter output. The destination could be a serial, parallel, or disk. The following command is used to create files for different items to be plotted. They cannot be plotted from this menu.

$$\boxed{\text{QUIT} \rightarrow \text{Plot} \rightarrow \text{Destination} \rightarrow \text{Hard Copy}}$$

The hard copy of these files is plotted by using the **Print PCB** tool located in (PCB_MENU) under Reporters.

QUIT Report will generate a variety of reports about a PCB layout.

QUIT → Report → Netlist will generate a netlist file from your layout file. This feature is quite useful because your layout file may have some additional module and net information that was not in your original schematic.

QUIT Suspend to DOS will let you temporarily go to DOS mode. You can run DOS commands normally and resume your work without losing any information from your layout. When you are in DOS mode, a double arrow prompt will remind you that you have OrCAD running in the background. To get back to your layout, just type as follows:C\>>Exit <Enter>.

QUIT Update File will save the modified layout that is currently on the screen.

QUIT Write to File will prompt for a new filename of the layout and, if provided, will save the layout under that new name.

8-4.11 Routing

ROUTING has many commands. Using this set of tools, you can do manual and autorouting of a PC board layout. Figure 8–7 shows the **Routing Begin** submenu. **ROUTING Begin** is used to draw tracks on the current active layer of the layout. **Begin** will start a track or continue a track in a new direction. You must begin a track from a pad. Two pads cannot be connected with a track either manually or using the autorouter unless they have the same netname. Note this condition carefully because you may face it and not know why it is happening.

If you fetch a module or a single pad from the PCB library, the module and the pad do not have a netname. Modules that you downloaded from the PCB library by using the netlist file have netnames. So connecting the pads that have a netname to those pads that do not have the same netname will not be possible. To connect a single pad or the pads of the module you fetched to the other existing module with routes, you must provide the appropriate netname to each new pad. If there is a situation in which you have to connect two pads with different netnames, you need to set the **Design Rule Check Routing (DRC)** located under the SET submenu to **No**. Set the DRC back to its **YES** condition after you make the connection.

ROUTING Other is used to place a **Via** at the cursor location and switch to the other working layer. The type of via placed depends on the via type selected under **SET**. Before executing this command, make sure that the correct via type is set. There are three types of via: Through, Buried, and Blind.

ROUTING Inquire is used to find the information about a pad, track, etc. For example, during routing it often becomes necessary to find the netname of a pad or width of a route. Using this command makes this process easy. Figure 8–8 shows the submenu.

ROUTING Show will display the unit of measurement. Figure 8–9 contains this submenu.

ROUTING Highlight is used to highlight the track of a net. This means that if you place the mouse pointer at a net point and select **ROUTING → Highlight,** it will highlight all the tracks in the active layer under one netname. This will help you identify the entire net without any extra effort. Figure 8–10 shows the submenu.

ROUTING Width is used to set width of Tracks and Vias on your current layout. Figure 8–11 shows the submenu.

ROUTING Layer Pairs is used to define the two layers of your board that will be routed. OrCAD/PCB software allows you to route up to two layers of the board at a time. Your board may have several layers (up to sixteen), but you can route only two layers at once.

ROUTING Netlist will let you configure the routing parameters for the autorouter. Figures 8–12a through 8–12d are the submenus for the **ROUTING Netlist** options.

ROUTING Netlist Net Pattern will let you define the net patterns for the autorouting tool. There are two net patterns: Tree and Chain.

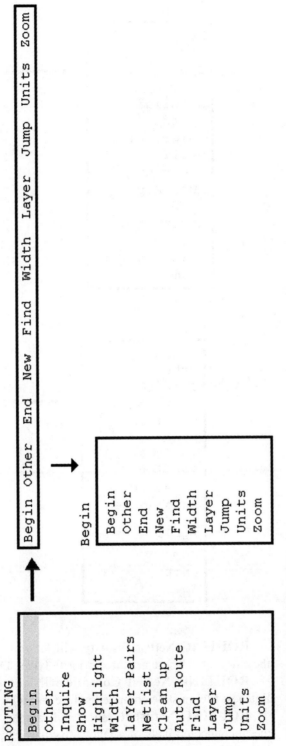

Figure 8-7. ROUTING Begin Submenu

Figure 8-8. ROUTING Inquire Submenu

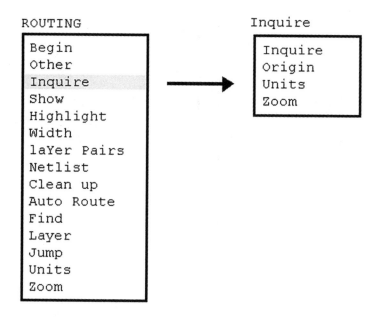

Figure 8-9. ROUTING Show Submenu

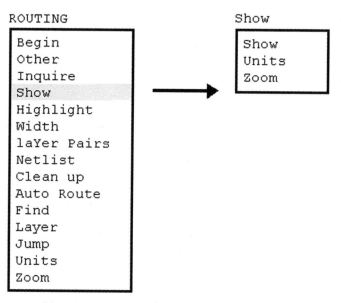

ROUTING Netlist Priority will let you select the routing priority for the autorouter. There are three priority types: Default, Short, and Long

ROUTING Netlist Compile will compile the Ratsnest file. The Ratsnest file must be compiled before it can be used.

ROUTING Netlist Ratsnest will display the Ratsnest vectors for all the unconnected pads of a layout. By using this command you can see the Ratsnests for all unconnected pads or one pad, module, or net.

ROUTING Netlist Vector will display the set of vectors that will help you determine the optimum position for a module. By using the command you may display the vector for one module or for all modules.

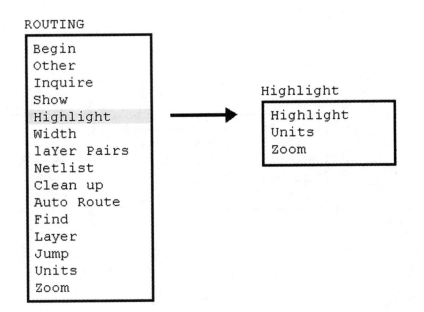

Figure 8-10. ROUTING Highlight Submenu

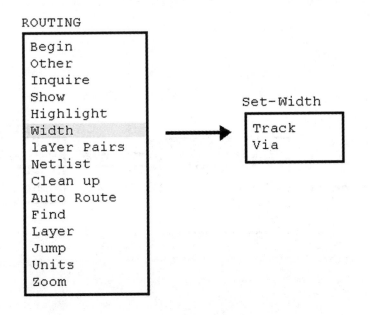

Figure 8-11. ROUTING Width Submenu

ROUTING Clean Up will erase all tracks that do not end on a pad or another track. Figure 8-13 shows the submenu.

ROUTING Auto Route will automatically place routes for unconnected nets. The routes that already exist will not be affected by the command. The submenu for **Auto Route** is shown in Figure 8-14. By using this command, it is also possible to route only one pad, module, net, or block of a layout. Along the bottom of the **Auto Route** screen much information is displayed: strategy, number of nets, number of unconnected routes, etc. You can alter some of the attributes while routing by accessing them normally from the submenus.

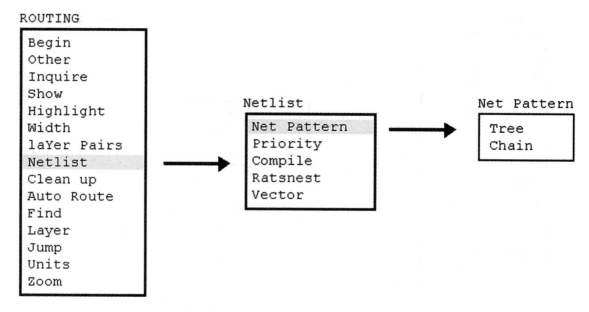

Figure 8–12a. ROUTING Netlist Net Pattern Submenu

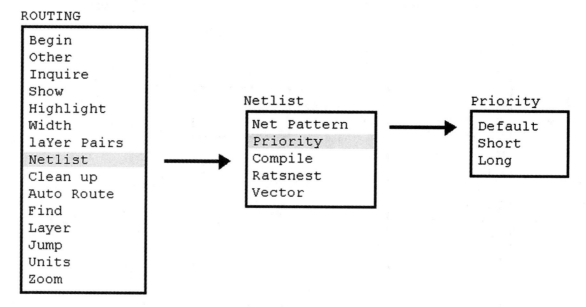

Figure 8–12b. ROUTING Netlist Priority Submenu

ROUTING Auto Route Pad will route only the selected pad of a layout.

ROUTING Auto Route Module will route only the selected module of a layout.

ROUTING Auto Route Block will route only the selected block of a layout.

ROUTING Auto Route One will place one route of a net as having the highest priority.

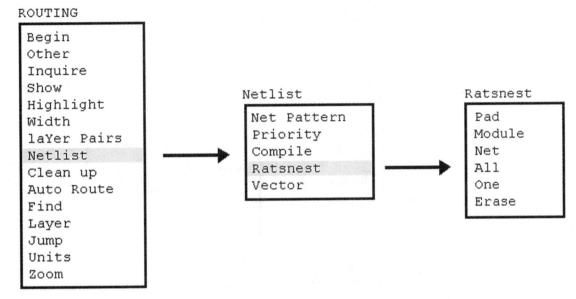

Figure 8–12c. ROUTING Netlist Ratsnet Submenu

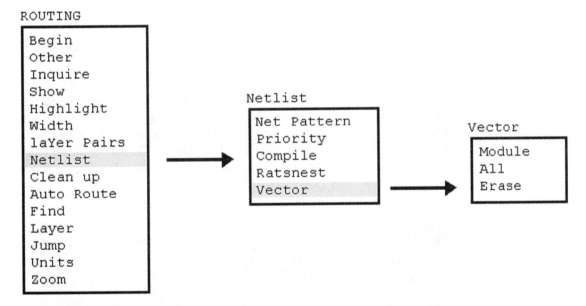

Figure 8–12d. ROUTING Netlist Vector Submenu

ROUTING Auto Route All will route all the unconnected nets of a layout. The command will place routes on the board according to the defined Strategy and Design Conditions. For a large board, this may take quite some time. For some boards, it may be necessary to define different characteristics for a net or group of nets. These characteristics are track width, Net Pattern, Isolation, Strategy, etc. They can be done in two different ways. You may set them before routing or define them under Net Conditions. Characteristics of a net defined under Net Conditions will always take

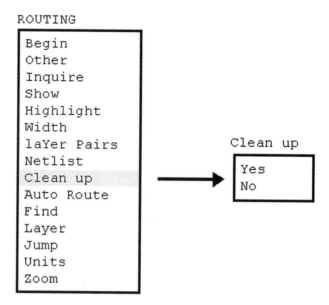

Figure 8-13. ROUTING Clean Up Submenu

precedence over characteristics defined under either Design Conditions or the Auto Route menu.

ROUTING Auto Route Strategy will let you select the routing strategy for the board. Routing strategy can also be defined under Design Conditions and Net Conditions. Design Conditions and Net Conditions are located under PCB Configuration. A strategy defined under Net Conditions always takes precedence over a strategy from Design Conditions or set from the Auto Route menu. A strategy defined from the Auto Route menu will take precedence over Design Conditions. There are six different types of routing strategy that can be defined:

Normal
Flexible
Extensive
Power
90 Degree
No Via

Each of these strategies routes a board differently. For example, the **No Via** strategy will not put via on a double-sided board. However, there is no guarantee that the entire board can be routed by using only one of these strategies. One of these strategies may work better than the others in a particular situation. This means that you may route part of the board with one strategy and the other parts with other strategies.

ROUTING Auto Route Unroute will delete the selected route from the layout.

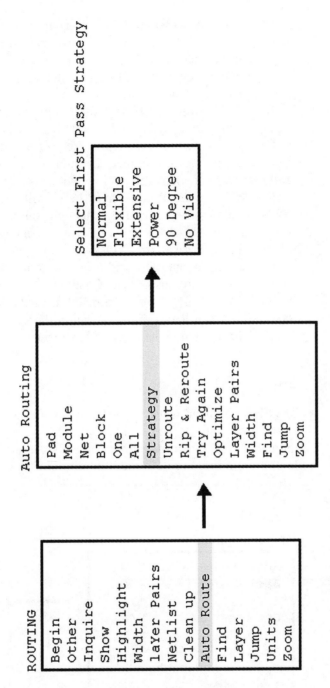

Figure 8–14 ROUTING Auto Route Submenu

ROUTING Auto Route Rip & Reroute will rip up all tracks and their associated nets that fall within a two-grid-unit radius of the source and target pads of the unconnected routes, and attempt to place all the routes starting with unconnected routes by using the strategy defined as the second pass under Design Conditions.

ROUTING Auto Route Try Again will attempt to place routes for the unconnectable nets by using any new strategies you have selected.

ROUTING Auto Route Optimize will reroute a track or a group of tracks and attempt to minimize the number of vias and the track length. **Optimize** will not attempt to put in previously unconnected routes.

ROUTING Auto Route Layer Pairs will specify the two layers that will be routed by the Autorouter. Although your board may have more than two layers, OrCAD allows only two layers to route at one time.

ROUTING Auto Route Width will specify the width of tracks and vias for the current layout. The width of tracks and via can also be selected under Design Conditions and Net Conditions.

ROUTING Auto Route Find, ROUTING Auto Route Jump, and **ROUTING Auto Route Zoom** function the same way as **FIND, JUMP,** and **ZOOM,** located in the PCB_MAIN menu.

8-4.12 Set

The **SET** command is used to define a number of parameters for the board layout environment. The submenus for **SET** are shown in Figures 8–15a through 8–15h. Each of these parameters can be invoked and set for the layout operation. For example, the number of layers for the PCB must be

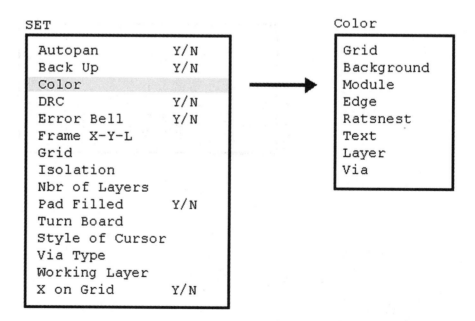

Figure 8–15a. SET Color Submenu

8-4. Common Functions of the PCB Tool Set

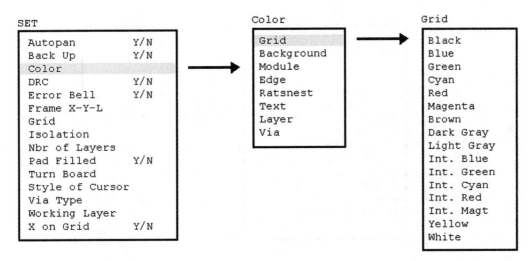

Figure 8-15c. SET Color Module Submenu

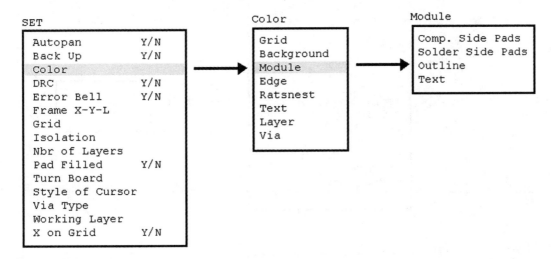

Figure 8-15b. SET Color Grid Submenu

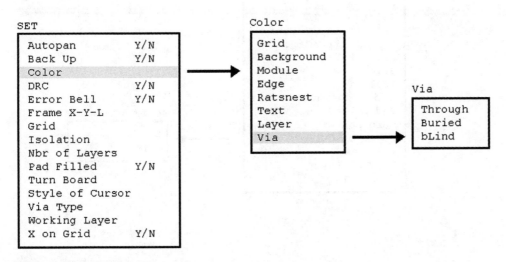

Figure 8-15d. SET Color Via Submenu

Figure 8–15e. SET Grid Submenu

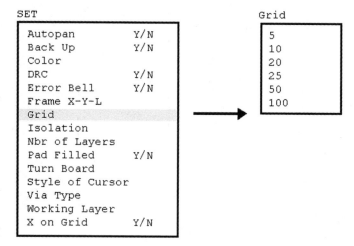

Figure 8–15f. SET Isolation Submenu

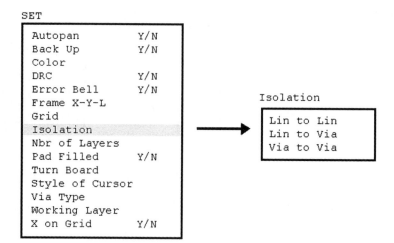

Figure 8–15g. SET Style of Cursor Submenu

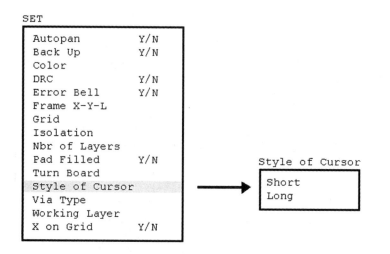

8–4. Common Functions of the PCB Tool Set

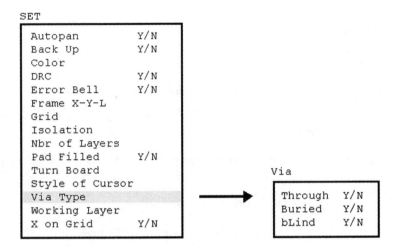

Figure 8–15h. SET Via Type Submenu

set before routing starts. OrCAD supports a maximum of sixteen layers per board. To set the number of layers to two for a design, click on **SET→Nbr of Layers,** then scroll down to number 2, and click again. From this time on, the board will be routed with two layers. This means that the connection from one net to the other may be done partly in Layer 1 and partly in Layer 2.

8–4.13 Tag

The **TAG** command is used to jump from one part of the board layout to another part by identifying positions on the board by tags. You can tag eight positions, **A** through **H,** on a board and jump from one to another by specifying the tag name. Tags will remain set only during the editing sessions. Exiting OrCAD/PCB or loading a different layout file will destroy the tags.

8–4.14 Units

The **UNITS** command is used to set the units of measurement used during the layout design. You may specify the units of measurement in either inches or millimeters. The default setting is in inches.

8–4.15 Zoom

The **ZOOM** command is used to view the PCB layout at varying levels of detail. Click on **ZOOM** to get the submenu, and by selecting **In, Out, Auto,** or **Scale** you may go to a convenient viewing depth of the board.

QUESTIONS

1. What is the fundamental difference between a PCB layout and a schematic?

2. Which type of board is less expensive, single-sided or double-sided?

3. Generally, what type of board (single- or double-sided) is comparatively difficult to route?

4. What tool should you use to modify the pads of a PCB module from round to oval?

5. Why are oval-shaped pads sometimes desirable?

6. Write the command sequence for loading a previously drawn board file.

7. Write the command sequence for generating and printing a PC board artwork layout.

8. You are trying to autoroute a PC board, and the software prompts you *not enough memory*. What could you do to solve this problem?

9. How are Design Conditions and Net Conditions used?

10. If you define a track width for a net differently under Design Conditions and Net Conditions, which one will be used by the autorouter during routing?

11. Write the command sequence for moving a module from one place to another on the board.

12. Write the command sequence by which a module can be moved by calling its reference designator.

13. If you want to route your layout without any via, what routing strategy will you choose?

14. What is the maximum number of board layers that OrCAD supports?

15. Write the command that is used to jump quickly from one part of the board to another.

16. What is the maximum number of layers that can be autorouted at one time?

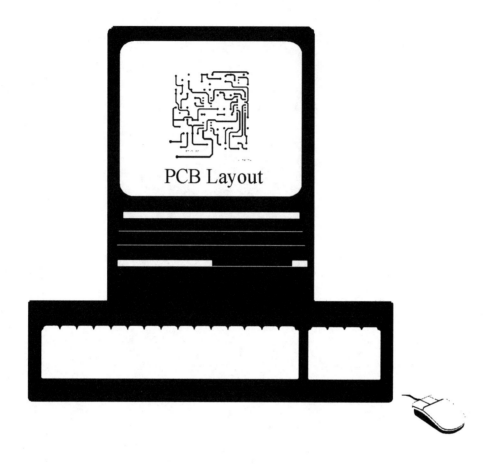

PCB Layout Design

9

COMPUTER-AIDED TECHNIQUES FOR PCB LAYOUT DESIGN

9–1. THE PC BOARD LAYOUT DESIGN PROCESS

The PC Board Layout design process using the OrCAD/PCB tool set has ten steps:

1. Set up design environment
2. Configure PC Board Layout Tools
3. Invoke PC board worksheet
4. Set up the PCB design environment
5. Draw a board edge
6. Use netlist to unload modules from the PCB Library
7. Load module from PCB Library
8. Arrange modules on the board
9. Place text and special zone
10. Route board

9–1.1 Setting Up the Design Environment

The design environment must be set using the Design Management Tools so that the netlist file, the schematic file, etc., can be easily accessed and the board file, when generated, can be saved in an appropriate subdirectory. Setting up the design environment is important, because it will keep the files in order. The following procedure can be used:

1. From the ESP_MENU, click on the **Design Management Tools** icon, and the Design Management tool set menu (DM_MENU) will appear on the screen.

Objectives

After completing this chapter, you should be able to

1. Describe the steps of the master artwork design process
2. Set up the design environment
3. Configure the design condition of the PC Board Layout tool set
4. Create board edge and download modules using a netlist file
5. Download modules from the PCB library
6. Arrange modules, place text, and place zones on board
7. Do manual and automatic routing of a board
8. Set strategy for autorouting and optimize routes
9. Inquire and edit tracks, vias, and nets
10. Route unconnected pads
11. Print hard copy of master artwork, drill template, component legend, and solder mask

Objectives (cont.)

12. Create soft copy of all the above files and soft copy of drill files for a CNC machine

2. Since you are ready for board layout, you must have created the design name directory. The design directory will be in the left list box among the design names. Locate the design name for which you are about to begin the board layout.

3. Left-click the mouse on the design name, and it will be highlighted. Return to the ESP_MENU menu by left-clicking on the OK icon. The design environment is set, and the TEMPLATE design name will be replaced by the chosen design name.

9-1.2 Configuring PC Board Layout Tools

The basic configuration of PC Board Layout Tools such as Driver Options, Board File Options, Serial Options, Color Table, and Pen Carrousel should already be complete. However, configuration options such as Memory Allocation, Design Conditions, and Net Conditions are specific to a design. Therefore, these options can be configured now. Your instructor, supervisor, or person for whom you are designing the PC board may have given you some limitations. These limitations could be allowed Number of Layers, minimum Track Width, minimum Track-to-Track Isolation, Pad Diameter, Via Diameter, or Via-to-Via isolation. These limitations often are driven by the product for which you are designing the PC board. They may also come from the fabrication facility your school or company has. You may enter this information under Design Conditions. These limitations are your basic design rules. In other words, you want to follow them for most of the board. You will be able to change the design conditions as you route the board, but this is the place to lay down general design rules. The general design rules I like to follow for the projects described earlier in the book are provided here. Enter the following information under Design Conditions:

- Minimum Route Width = 0.025"
- Pad Diameter = 0.055"
- Via Diameter = 0.045"
- Pad Drill Size = 0.039"
- Via Drill Size = 0.039"
- Solder Mask Guard = 0.020"
- Isolation Track-to-Track = 0.025"
- Isolation Via-to-Via = 0.020"
- Minimum Route Width Between Pins = 0.019" (To comply with this rule, you may have to manually change the width of the segment passing between the pins.)

Number of Layers, Text Dimension, Routing Grid, Working Layer, Strategy, and Net Pattern are best configured during the routing process. Set short or long cursor according to your preference. The short one is more convenient for me. However, when drawing the board edge, the long cursor is more convenient.

If any special directives are provided by the circuit designer or if you want to provide special design rules for any connection or net, you should enter them under **Net Conditions.** To do this you need to know the netname of the connection. Track Width for power and ground nets can be entered here. Enter the netname for power and ground, and enter the following design rules under **Net Conditions:**

- Track Width = 0.050″
- Isolation Track-to-Track = 0.050″
- Isolation Via-to-Via = 0.025″

9–1.3 Invoking the PC Board Worksheet

PC board master artwork designing takes place on the PC board worksheet. This is located under the **Route Board** tools. To invoke the worksheet, follow the procedure below:

1. From the ESP_MENU, left-click on the **PC Board Layout Tools** icon. The PC Board Layout tool set menu will appear.

2. Click on **Route,** located under **Editors,** and the OrCAD/PCB II configuration screen will appear. Assuming that you have configured the PC Board Layout Tools earlier, type **U** (for Update), followed by **R** (for Run). A welcome screen with the OrCAD logo will appear.

3. Pressing the **Enter** key once will bring the prompt for *Load Board.* Since this is the first time, there is no board file, so press the **ESC** or **Enter** key. The PC board worksheet, along with the main menu, will appear on the screen.

9–1.4 Setting the Board Layout Environment

The PCB design environment must be set. All the environment parameters have default settings; however, some of the default settings may not be suitable for the specific artwork you are planning to create. The default value for number of layers is two. This value may not be suitable for your design. Once you start your artwork design, some of the design environment parameters are time-consuming to change and sometimes impossible to change. Therefore, always set the design environment before starting on the design.

From the PCB_MAIN menu, located at the upper left corner of the worksheet, click on the option **SET,** and a submenu will appear. Set the design environment to the following values:

- Autopan ➜ Y (Yes)
- Back up ➜ Y (Yes)
- Color ➜ keep the default values

- DRC (Design Rules Check) → Y (Yes)
- Error Bell → Y (Yes)
- Frame X-Y-L → Do not invoke
- Grid → Set at 0.050 inch
- Isolation → Keep all the default values
- Pad Filled → N (NO)
- Turn Board → N (NO)
- Style of Cursor → Short
- Via Type → Through
- Working Layer → 1
- X on Grid → Y (YES)

Grid. Grid points on the worksheet are used to ensure easy and correct spacing of components. Various values for the grid can be set. However, 0.050-inch grid spacing is the optimum for most artwork design. If you set X on Grid option to **Yes,** the cursor will stay on grid only. You cannot place the cursor between two grid points when the grid option is set. Although this appears inflexible, it allows convenient placing of modules on the worksheet. Keep this option at **Yes.**

Layers of PC board. PC boards can be single, double, or multilayer. However, multilayer is uncommon in a school laboratory or in a cost-sensitive environment. Multilayer boards are generally used for high-density, high-quality, and compact product applications. For a single-sided PC board, components are mounted on one side and routes on another side. For a double-sided board, components are mounted on one side and routes on both sides. Thus by convention, the side where the components are mounted is called the component side of the board and the other side is called the route side of the board. OrCAD/PCB numbers the layers. Layer 1 is the route side, and Layer 2 is the component side. On the computer screen, OrCAD/PCB always looks at the board artwork from the component side of the board. This means while you are looking at Layer 1 on the computer screen you are looking through Layer 2 of the board. This is an important fact to remember.

Frame X-Y-L. For schematic design tools, the function of a part is important. For PCB tools, the dimension of a part is important. Therefore, the PCB worksheet is provided with a grid reference scale, located at the upper right corner of the screen. X and Y values are the current horizontal (X) and vertical (Y) position of the cursor in inches. The 0-0 point of the grid is the upper left corner of the worksheet frame. The number beside **L** represents the current working layer. If you select the **Layer** option from the PCB_MAIN menu and select Layer 1, the color of the grid reference box will change to copper. This means that the current working layer is the route side of the board. If you change the layer to 2, it will be green, and this means that the current working layer is the component side of the board.

Isolation. Isolation is the defined value of track to track, track to via, and via to via spacing. These specified values are obeyed by the router tool. Product requirements and the PC board fabrication facility are the determining factors for these physical spacing limitations. The values of **Line to Line, Line to Via,** and **Via to Via** spacing set in section 9–1.2 are often adequate for a laboratory-based fabrication facility. Therefore, for the time being, keep these values for the spacing.

9–1.5 Drawing a Board Edge

The board edge defines the physical size of the PC board. PCB tools use the board edge as the limit for the routes. Since the physical size of a PC board depends on the number and sizes of the components, it is always necessary to estimate the size of a PC board. As a general rule of thumb, components should cover only 25% of the total board area. Different PC board manufacturers may have different density standards. However, this guideline is good enough for determining the size of a PC board for a school laboratory project. For a manufactured product, there are other critical factors that need to be considered. After routing or at any time during routing, you may choose to change the size of a board by deleting and redrawing its edge. Let us create the edge of a board whose size is $3'' \times 6''$. Use the following steps to create its edge:

1. From the PCB_MENU, click on **PLACE,** and a submenu for **PLACE** will appear.

2. Click on **Edge,** and a submenu for **PLACE Edge** will appear. Select **Begin** from the step menu located at the upper left corner. This action will change the step menu into a bar menu. Now you can move your mouse pointer.

3. Place the mouse pointer at the $X = 1''$, $Y = 1''$ position. To determine the coordinate, you need to look at the coordinate display box (grid reference box) located at the upper right corner of the screen. Now left-click the mouse and select **Begin.** Drag the mouse straight down until the Y value is $4''$ and the X value is still $1''$. A vertical line will appear whose exact length is $3''$. This is one of the $3''$ sides of your board.

4. Left-click the mouse and select **Begin** again. Drag the mouse to the right until the X value is $7''$ and the Y value is $4''$. This is one of the $6''$ sides of the board.

5. Left-click the mouse once and select **Begin.** Drag the mouse straight up until the Y value becomes $1''$ and the X value is $7''$.

6. Left-click the mouse once and select **Begin.** Drag the mouse until both the X and Y values equal $1''$. Left-click the mouse once, and this time select **End.** The above procedure has created a board edge of $3'' \times 6''$ in size.

Delete a Board Edge. If you make a mistake, or wish to resize the board, the board edge can be deleted by using the following procedure:

1. Keep on right-clicking the mouse until the PCB_MAIN menu appears.
2. Select **Delete** and select **Object** from the **Delete** submenu.
3. Place the mouse pointer on the edge you want to delete, and left-click the mouse button.

9–1.6 Using a Netlist to Unload a Module from the PCB Library

A netlist file contains all the logical connections among parts and the type (physical dimension) of module suitable for the schematic parts. Information contained in the netlist file is used by the OrCAD/PCB tools to unload appropriate modules from the PCB module library and to convert the logical connections among parts into physical connections called tracts and nets. Also, netlist information is used to generate ratsnest and force vectors to optimize the placement of modules on a board. A netlist file for a schematic is generated by OrCAD/SDT tools. The following procedure will download modules from the PCB module library and place them on the specified area on the board:

1. From PCB_MAIN menu, select Quit, and a submenu will appear.
2. Left-click on **Initialize,** and its submenu will appear.
3. Left-click on **Use Netlist,** and a menu bar will appear at the top of the screen. Position the mouse pointer close to the upper left corner of the edge of your board (within the board edge) and left-click the mouse again. The bar menu will convert to a step menu.
4. Select **Begin,** and drag the mouse down and toward the right to create the area where the modules will be downloaded.
5. Left-click the mouse and select **End** from the menu. A *Read Net File?* prompt will appear at the top. Provide the name of the netlist file (with NET extension) for which you are creating the board file. The board file will generally have the same file name, with a BRD extension.
6. OrCAD/PCB tools will download modules from the PCB module library appropriate to each device of the schematic. In some cases, when a number of devices are packaged in one module, PCB will download the appropriate module for them also.
7. If for some reason an appropriate module is not found in the PCB library, the tool will ring an error bell with a prompt *File Not Found.* Note the part reference number and part value for each library file not found in the PCB library. Press **Enter** if you still want to download the rest of the modules on the worksheet.

Generally, there are three reasons for the *File Not Found* message to appear:

1. The package or module number you specified in Part Field 1 is not available in the PCB library.

2. The package or module number you specified in Part Field 1 has a typographical error. A common error is the confusion between the number zero and the letter **O.**

3. The netlist file does not have the module or package number for the part. The reason for an inappropriate netlist file is the wrong Key Field Configuration.

Find the source of error, eliminate it, and try to download the modules again. In the case of an inappropriate netlist file, you have to go back to the SDT tool and generate the netlist file again.

9–1.7 Loading Modules from the PCB Library

Often it is necessary to fetch modules directly from the PCB library. They are not part of the schematic drawing, but they are needed for power, ground, and off-board parts connection. The following procedure will fetch any module from the PCB library to the worksheet:

1. From the PCB_MAIN menu, left-click on **PLACE** and the submenu for **PLACE** will appear.

2. Left-click on **Module.**

3. Left-click on **Load** from the **Module** submenu. The software will prompt for the library filename for the module as *Load Library File*. Provide the library file name and left-click the mouse; the module will appear on the worksheet.

If, however, the library filename is not known exactly, left-click the mouse once, and the software will let you browse through the module names. If you find the desired one, click the mouse on it, and the module will appear on the worksheet. You may also browse through the library by selecting **Browse** from the **PLACE Module** submenu shown in Figure 9–1.

Modify module name. Often it is necessary to give an appropriate name to the fetched module in relationship to your circuit. The following procedure will let you modify the name of a library module:

1. From the PCB_MAIN menu, click on **QUIT** and click on **Library** from **QUIT.**

2. Click on **Quit** from the **Library** submenu. The **Quit Library** submenu will appear.

3. Click on **Load** from the **QUIT Library Quit** submenu shown in Figure 9–2a, and the software will prompt for *Load Library Module?*.

Chapter 9 / Computer-Aided Techniques for PCB Layout Design

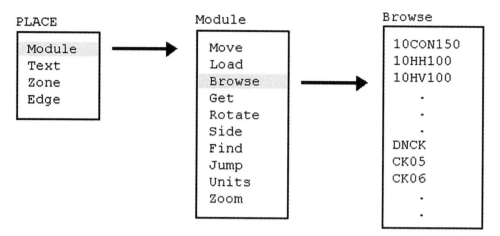

Figure 9-1. PLACE Module Browse Submenu

Provide the name of the library module you want to modify, and left-click once. The library module will appear on the screen. Remember, this is not the worksheet of your board. If you are not sure about the module name, click once to browse through the module names.

4. Right-click the mouse once and select **Name** from the **QUIT Library** submenu. The **Edit Name** submenu shown in Figure 9–2b will appear.

5. Select **Place** for the **Edit Name** submenu, and provide the desired name at the prompt. Click once, and the new name will appear.

6. Place the mouse pointer on the name you want to delete, and select **Delete** from the **Edit Name** submenu.

7. Right-click the mouse once and then click on **Quit.**

8. Left-click the mouse on **Write** from **QUIT Library Quit** submenu. Provide a name for your new module. (**Caution:** Never click on Update from the **Quit Library** submenu. This will destroy the original library parts. Similarly, do not save the new part under the same module name. Provide another appropriate name for the modified module and save it under the Custom library.)

There is another way to modify the name associated with each module. Use the following procedure:

1. Put the mouse pointer on the module value (for a single pad, it is MHOLE 1) or module reference designator (such as M ***), and click on **EDIT** from PCB_MAIN. The module value or the reference designator will be at the pointer for you to move. You may need to position the mouse pointer carefully on the module value or reference designator for this operation. Once you have it on your pointer, move it off the board edge.

2. Use **PLACE → Text** to name the module appropriately.

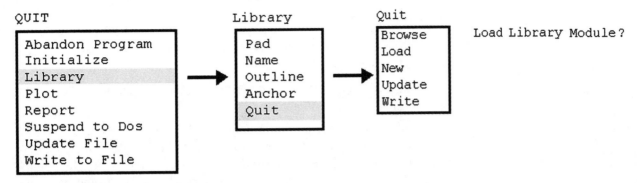

Figure 9–2a. QUIT Library Quit Submenu

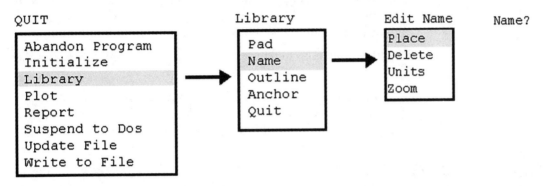

Figure 9–2b. QUIT Library Edit Name Submenu

9–1.8 Arranging Modules on the Board

Once all the modules are downloaded, they must be arranged logically on the board. During the process of physical arrangement, several factors need to be considered by the designer. The three most important factors that determine the relative placement of modules are as follows:

1. The number of connections between two modules
2. The length of tracks between two modules
3. The architectural beauty of the board

The length of tracks between modules can be optimized by using ratsnest and force vectors. However, choosing the position of a module depending on the number of connections among its neighboring modules is strictly a decision of the designer. Architectural beauty of the board also depends on the designer.

Arranging the modules on the board is a matter of moving them from one part of the board to another. Using the following two commands, located under the **PLACE Module** submenu, a module can be moved using: **Move** and **Get.** The following procedures will show how these commands work.

Move command

1. From the PCB_MAIN menu, left-click on **PLACE** and left-click **Module** from the **PLACE** submenu.
2. Left-click on **Move** from the **Module** submenu. The menu bar shown in Figure 9–3a will appear.
3. Position the mouse pointer on the module to be moved and left-click the mouse once. The selected module and its connections with other modules will be transformed into outline mode and white lines, respectively. The menu bar at the top will change to the bar menu shown in Figure 9–3b.
4. As you move a module, white lines which show connections with other modules will give you a rough idea about where to place it with respect to other modules on the board.
5. Drag the module by moving the mouse at the desired location on the board.
6. Left-click the mouse once to place the module or use the first letter of each option to execute the command. As soon as you place the module, the menu bar will change to the one shown in Figure 9–3a.

Get command

1. From the PCB_MAIN menu, left-click on **PLACE** and left-click **Module** from the **PLACE** submenu shown in Figure 9–4a.
2. A left click on **Get** from the **Module** submenu will display the menu bar shown in the figure.
3. Left-click the mouse once and a Get? prompt will appear. Look at the hard copy of the schematic and provide the reference designator of the module you want to relocate on the board.
4. You may also choose the reference designator from the list of modules. If you left-click the mouse once at the Get? prompt, the module that you have already placed on the board will display an asterisk on the right. This is a reminder for you of which modules have yet to be placed properly on the board.

```
| Move    Find    Jump    Origin    Unit    Zoom    Escape |
```

Figure 9–3a. Bar Menu for PLACE Module

```
| Place    Rotate    Side    Find    Jump    Origin    Unit    Zoom    Escape |
```

Figure 9–3b. Bar Menu for PLACE Module Move

5. As you select a module, the menu bar at the top will change to the one shown in Figure 9-4b.

6. As you drag a module by moving the mouse, white lines that show connections with other modules will give you a rough idea about where to place the module with respect to other modules on the board.

7. Drag the module to the appropriate position on the board. Left-click the mouse once to place the module or use the first letter of each option to execute the command. As soon as you place the module, the menu bar will change to the **Get** mode.

8. In this way, move or drag each module to the desired place on the board and arrange them according to the three rules mentioned above. The module arrangement that you are making on the board is not the final one; you may need to rearrange the modules again during the board routing process.

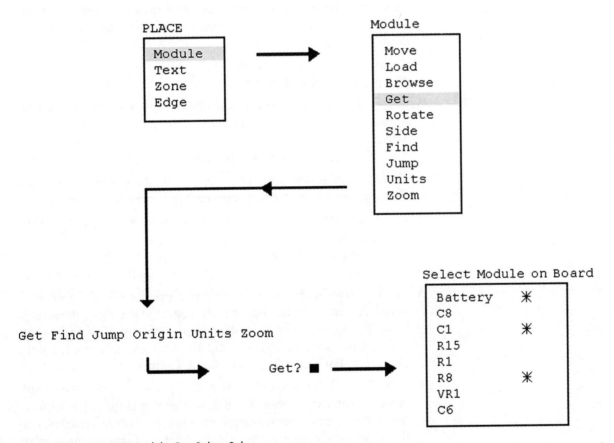

Figure 9-4a. PLACE Module Get Select Submenu

Figure 9-4b. PLACE Module Get Submenu

9-1.9 Placing Text and Special Zones

Placing text. Text should be written in copper on each board layer for the purpose of identifying them during the fabrication process. Tracks cannot intersect with this kind of text because this text is considered as another track. On a double-sided board, this is true for both sides. However, text which is written in silkscreen will not interfere with tracks. The text in copper should be written on both sides of the board before the routing process. Use the following procedure:

1. From the PCB_MAIN menu, left-click the mouse on **PLACE.**

2. Left-click on **Text** from the **PLACE** submenu. The software will prompt for Text? Enter the short name of your project and the layer number (such as Layer 1 or Layer 2), and left-click the mouse once. The menu bar shown in Figure 9–5 will appear at the top of the screen.

3. For example, enter text as **Logic Probe Layer 1.** Move the mouse to position the text at the top or at the bottom of the board. If you place the text at the top or bottom of the board, the software will have less trouble in routing.

4. Enter **V** (vertical sizing of the text) from the keyboard. The vertical sizing parameter will appear at the bottom of the screen.

5. Move either the mouse or the up/down arrow keys (↓↑) to set the vertical size of the text. Watch the sizing parameter change at the bottom. For example, set the size to 0.085 inches. When the vertical size of the text attains the desired value, left-click the mouse once.

6. Enter **H** (horizontal sizing of the text) from the keyboard. The horizontal sizing parameter will appear at the bottom of the screen.

7. Move either the mouse or the up/down arrow keys (↓↑) to set the horizontal size of the text. Watch the sizing parameter change at the bottom. For example, set the size to 0.075 inches. When the horizontal size of the text attains the desired value, left-click the mouse once.

8. Enter **T** (type of text, written in either copper or silk) from the keyboard. A menu with two options will appear at the top of the screen. Choose **Copper** if you want the text to appear with the master artwork. Choose **Silk** if you want the text to appear with your silkscreen. For example, your choice is **Copper.**

9. Enter **L** (the layer where the text will be placed) from the keyboard. A menu with two options, **1** and **2,** will appear at the upper left corner of the screen. The working layer number is always displayed at the upper right corner in the grid display box. Click on the number

```
PLACE  Rotate  Mirror  Vert  Horiz  Type  Find  Jump  Layer  Origin  Unit  Zoom
```

Figure 9–5. Bar Menu for PLACE Text

of the layer where you desire to place the text. For example, your choice is **1**. If the grid display box had Layer 2, it would change to 1 and vice versa.

10. Since your choice of placing text is Layer 1, the text needs to be mirror imaged. This mirror-imaged text on the route side will help you during the fabrication process. Enter **I** (mirror image the text) from the keyboard. The text will be mirror imaged. Reposition the text by moving the mouse, and when the text is in the appropriate place, click the mouse once. The text will be positioned. If the default copper color for Layer 1 is brick red and Layer 2 is green, the text you have just placed will be brick red.

You may place text in copper or silk for power, ground, and other purposes at any time following the above procedure.

Placement of special zones. Special zones are the restricted areas of the layout. These areas are created in the layout plan of the board for various purposes—for example, a mounting screw, a hole for something to pass through, or a ground or power plane for a component needing a heat sink. The zones can be placed on either side of the board. There are three types: Copper, Forbidden, and No Via. All of these zones are layer dependent. This means that a **Copper** zone in Layer 1 will not affect the routes in Layer 2; routes in Layer 2 will pass through Layer 1 zones. Similarly, a **Forbidden** zone in Layer 2 will not affect the routes in Layer 1; routes in Layer 1 will pass through the Layer 2 zones. However, it is not the same for a **No Via** zone. Since vias are through-holes, a via on either layer will not be placed in a **No Via** zone. But a route may pass through a **No Via** zone. If you want a zone where you do not want anything on either side, you need to create a Forbidden zone of the same size on both layers at the same place on the board.

A Copper zone is filled with copper (brick-red color), and, depending on its layer, routes on that layer will not pass through that zone. However, routes on the other layer may pass through the zone. The Copper zone in Layer 1 is brick red and in Layer 2 is green.

A Forbidden zone is shown as a crosshatched pattern, and, depending on its layer, it will have the appropriate color.

A No Via zone looks like graph paper, and, depending on its layer, it will have the appropriate color.

The following procedure will place a special zone on a board:

1. From the PCB_MAIN menu, left-click the mouse on **PLACE**.

2. Left-click on **Zone** from the **PLACE** submenu. The menu bar shown in Figure 9–6 will appear.

3. Position the mouse pointer at the upper left corner of your prospective special zone. Enter **B** (begin drawing) from the keyboard or click once to choose from the step menu. A left click and drag of the mouse will let you draw a box of desired size for the special zone.

```
Begin  Seed  Type  Width  Find  Layer  Jump  Origin  Unit  Zoom
```

Figure 9-6. Bar Menu for PLACE Zone

4. Enter **T** (type of zone) from the keyboard and choose the type of zone from the submenu. For example, you have chosen **Copper**. Left-click the mouse once to go to the next step.

5. Enter **L** (choice of layer) from the keyboard and choose the layer number where you want to put your special zone. For example, you have chosen Layer 1. Click the mouse once to go to the next step.

6. Enter **S** from the keyboard while keeping the mouse pointer inside the special zone area you have just drawn. The entire special zone will turn brick red. You have created a Copper zone in Layer 1. When you autoroute the board, the software will not put any Layer 1 tracks through this zone. This zone is forbidden for Layer 1 routes. However, Layer 2 routes (which are on the other side of the board) may pass through this zone.

9-1.10 Routing a Board

The final task of designing artwork is to connect the modules. This process of converting logical connections among parts into physical ones is called routing. Each pad is assigned a net name. When two or more pads are connected together, they are given the same netname. In addition to a lot of other information, the netlist file contains all the netnames and their connection scheme. During the process of routing, the connection from one pad to the other is being converted to physical tracks. These tracks cannot intersect with each other, making the routing process not so straightforward.

To avoid interference between tracks, depending on the number of nets, boards can be single sided, double sided, and multilayer. From single layer to multilayer, the board is progressively more difficult to design and more expensive to make. However, a single-layer board is more difficult to route than a double-layer board. The chances of finding a noninterfering path for a track on a single-layer board compared to a double-layer board is much lower. A double-sided board presents an optimum situation for both routing and fabrication. On a double-sided board, when the autorouting tool cannot find a noninterfering path for a track on one side of the board, it will go to the other side of the board and find a noninterfering path. At the point where this changeover takes place, tracks on both sides need to be connected through the board by using a conducting path. This conducting path from one layer to another is called a **via.** If you are routing the board by using the **No Via** strategy, the route tool may also find this conducting path through a lead of a component.

9-1. The PC Board Layout Design Process

In OrCAD/PCB II, routing can be done manually or automatically. Any combination of automatic and manual routing can be done to route a board. In addition to many common features, the PCB II automatic routing tool has three more advanced features:

1. Optimizing Module Placement
2. Optimizing Routes
3. Strategies for Automatic Routing

Before you proceed with the routing process, you must know how to do the following operations:

- Compile a netlist file
- Optimize module placement
- Optimize routes
- Inquire about track width, length, netname, etc.
- Edit a netname
- Delete an object or a block

Compile a netlist. Compiling the netlist file means extracting the connectivity information from the file for use by the PCB software. Remember to set the design environment for your design before you start compiling. To compile the netlist file, use the following steps:

1. From the PCB_MAIN menu, click on **ROUTING**. The **ROUTING** submenu shown in Figure 9–7 will appear.
2. Click on **Netlist** from the submenu. The **Netlist** submenu will be on the screen.
3. Click on **Compile** from the **Netlist** submenu. The software will neither prompt you about the completion of the compiling process nor give you any indication of it, but the compilation is complete. If you wish you may click on **Compile** again.

Set strategies for automatic routing. A strategy tells the software how to route a net during an autorouting process. There are six strategies: **Normal, Flexible, Extensive, Power, 90 Degree,** and **No Via.** Each of these strategies has a specific meaning to the autorouter.

- The **Normal** strategy discourages the autorouter tool from making 45° connections to tracks and forbids them entirely on pads. Pad connections are made according to their orientation.
- The **Flexible** strategy lets the autorouter make 45° connections to pads, if the connection cannot be made according to the pad orientation.
- The **Extensive** strategy lets the autorouter make 45° connections anywhere in the layout.

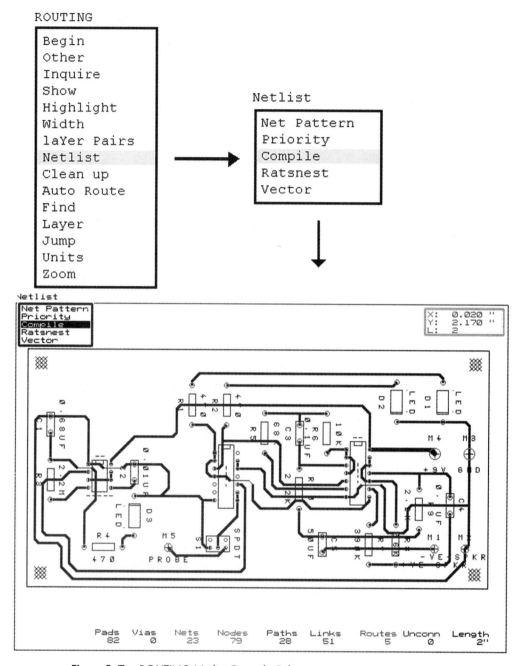

Figure 9-7. ROUTING Netlist Compile Submenu

- The **Power** strategy allows the autorouter to make connections to existing tracks. The strategy is useful for routing power and ground nets.

- The **90 Degree** strategy allows the autorouter to make only 90° connections to pads and tracks.

- The **No Via** strategy forces the autorouter to place routes completely on one working layer or the other, without any via.

These strategies can be specified in three different ways: under **Design Conditions**, using the **SET** command from the PCB_MAIN menu, and under **Net Conditions**. The strategies specified under **Net Conditions** take precedence over either **Design Conditions** or the **SET** command. The **SET** command takes precedence over **Design Conditions**. This means when the autorouter routes a net and the netname is provided under **Net Conditions**, the strategy specified under **Net Conditions** is used for routing. If the router cannot find any strategy under it, the autorouter looks for the strategy set by the **SET** menu. If no strategy is available here, it routes the net by using the strategy specified under **Design Conditions**. You can specify different strategies, route width, isolation, etc., for some specific nets by specifying their netname under **Net Conditions**. **Net Conditions** is located under the PC Board Layout configuration screen.

Optimizing module placement.

The **Optimizing** tool provided by PCB II can be used to optimize the placement of modules on the board. The purpose of this optimization is to minimize the total length of tracks. To minimize the total length of tracks, the optimum position of the modules with respect to one another needs to be determined. The tool uses a **Force Vector** and **Ratsnest Vectors** for each module to indicate the direction of optimum position of the module. The **Force Vectors** represent a mathematically weighted average of the individual connections between the pins for some or all of the modules. Each of these **Force Vectors** is dependent on the position of all the modules to which it is connected. Therefore, the movement of a module will change the **Force Vectors**. The **Ratsnest Vectors** are straight connections between two or more pads. The following procedure will let you optimize the module placement on a board by using the **Force Vectors:**

1. From the PCB_MAIN menu, left-click on **ROUTING.**
2. Left-click on **Netlist** from the **ROUTING** submenu.
3. Left-click on **Vector** from the **Netlist** submenu.
4. Left-click on **All** from the **Vector** submenu. The software will display each module on the board with a yellow straight line and a small diamond-shaped box at one end. These lines are called **Force Vectors.** The diamond indicates the center of gravity of the module. It also indicates the direction an individual module needs to be moved for its optimum position. Each of these **Force Vectors** is dependent on the position of all modules to which it is connected.
5. Go to the **Netlist** submenu and left-click on **Ratsnest.**
6. Left-click on **All** from the **Ratsnest** submenu. Straight line connections from pad to pad of each module will now be displayed in addition to **Force Vectors.** These straight lines are the actual connections among modules and will be converted into routes. You may take advantage of both **Force Vectors** and **Ratsnest Vectors** when positioning a module.

7. Move the modules on the board such that the diamonds of most of the **Force Vectors** are within the boundary of the module. Since the movement of a module will affect other **Force Vectors,** you may have to rearrange the modules several times for the optimum position of the modules.

Optimizing module placement using the above procedure is not perfect, but that is a relative matter. Figures 9–8a and 9–8b show a board before the optimization, and Figures 9–8c and 9–8d show the board after placing the modules at their optimum position. Once the modules are placed at their optimized position on the board, the board is ready for routing. During the process of routing, the modules may have to be repositioned slightly again.

Optimizing routes. The total number of routes and their length in a PC board layout can be reduced by using the **Optimize** tool. The tool does not make any attempt to place previously unconnected routes; instead, it rips and reroutes existing tracks a little bit differently to reduce their length. A board that has already been routed or partially routed can be optimized by using the following procedure:

1. Click on **Routing** from the PCB_MAIN menu, and the **Routing** submenu will appear.
2. Click on **Auto Route** from the submenu, and the **Auto Route** submenu will appear.
3. Click on **Optimize** from the submenu shown in Figure 9–9.
4. Select your choice of whatever you want to optimize from the **Optimize** submenu:
   - **Track** will optimize a selected track.
   - **Module** will optimize all the nets associated with a module.
   - **Net** will optimize all the tracks associated with a net.
   - **Block** will optimize all the nets within the specified block.
   - **All** will optimize the entire board.
5. Position the mouse pointer according to your choice. For example, if you have selected **Block,** the software will display another menu for you to draw a block on the layout. After drawing the block enclosing the area you want to optimize, click on **Block** again. The software will rip up the nets within the block and reroute them. If, however, you have selected **All,** the software will immediately rip up all the nets and reroute them one by one.

Optimizing reduces the total length of the route and also reduces the number of routes. The software also lets you select another strategy and select layer pairs other than the current ones.

9–1. The PC Board Layout Design Process   327

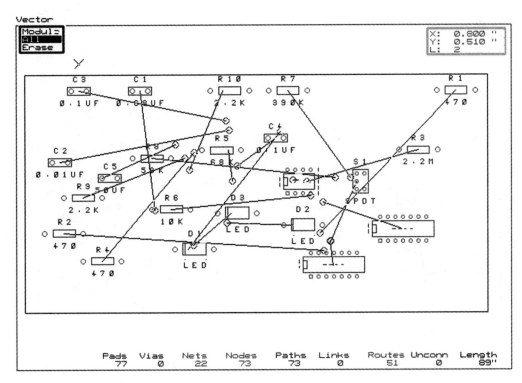

Figure 9–8a. Board with Force Vectors before Optimization

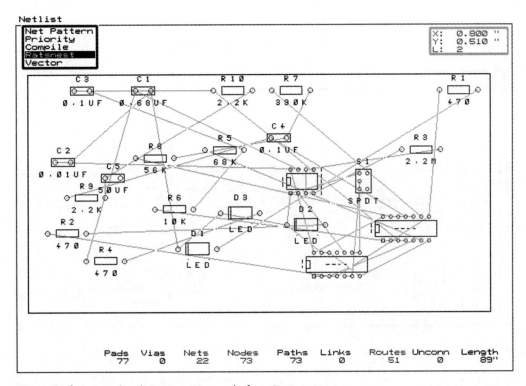

Figure 9–8b. Board with Ratsnest Vectors before Optimization

328 Chapter 9 / Computer-Aided Techniques for PCB Layout Design

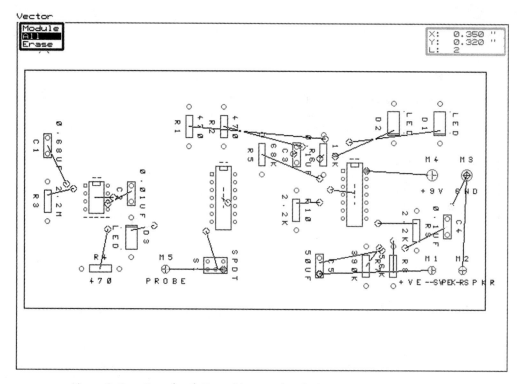

Figure 9–8c. Board with Force Vectors after Optimization

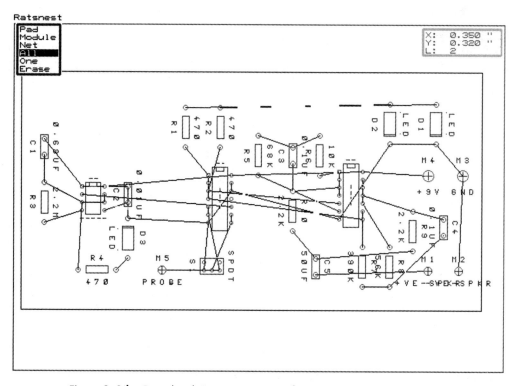

Figure 9–8d. Board with Ratsnest Vectors after Optimization

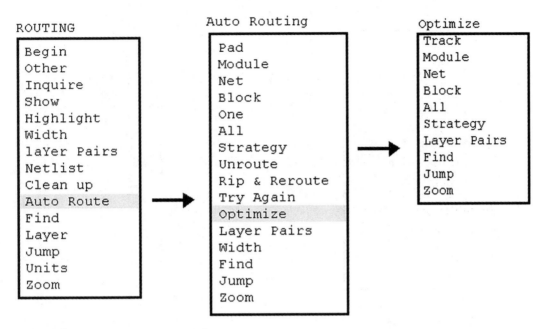

Figure 9-9. ROUTING Optimize Submenu

Inquire. On a PCB worksheet, it is difficult to know about netname, track width, track length, etc., without the help of the **Inquire** command. The command can be accessed from the **ROUTING** submenu. To inquire about some object, place the mouse pointer on that object and click on **Inquire** from the **ROUTING** submenu. The **Inquire** submenu will appear. Click on **Inquire** again from the submenu, and the result will appear at the bottom of the screen. The following are examples of **Inquire** used on four different types of objects of a PC board:

Inquire About a Track:

Length	Track	Width	Layer	Netname
1.392"		0.008"	2	N00001

Inquire About a Pad:

Pad	Vert	Horiz	Netname	Drill
1	0.075"	0.075"	V_{DD}	0.045

Inquire About a Via:

Via	Width	Layer	Netname
Through	0.045	1/2	N00032

Inquire About a Module:

Module	Designator	Part
	U7	7420

Also, Figures 9–10a through 9–10c show the picture of the actual screens when **Inquire** is used.

Chapter 9 / Computer-Aided Techniques for PCB Layout Design

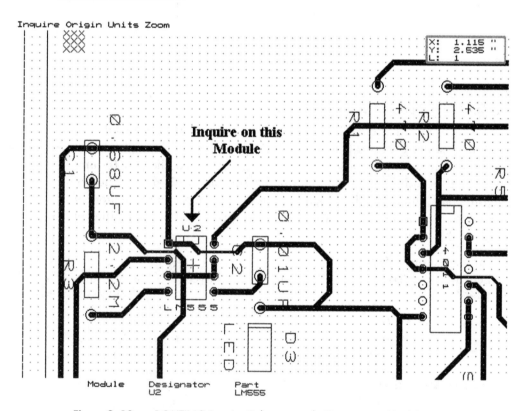

Figure 9–10a. ROUTING Inquire Submenu with Cursor on a Module

9-1. The PC Board Layout Design Process

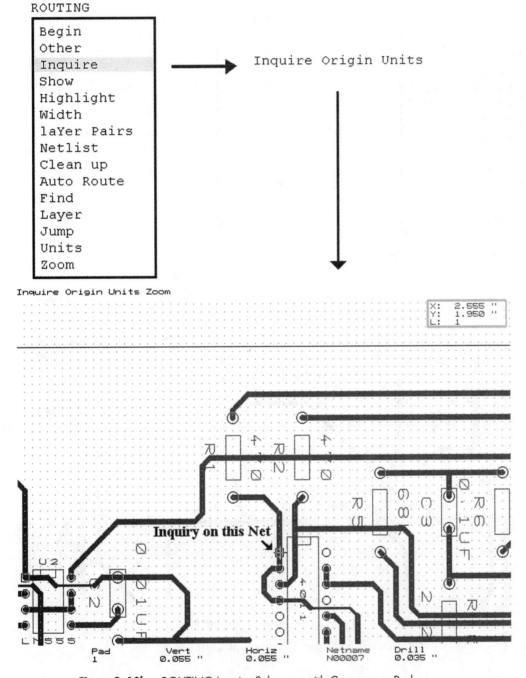

Figure 9-10b. ROUTING Inquire Submenu with Cursor on a Pad

332 Chapter 9 / Computer-Aided Techniques for PCB Layout Design

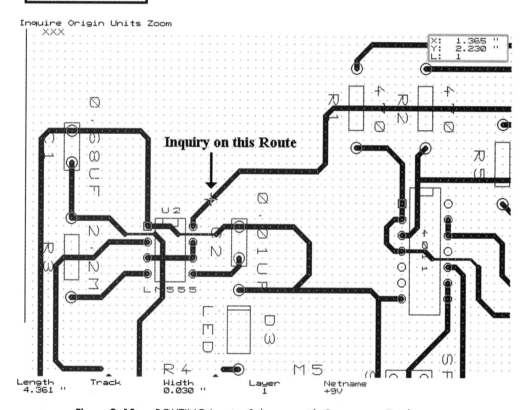

Figure 9-10c. ROUTING Inquire Submenu with Cursor on a Track

Editing netname. Modules that are fetched directly from the PCB library onto a board do not have any logical connections to other modules. They are not a part of the netlist file. In order to route them with other modules on the board, pins of these modules must have the appropriate netname. Pins that are connected together are assigned the same netname by the netlist tool located under the SDT tool set. To create a route either manually or automatically, pins must have the same netname. To give a netname to a pin or pins of a module whose net information is not in the netlist file, it/they must be edited. This means you must know the netname of the pin/pins to which you will route the pins of a newly fetched module from the PCB library. Only when the appropriate netname is assigned can they be routed either manually or automatically to other modules. Generally, netname editing is required only for the pins of the modules you fetched from the PCB library. To find the netname of a pad that already exists, the easy way is to use **Inquire** from the **ROUTING** submenu. Position the mouse pointer on the pad for which a netname is desired, and click once. Then left-click on **Inquire** from the **Inquire** submenu. The following procedure will assign a netname to a pin:

1. Invoke the PC board on the worksheet for which you want to edit the netname of pins.

2. Position the mouse pointer on the item you want to edit from the appropriate layer, and left-click on **EDIT** from the PCB_MAIN menu. The **EDIT** submenu shown in Figure 9–11a or Figure 9–11b will appear. Figure 9–11a appears if you select only an item, and Figure 9–11b appears if you have more than one item selected (a pad and a track) by the mouse pointer.

3. Click on **Pad** from the **Select Item** submenu. Information about the net shown in Figure 9–11b will appear at the bottom of the screen. Generally, nets can be edited from both sides of the board. This means that you can select from either layer. However, for routes and silkscreen items you may have to go from one layer to the other. You can take your mouse pointer from Layer 1 to Layer 2 by using the **Layer** option from the **EDIT** submenu and selecting **2** from the **Layer** submenu. The scale box at the upper left corner should show **L:2**, indicating Layer 2. Switching between Layer 1 and Layer 2 can also be done by left-clicking on **Other,** located in the **ROUTING** submenu.

4. A left click on **Netname** will bring about a highlighted prompt at the bottom of the screen. If the pin already has a netname and you are changing it to a different netname, the current netname will appear at the prompt. You may overwrite or enter a new netname that is exactly the same as one of the pins it will be connected to.

Following the above procedure, a netname can be assigned to pins that do not have any netname and pins whose netname needs to be altered. Also, other attributes of a Pad can be altered.

Chapter 9 / Computer-Aided Techniques for PCB Layout Design

Editing a track or a via. Editing a **Track** is similar to editing a **Netname.** To edit a **Track** or a **Via,** you may use the following procedure:

1. Position the mouse pointer in the correct layer onto a track, net, or via, and invoke **EDIT** from the PCB_MAIN menu.

2. Select **Width** from the **Route Width** submenu shown in Figure 9–11c. Select **Track** or **Via,** depending on the item you wish to edit. If you have selected **Track, Track** at the bottom of the screen will be highlighted and you can set the track width. If you have selected **Via,** then **Via** at the bottom of the screen will be highlighted and you can set the diameter of the via. Skip step 3 and go to step 4. (Nothing has changed on your layout yet. If you do not like it, right-click the mouse several times and start the process again by selecting **EDIT.**)

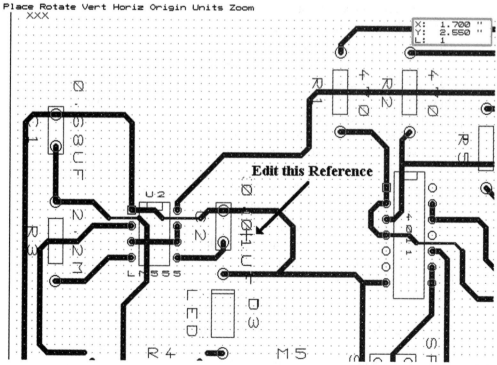

Figure 9–11a. ROUTING Edit Submenu with Cursor on a Reference Designator

9-1. The PC Board Layout Design Process

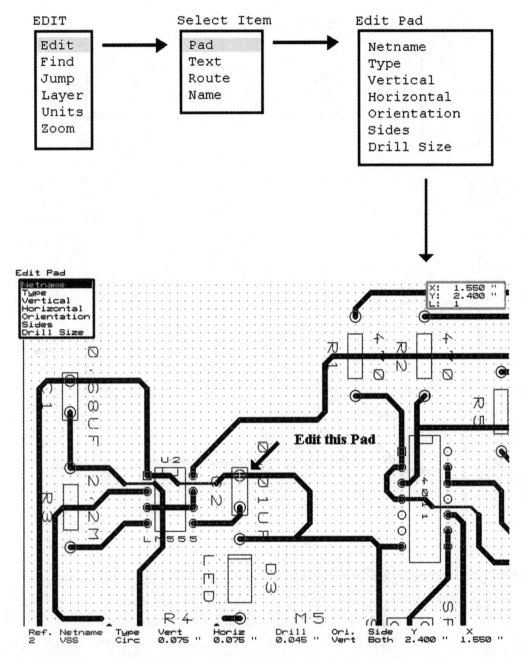

Figure 9-11b. ROUTING Edit Submenu with Cursor on a Pad and a Track

3. If, however, the submenu shown in Figure 9-11b appears, select **Route** from the **Select Item** submenu, and **Route Width** submenu will appear. Click on **Width** from the submenu and select **Track** from the **Set Width** submenu. **Track** at the bottom of the screen will be highlighted, and you can change the track width.

4. Right-click the mouse once, and the **Route Width** submenu will appear. Click on **Segment**, **Track**, or **Net**. Depending on your selection, you will see the change in your route.

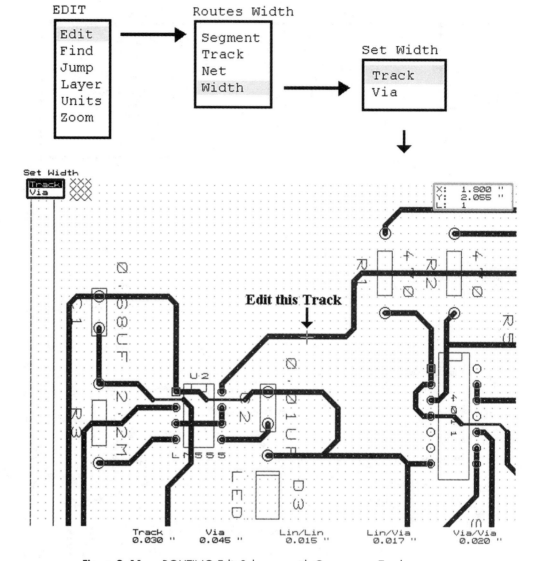

Figure 9–11c. ROUTING Edit Submenu with Cursor on a Track

- **Segment:** will change only the selected part of the track
- **Track:** will change the entire track from one pad to another pad that may have several segments
- **Net:** will change the track width of the entire net that may have several tracks

Delete an object or a block

1. To delete an object, place the mouse pointer on that object and click on **DELETE** from the PCB_MAIN menu. The **DELETE** submenu will appear.
2. Click on **Object** to display the **DELETE** bar menu.

3. Place the mouse pointer on the object to be deleted and click on it. Depending on the type of object, the software will prompt appropriately to delete that object. For example, if you place the mouse pointer on two items at the same time (a track and a net), the **DELETE Object Delete** submenu will prompt you to delete the desired object.

4. Click on **Segment** if you want to delete only that portion of the track on which your mouse pointer is positioned. If you want to delete the entire track, click on **Track.** If, however, you want to delete the entire net, click on **Net.**

5. If you want to delete a portion of a board, click on **Block** instead of **Object** from the **DELETE** submenu. The **DELETE Block** submenu will appear.

6. Click on **Begin,** and the Begin bar menu will appear. Position the mouse pointer at the upper left corner of the area you would like to delete.

7. Left-click once, click on **Begin,** and drag the mouse pointer down and right to draw a box around the area to be deleted.

8. Click once and click on **End.** Another DELETE submenu will appear. Click on the type of item you want to delete from that box. Several right clicks and a left click will bring the PCB_MAIN menu back on the screen.

Manual routing. When designing artwork for a PC board, there are usually some connections in the layout that you would like to do in a specific way—especially those connections for which you need to control the path of the track. The routing of these needs to be done manually. The following procedure will convert a logical connection between two pins to a physical track and route the track through the desired places on the board.

1. From the PCB_MAIN menu, left-click on **ROUTING.**

2. Left-click on **Layer** from the **ROUTING** submenu, and select **1** or **2,** depending on which side you want to create your route. For a double-sided board, the default will be Layer 1.

3. Left-click on **Width** from the **ROUTING** submenu. The present settings of track width and via diameter will appear at the bottom of the screen. The settings may be given in inches or centimeters, depending on the unit selection. The default value of the track width and via diameter is 0.015″ and 0.045″, respectively. Left-click once, and the **Set Width** submenu will appear.

4. Left click on **Track,** and the track width dimension at the bottom of the screen will be highlighted. The track width can now be changed either by moving the mouse up and down or with the up and down arrow keys. Use the up and down arrow keys as you get closer to the

set value. Generally, tracks connecting main power and ground are wider than other tracks. Select the desired value of the track width.

5. Right-click the mouse once, and the **Set Width** submenu will come back on the screen. This time left-click on **Via.** The Via dimension at the bottom of the screen will be highlighted. Now the via diameter can be changed by the mouse or the up and down arrow keys. Set the via diameter to a desired value.

6. Right-click the mouse twice and left-click once to get back to the **ROUTING** submenu.

7. Left-click on **Begin** from the **ROUTING** submenu. The **Begin** bar menu will appear at the top of the screen. Position the mouse pointer where you want to begin your track. Generally, it should be a pad. Click once to change the bar menu to a pull-down menu.

8. Click on **Begin** again, and drag the mouse to draw a segment of a track. If you need to turn the track to go in another direction, right-click the mouse once to get back to the **Begin** submenu. Then left-click from the **Begin** submenu either on **New** or **Begin,** and drag the mouse to draw another segment of the track.

9. If you cannot find a nonintersecting path for a track, go to the other side of the board and draw the track. To go to the other side, click on **Other** from the **Begin** submenu. A via of currently selected size will be created, and the track will transfer to the other side of the board. If vias are not allowed, the track can go to the other side through a component pad belonging to the same netname. The track can run on one side as long as it is required and can be transferred to the other at any time through a via or a component pad belonging to the same net.

10. To end or begin another track, either click on **New** or **End** from the **Begin** submenu.

Auto routing. The Auto Routing tool can route all or only the selected unrouted nets of a board. Paths or nets that have already been routed are not affected by this process. If there are no manual routes on the board, it is very easy to erase all the routes and route the entire board again by using the Auto Route tool. After auto routing a board, if several unconnected paths or routes remain, there are several options for that board:

1. Erase all routes and route again with a different routing strategy.
2. Erase all routes and rearrange the position of the modules and routes again.
3. Route the unconnected paths manually.
4. Use the **Rip and Reroute** command.
5. Use the **Try Again** command.

9-1. The PC Board Layout Design Process

The following procedure will route a board by using the **Auto Route** tool.

1. From the PCB_MAIN menu, left-click on **ROUTING.** Then click on **Auto Route** from the **ROUTING** submenu. This will bring up the **Auto Route** submenu.

2. Left-click the mouse on **Width** from the **Auto Route** submenu.

3. Click on **Track** from the **Set Width** submenu. Width value at the bottom of the screen will be highlighted. Set the desired value of the track width by using the mouse or the up/down arrow keys. The default value of the track width is 0.015 inches. If you do not set the track width by using the **Set Width** command, OrCAD/PCB will use the track width value from the **Design Conditions** defined in the PCB configuration file. However, the **Set Width** command will not override the **Net Conditions** specified in the configuration menu.

4. Left-click the mouse once, and the **Set Width** submenu will appear. This time set the via diameter by using the same procedure as for track width. The default value for the via diameter is 0.045 inches.

5. Right-click on the mouse several times and left-click once to display the PCB_MAIN menu.

6. A left click on **SET** will bring up the **SET** submenu.

7. A left click on **Isolation** will bring up the **Isolation** submenu.

8. Left-click on **Lin to Lin,** and the line to line isolation value at the bottom of the screen will be highlighted. Change the value to the desired number by using either the mouse or the up/down arrow keys.

9. Left-clicking the mouse once will bring the **Isolation** menu back. Following the above procedure, set the desired value for the **Lin to Via** and **Via to Via** values. If you do not set these values here, they will be taken from the values defined in the Design Conditions in the PCB configuration file.

10. From the **SET** submenu, click on **Via Type,** and from the **Select Via Type** submenu, set **Through = Yes, Buried = No,** and **Blind = No.** If you do not set these values here, they will be taken from the Design Conditions.

11. From the **SET** submenu, left-click on **Nbr of Layers** and select the total number of layers for your board. If you do not set the number of layers for the board here, it will be the same as specified in the Design Conditions.

12. If your board is more than two layers, click on **Layer Pairs** from the ROUTING submenu. Follow the prompt to select the layer pair you want to route first. The Auto Router can route only one pair at a time.

13. From the **Routing** submenu, left-click on **Netlist,** and the submenu for Netlist will appear.

14. Left-click on **Net Pattern,** and set either **Tree** or **Chain** from the **Select Either Net Pattern** submenu. If you do not set a net pattern, it will be taken from the Design Conditions.

15. From the **Auto Routing** submenu, left-click on **Strategy** and the submenu for strategy will appear. Click on the desired strategy from the submenu for Pass 1, and the software will prompt for the Pass 2 strategy. Click on the desired strategy for Pass 2. If you do not select the strategy here, both Pass 1 and Pass 2 strategies will be taken from the Design Conditions.

16. If you want a certain net or nets to have different track width, via diameter, isolation, strategy, and net pattern, then they need to be set under **Net Conditions.** Remember, once you set these parameters for a net, they will override any setting made under **Design Conditions** and using **Set** commands.

17. Now the environment for the Auto Router is completely set. The Auto Router can be invoked to route a pad, a module, a net, a desired area on the board, just one route, or all the unconnected routes.

18. From the **Auto Routing** submenu, click on the desired option to route the board. If you left-click on **All,** the software will start routing all the unconnected routes one after another. After completing the routing, the software will report the information and display them at the bottom of the screen. Unconnected routes and the associated pads will be shown by ratsnest vectors.

The unconnected routes on the board can be routed by using the following procedure:

1. Use the **Try Again** command.
2. Reroute with a different strategy and use the **Try again** command.
3. Rearrange modules using **Ratsnest** and **Force Vectors** and use the **Reroute** and **Try Again** commands.
4. Use **Rip & Route** and the **Try Again** command.
5. Change the pads of modules from round to oval.
6. Increase the board size.
7. Reduce the route size.

Variables that affect routing.

- Number of modules and their sizes
- Number of nets and route width
- Routing strategy
- Number of layers
- Size of the board

9-1. The PC Board Layout Design Process

The try again command. If there are only a few (maximum five) unconnected routes, use the **Try Again** command with a different routing strategy. This is the quickest way to complete the routing. If this does not work, go to the next method.

Reroute with different strategy and use try again command

1. From the PCB_MAIN menu, left-click on **QUIT.**
2. Left-click on **Initialize** from the **QUIT** submenu.
3. Finally, left-click on **Erase Routes** from the **Initialize** submenu. Do not choose **Clear Board** because it will wipe out everything from your board, including the modules. **Erase Routes** will clear all the routes from the board, leaving only the modules at their same position. Now the board is ready for routing again with a different strategy.
4. Left-click on **Strategy** from the **Auto Routing** submenu, and select other strategies for Pass 1 and Pass 2.
5. Left-click on **All** from the **Auto Routing** submenu.
6. This time the number of unconnected routes should be fewer than the previous case. If the board has only a few unconnected routes, you may use the **Try Again** command to complete the routing. If it does not work, go to the next procedure.

Rearrange modules using ratsnest and force vectors and use reroute and the try again command.

1. Erase all routes from the board by using **Quit → Initialize → Erase Routes.**
2. Left-click on **Netlist** from the **ROUTING** submenu. Then left-click on **Vector** from the **Netlist** submenu.
3. A click on **All** will bring all the Force Vectors for the unconnected routes onto the worksheet.
4. From the **Netlist** submenu, click on **Ratsnest** and then click on **All** from the **Ratsnest** submenu. This action will bring all the Ratsnest Vectors onto the worksheet.
5. Using the Ratsnest and Force Vectors, rearrange the modules on the board as efficiently as possible. For this purpose, follow the procedure described earlier in this chapter.
6. Auto-route the entire board and find out the number of unconnected routes.
7. If you have quite a few unconnected routes, you may choose to erase all routes, rearrange the modules on the board a little more efficiently, and reroute the board again by using the Auto Router. You may also choose the next option.

Use rip & route and try again command

1. Route the board normally, using the Auto Router.
2. Left-click on **Rip & Route** from the **Auto Route** submenu, and the step menu will change to a bar menu. Position the mouse pointer on one of the unconnected routes, and left-click once. The software will rip all tracks and their associated nets that fall within a two-grid-unit radius of the source and target pads of the unconnected routes. Then, starting with the unconnected routes, the Auto Router attempts to place all the routes by using the strategy indicated in the PCB configuration file.
3. At this time, if your board still has a few unconnected routes, use the **Try Again** command.

Change pads of modules from round to oval. This method may work for your board because it allows the route to connect to the pad from various angles. Change pads of some of the modules from their usual round shape to oval. Reroute the board again. This time, try all five different strategies.

Increase board size. If there are no other restrictions, increase the board size and reroute the board.

Reduce the route width. If your board design criteria permit, reduce the route width only a little bit and reroute the board.

If all the above procedures cannot route your board completely, do not get frustrated. There are still several dozen things you can do to route your board. There are logical techniques but no magical and automatic ways to route a board. This is why people pay money to have a PC board designed. The following are a few more options:

1. Rearrange the modules on the board more efficiently by using Ratsnest and Force Vectors.
2. Reduce the size of the tracks. While reducing the width, be mindful of the manufacturing environment available at your school laboratory. Otherwise, you may end up with a board with broken tracks. Boards with thinner tracks are progressively difficult to make.
3. Route the board with a track width of 0.015" and then increase it wherever possible by manually editing the track width.
4. Change the pads of all modules to oval instead of round.
5. Increase the board size and rearrange the modules.

If the board size is restricted by the product for which the board is being produced, increase the number of layers of the board. For example, if you have a single-sided board, increase it to double-sided. If you have double-sided, increase it to four-layer and so on. However, the price differential between a single-layer board and a double-layer board is much less than that of a double-layer board and a four-layer board. Also, it is difficult to make a four-layer board in a school laboratory.

9-2. PCB MASTER ARTWORK PRINTING

Once you have designed the artwork, it is not difficult to print a hard copy. A hard copy of the artwork is used to produce a film mask for fabricating the PC board. To fabricate a PC board, among many other things, the resolution of the artwork plays a very important role. The better the accuracy of the film mask, the higher the quality of the fabricated board. By using photoplotters, the accuracy of the hard copy of the artwork usually goes up to 1 mil (0.001″). High-resolution photoplotters are quite expensive. For high-resolution output, print artwork directly on a film mask. Names and addresses of companies that provide photoplotting service are provided in Appendix C. However, for school laboratory projects, high-resolution plotting is not necessary. A laser printer of 600 × 600 dpi will be enough.

Hard-copy generation of the artwork in OrCAD is a two-step process. First, you have to create a printer file using the **Route Board** tools and then print hard copy by using the **Print PCB** tool located under **Reporters** in PCB_MENU. The following procedure will generate plot files for each layer of a double-layer board for the HP Laserjet IV printer.

1. After setting the design environment for your design, click on the **Route Board** icon and invoke **Configure Layout Tools** from the Transit Menu. Under **Driver Options,** scroll down to **Printer Drivers** and select the appropriate printer driver for which you are generating printer files. In this case, select the printer driver for HP Laserjet IV. Exit the configuration by clicking on the OK icon.

2. Invoke **Route Board** by clicking on the icon and selecting **Execute** from the Transit Menu. Your board file should be on the screen now. If not, click again and select from the file list.

3. Click on **QUIT** from the PCB_MAIN menu. The **QUIT** submenu will appear. Select **Plot** from the submenu, and the **Plot** submenu will appear.

4. Click on **Item to Plot** from the **Plot** submenu. The **Item to Plot** submenu will appear.

5. Select **Layer** and then **Layer 1** from the **Layer** submenu. Select **All** from the **Board** submenu.

6. Select **Filled** and then select **Auto Sel.**

7. Right-click the mouse till you get back to the **Item to Plot** submenu, and click on **Destination.**

8. Click on **Hard Copy** from the **Destination** submenu. A prompt will appear at the top of the screen: *Write Drill Tool File?* Answer this prompt with LAYER1.TOL. Another prompt will appear at the top

of the screen: *Save Printer Data Base?* Answer this prompt with LAYER1.PRN. The software will create both the files. The LAYER1.TOL file contains a list of all the tools needed to drill that layer of the board. The LAYER2.PRN file contains the binary data base for the **Print PCB** tool to print the artwork.

9. Select **Item to Plot** from the **Plot** submenu. Click on **Layer** and then on **Layer 2.**

10. Select **All** from the **Board** submenu. Select **Filled** and then **Auto Sel.**

11. Right-click the mouse until the **Item to Plot** submenu appears.

12. Select **Destination** and select **Hard Copy;** two prompts, one after another, will appear. Answer them as follows: *Write Drill Tool File?* should be answered with LAYER2.TOL. *Save Printer Data Base?* should be answered with LAYER2.PRN.

Now both the drill and master artwork files are available to be printed. The drill template is available as the LAYER1.TOL file, and master artwork for Layer 1 and Layer 2 is available as LAYER1.PRN and LAYER2.PRN, respectively.

The second part of this procedure is to actually print a hard copy of the master artwork (LAYER1.PRN and LAYER2.PRN) and hard copy of the drill template by using the **Print PCB** tool:

1. Click on the **Print PCB** icon from PCB_MENU.

2. Click on **Local Configuration** from the Transit Menu of the **Print PCB** tool. The **Local Configuration** of the **Print PCB** tool shown in Figure 9–12 will appear on the screen.

3. Under **File options,** provide the name of the appropriate source file (in this case, LAYER1.PRN or LAYER2.PRN) with a PRN extension.

4. Click on the bubble for *Send Output to the Printer.*

5. Under **Processing Options,** click on the bubble *Produce an Artwork* and *Show Drill Holes.*

6. If an inverted image of the artwork is necessary, then click on the button for it.

7. If you need to print the artwork at a scale factor other than 1, click on the button and provide the scale factor in the dialogue box associated with it.

8. Exit the configuration by clicking on the OK icon.

9. Click on the **Print PCB** icon again, and this time select **Execute** from the Transit Menu. The software will send the output to the printer.

Both the inverted image and the scaling of the artwork are requirements of the fabrication process. Some manufacturers of PC boards require inverted images for their process, and some require double- or quadruple-size artwork.

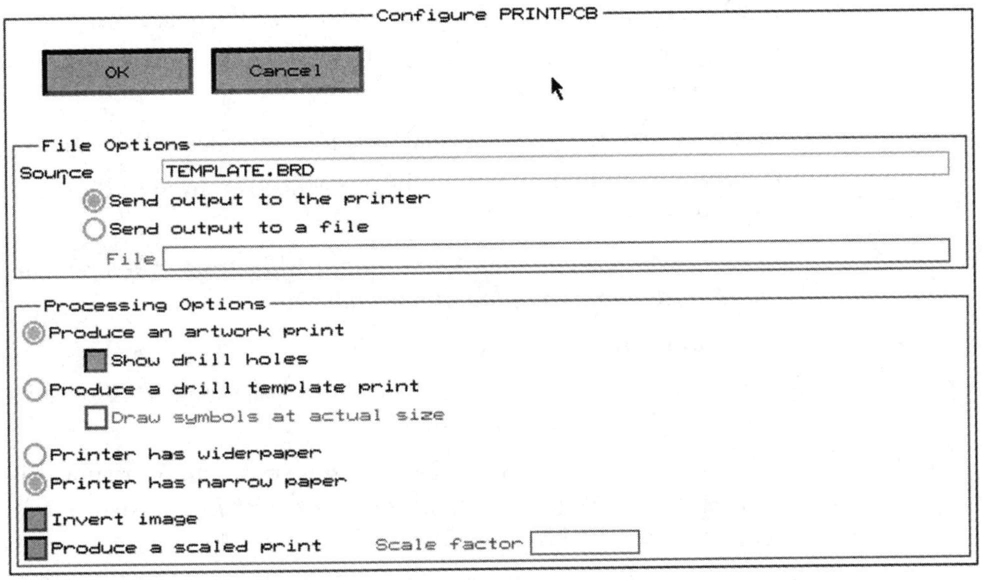

9–3. Printing Component Legend and Solder Mask

Figure 9–12. Local Configuration Screen of Print PCB (Courtesy of OrCAD Inc.)

Printing the drill template. To print the drill template, you should use the same procedure above except for the following:

1. Under **File Options,** provide the name of the source file (LAYER1.TOL) with a TOL extension
2. Under **Processing Options,** select the bubble for *Print Drill Template*.

By using the **Print PCB** tool, either a drill file or a layer of artwork can be printed, but only one item at a time. You may have to use the tool several times to print all the required files.

9–3. PRINTING COMPONENT LEGEND AND SOLDER MASK

9–3.1 Component Legend

The component legend is printed on the board to aid testing and field service personnel in identifying a component on an assembled board. This legend also helps the PC board assembly process. The legend can be printed on the board in many different ways. Silkscreen printing is the most popular way to do this. However, the process is cost effective only if the board is mass produced. To produce a silkscreen stencil, some specific information about the board, termed *silkscreen information,* needs to be extracted, and a hard copy of it needs to be produced. The following procedure will extract silkscreen information from a board file and print a hard copy of it:

1. Set the environment for your design and invoke the board file.
2. Click on **QUIT** from the PCB_MAIN menu. The **QUIT** submenu will appear.
3. Select **Plot** from the submenu, and the **Plot** submenu will appear.
4. Click on **Item to Plot** from the **Plot** submenu. The **Item to Plot** submenu will appear.
5. Select **Silkscreen** and then **Component Side** from the submenu. Select **All** from the **Component Side** submenu.
6. Select **Auto Sel.**
7. Right-click the mouse until you get back to the **Item to Plot** submenu, and click on **Destination.**
8. Click on **Hard Copy** from the **Destination** submenu. A prompt will appear at the top of the screen: *Save Printer Data Base?* Answer this prompt with Silk1.PRN.
9. Perform the same operation for the solder side of the board and save as Silk2.PRN.

Now the silkscreens for both sides of the board are available to be printed. To print them, use the same procedure as when printing artwork, except provide the name of the file to be printed as Silk.PRN in place of LAYER1.PRN. The component and the solder-side silkscreen have to be printed separately.

9-3.2 Solder Mask

Solder Mask (a more appropriate name is **Solder Resists**) protects the conductor side of the board during the soldering process. It is a nonconducting coating and is employed after the board has been etched. The only areas that are exposed to the soldering are component terminals, pads, lands, holes, and test points where solder will alloy with the conductor pattern. Solder mask is always used exclusively to protect the selected part of the board during automatic soldering. To put solder mask on a board, a stencil pattern is required. A negative of the pattern is generated by OrCAD/PCB. The following procedure will print a hard copy of the solder mask for a layout:

1. Set the environment for your design and invoke the board file.
2. Click on **QUIT** from the PCB_MAIN menu. The **QUIT** submenu will appear.
3. Select **Plot** from the submenu, and the **Plot** submenu will appear.
4. Click on **Item to Plot** from the **Plot** submenu. The **Item to Plot** submenu will appear.
5. Select **Solder Mask** and then **Component Side** from the submenu.
6. Select **Auto Sel.**

7. Right-click the mouse until you get back to the **Item to Plot** submenu, and click on **Destination.**
8. Click on **Hard Copy** from the **Destination** submenu. A prompt will appear at the top of the screen: *Save Printer Data Base?* Answer this prompt with SMASK1.PRN.
9. Perform the same operation for the **Component Side** of the board and save as SMASK2.PRN.

In the **Solder Mask** submenu, there are three selections, one of which is **Guard.** If you select **Guard,** a decimal number in inches will be highlighted at the bottom of the screen. Keep this number at its default value of 0.020″. This requirement comes from the fabrication process. The larger the **Guard** value, the more area around the pad will be guarded from the solder resists. Figures 9–13a and 9–13b show Solder Mask for a board with two different **Guard** values.

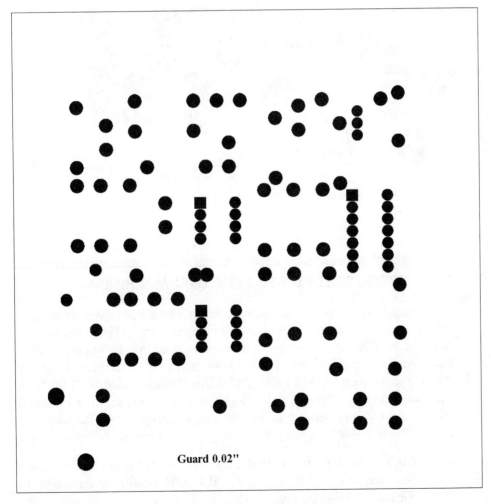

Figure 9–13a. Solder Mask Template with 0.02″ Guard

Chapter 9 / Computer-Aided Techniques for PCB Layout Design

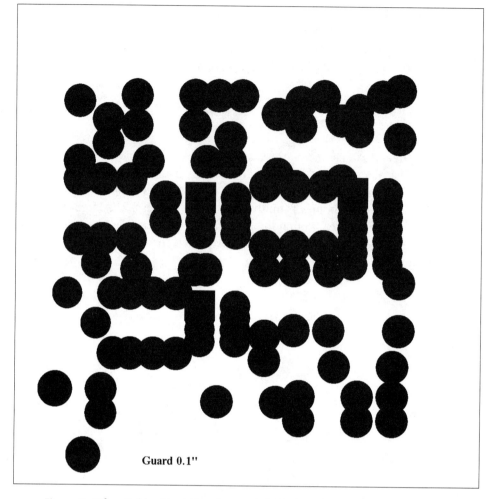

Figure 9–13b. Solder Mask Template with 0.1" Guard

9–4. CREATE DRILL FILES FOR CNC DRILL MACHINE

Commercial drilling of PC boards is performed by a high-speed, computerized, numerically controlled (CNC) drill machine. OrCAD/PCB generates a drill file for CNC machines in various industry-accepted formats, such as Excellon. This data is downloaded directly into a CNC machine. This drill data is the location of drill holes, their sizes, spindle speed, drill bit feed, and retraction rate. If you are planning to send your artwork to a PC board fabricator, you must send this file also. Unlike artwork, you have to send this drill data in soft form. To generate this file, perform the following steps:

1. Click on the **Create NC Drill File** icon. Select **Local Configuration** from the Transit Menu. The Local Configuration screen shown in Figure 9–14a will appear.

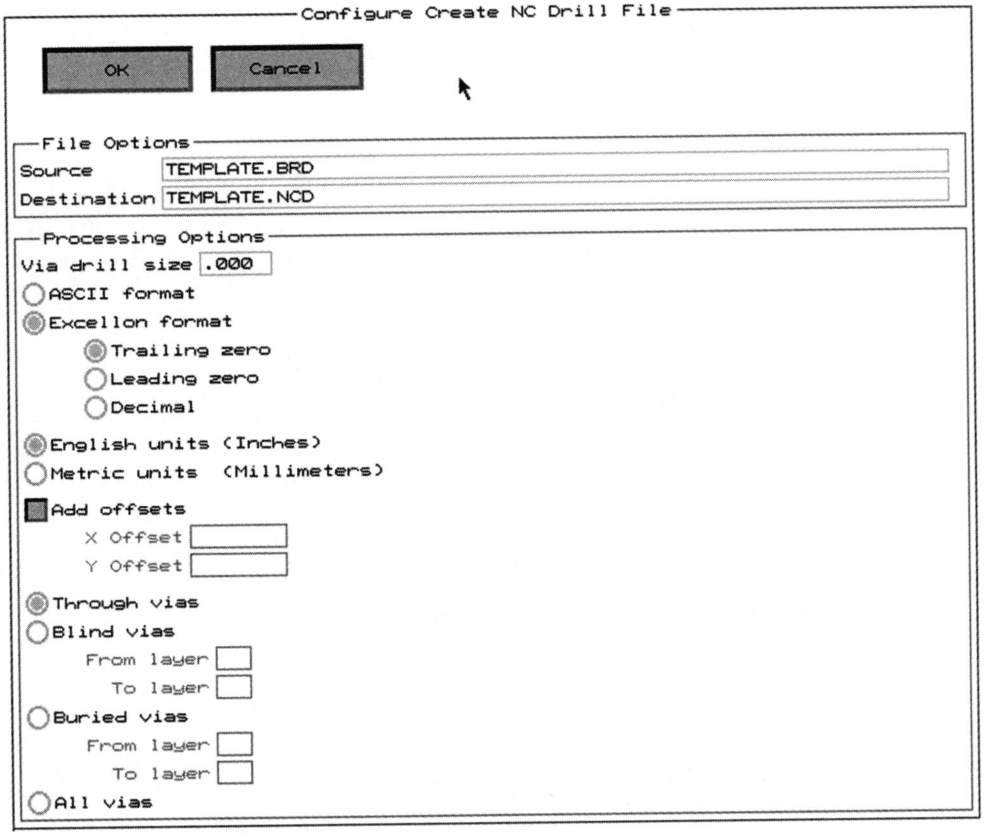

9–4. Create Drill Files for CNC Drill Machine 349

Figure 9–14a. Local Configuration Screen of Create NC Drill File (Courtesy of OrCAD Inc.)

2. In **File Options,** provide the source filename as **filename.BRD** and the destination filename as **filename.NCD.**

3. In **Processing Options,** provide Via Drill Size: 0.040 or other desired size. The via eyelets used for the projects have outer barrel diameters of 0.039" and 0.040". Select ASCII or Excellon format for the drill file by clicking on the respective button. Select **Through Vias** or make another selection, according to your requirements, by clicking on the appropriate button.

4. Exit the configuration by clicking on the OK icon.

5. Click on the **Create NC Drill File** icon again from the PCB_MENU, and this time select **Execute** from the Transit Menu. The OrCAD/PCB software will create the file and save it under **filename.NCD,** which was provided as the destination filename in the configuration.

If you have created the ASCII format of the file, it can be reviewed by using the M2EDIT editor. Figure 9–14b is the Drill Template for the Logic Probe Board.

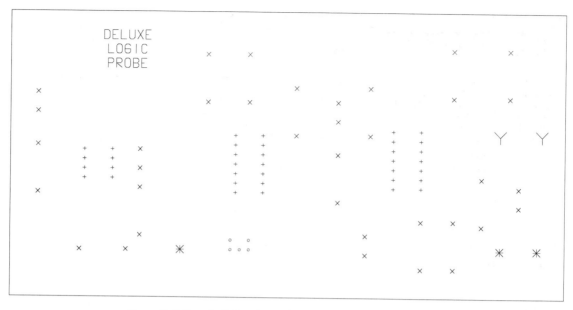

Figure 9–14b. Drill Template of Logic Probe

9–5. GENERATE MODULE REPORT

Often you like to have a list of all the modules used in the design, their X-Y location on the board, associated netnames, module values, reference designators, etc. OrCAD/PCB generates a Module Report that contains all of this information. The following steps will generate a module report for a design:

1. Click on the **Module Report** icon located under **Reporters** in the PCB_MENU. Select **Local Configuration** from the Transit Menu, and the local configuration screen shown in Figure 9–15a will appear.
2. In **File Options,** provide the source filename as **filename.BRD** and the destination filename as **filename.LOG.**
3. In **Processing Options,** select one of the report formats by clicking on the appropriate button.
4. Exit the configuration by clicking on the OK icon.
5. Click on the **Module Report** icon again from the PCB_MENU, and this time select **Execute** from the Transit Menu. The OrCAD/PCB software will create the file and save it under **filename.LOG,** which was provided as the destination filename in the configuration.

Module Report is produced in ASCII format, so you can review the file by using the M2EDIT editor. Figure 9–15b shows the Module Report for the Logic Probe board.

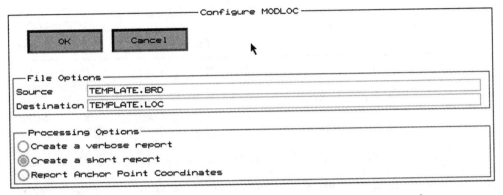

Figure 9–15a. Local Configuration Screen of Create Module Report (Courtesy of OrCAD Inc.)

9–6. THE PROJECTS

To design the board layout (artwork) for a project, the netlist file will be required. The netlist file is the interface between the OrCAD/SDT and OrCAD/PCB software.

9–6.1 The Deluxe Logic Probe

Set the design environment. The Design Environment must be set in order to start working with the project. The netlist file that was generated by SDT is located under the design name LOG_PRO. Invoke the Design Management Tools and set the environment to LOG_PRO.

Configure PC board layout tools. Invoke the PC Board Layout Tools configuration screen and enter the following design conditions for the Deluxe Logic Probe project.

- Minimum Route Width = 0.025″
- Pad Diameter = 0.055″
- Via Diameter = 0.045″
- Pad Drill Size = 0.039″
- Via Drill Size = 0.039″
- Solder Mask Guard = 0.020″
- Isolation Track-to-Track = 0.025″
- Isolation Via-to-Via = 0.020″

Invoke PC board worksheet. The board file does not exist at this time, because modules have yet to be downloaded from the PCB library. To download modules, a board worksheet must first be invoked. Follow the procedure described in Section 9–1.3.

Figure 9–15b. Module Report for Logic Probe

Module Refer.	Module Value	Module File Name	Orient	Pin	X	Y	Net Name
C2	0.01UF	CK06	(270)	2	1.510	2.170	VSS
				1	1.510	2.370	N00006
C3	0.1UF	CK06	(270)	2	3.660	1.670	N00014
				1	3.660	1.870	N00011
C4	0.1UF	CK06	(90)	2	5.610	2.770	N00019
				1	5.610	2.570	N00017
C1	0.68UF	CK06	(270)	2	0.410	1.570	VSS
				1	0.410	1.770	N00002
R6	10K	RC07	(270)	1	4.010	2.020	N00010
				2	4.010	1.520	N00014
R9	2.2K	RC07	(90)	1	5.210	2.470	N00017
				2	5.210	2.970	N00015
R10	2.2K	RC07	(90)	1	3.660	2.220	N00011
				2	3.660	2.720	N00015
R3	2.2M	RC07	(270)	1	0.410	2.620	N00001
				2	0.410	2.120	N00002
R7	390K	RC07	(90)	1	4.560	2.920	N00018
				2	4.560	3.420	N00019
U1	4011	14DIP300	(90)	1	2.560	2.020	N00007
				2	2.560	2.120	N00007
				3	2.560	2.220	N00008
				4	2.560	2.320	N00007
				5	2.560	2.420	PROBE
				6	2.560	2.520	PROBE
				7	2.560	2.620	VSS
				8	2.860	2.620	N00003
				9	2.860	2.520	N00003
				10	2.860	2.420	N00004
				11	2.860	2.320	
				12	2.860	2.220	VSS
				13	2.860	2.120	VSS
				14	2.860	2.020	+9V
U3	4011	14DIP300	(90)	1	4.260	1.970	N00010
				2	4.260	2.070	N00010
				3	4.260	2.170	N00011
				4	4.260	2.270	N00010
				5	4.260	2.370	N00008
				6	4.260	2.470	N00012
				7	4.260	2.570	VSS
				8	4.560	2.570	N00007
				9	4.560	2.470	N00018
				10	4.560	2.370	N00016

9-6. The Projects

Figure 9-15b. Continued

Module Refer.	Module Value	Module File Name	Orient	Pin	X	Net Y	Name
				11	4.560	2.270	N00017
				12	4.560	2.170	N00016
				13	4.560	2.070	N00016
				14	4.560	1.970	+9V
R1	470	RC07	(270)	1	2.260	1.670	N00007
				2	2.260	1.170	N00013
R2	470	RC07	(270)	1	2.710	1.670	N00008
				2	2.710	1.170	N00009
R4	470	RC07	(0)	1	0.860	3.220	N00002
				2	1.360	3.220	N00005
C5	50UF	CK06	(90)	2	3.960	3.270	N00020
				1	3.960	3.070	N00015
R8	56K	RC07	(90)	1	4.910	2.920	N00016
				2	4.910	3.420	N00019
R5	68K	RC07	(270)	1	3.210	2.020	N00012
				2	3.210	1.520	N00014
D1	LED	DO7	(270)	2	5.510	1.620	VSS
				1	5.510	1.120	N00013
D2	LED	DO7	(270)	2	4.910	1.620	VSS
				1	4.910	1.120	N00009
D3	LED	DO7	(90)	2	1.510	2.570	VSS
				1	1.510	3.070	N00005
U2	LM555	8DIP300	(90)	1	0.910	2.170	VSS
				2	0.910	2.270	N00004
				3	0.910	2.370	N00002
				4	0.910	2.470	N00001
				5	1.210	2.470	N00006
				6	1.210	2.370	N00002
				7	1.210	2.270	N00002
				8	1.210	2.170	+9V
S1	SPDT	RCKRSPDT	(270)		XXXX	2.510	3.120
					XXXX	2.710	3.120
				1	2.510	3.220	VSS
				2	2.610	3.220	N00003
				3	2.710	3.220	PROBE
M1	+VE_SPKR	MHOLE2	(0)	XXXX	5.410	3.220	N00020
M2	-VE_SPKR	MHOLE2	(0)	XXXX	5.810	3.220	VSS
M3	GND	MHOLE1	(0)	XXXX	5.860	2.020	VSS
M4	+9V	MHOLE1	(0)	XXXX	5.410	2.020	+9V
M5	PROBE	MHOLE2	(0)	XXXX	1.960	3.220	PROBE

Set board layout environment. It is always a good idea to set the board environment before downloading the modules. You can change any or all of the settings at any time before or during routing. For the Logic Probe, use the following setting:

- Autopan → Y (Yes)
- Back up → Y (Yes)
- Color → keep the default values
- DRC (Design Rules Check) → Y (Yes)
- Error Bell → Y (Yes)
- Frame X-Y-L → Do not invoke
- Grid → Set at 0.050 inch
- Isolation → Keep all the default values
- Pad Filled → N (NO)
- Turn Board → N (NO)
- Style of Cursor → Short
- Via Type → Through
- Working Layer → 1
- X on Grid → Y (YES)

Create board edge. The edge of the board should be created before downloading the modules from the PCB library. You may change the board edge later. Follow the procedure described in Section 9–1.5 to create a 6″×3″ board for the Logic Probe.

Use netlist to download modules from the PCB library. Use the procedure described in Section 9–1.6 to download modules from the PCB library. You should have 22 modules for the Logic Probe project. The modules are sitting on one another, and you will not be able to count them without separating them. If any module is not found in the PCB library, the software will indicate that during the downloading process by saying *File Not Found*.

Load modules from the PCB library. For the Logic Probe project, five single pads are required on the board to make connections to off-board parts. They are two for the piezo buzzer, one for the probe, one for power, and one for ground connections. The off-board power and ground connections need larger pads compared to other off-board parts. Therefore, using the procedure for modifying module name, download MHOLE2 as SPEAKER and PROBE, and MHOLE1 as +9V and GND. Figure 9–16a shows the Logic Probe board just after downloading the modules from the PCB library.

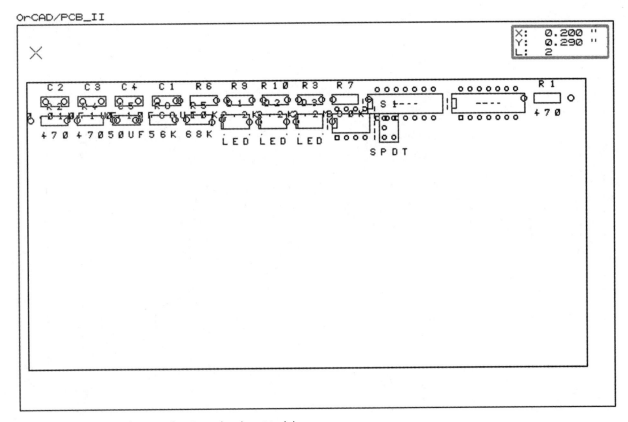

Figure 9–16a. Board View after Downloading Modules

Arrange modules on board. Essentially, module downloading must be complete and modules must be arranged on the board before routing can begin. Use the **Move** and **Get** commands to separate the modules from each other. Once they are separated, use **Compile Netlist** and display the Force Vectors and Ratsnest Vectors. Now move each module as closely as possible toward the diamond. The best approach is to place the module such that the diamond is within the boundary of the module. You must keep the board architecture in mind also. This is the hardest part of board design. Remember, this is not a perfect process and you as a designer still need to use common sense. Figure 9–16b is the Logic Probe board after the modules are arranged properly.

Place text in copper and special zones. If you want to add text and special zones, this is the time. I like to add text to Layer 1 to facilitate the fabrication process later on. Text and zones should be defined now so the Autorouter can route the board without using those places. Using the procedure described in Section 9–1.9, place text and zones as follows: Layer 1— Text in copper: Mirror image of "DELUXE LOGIC PROBE." Special zone: Add Forbidden Zone in all four corners of the board to hold mounting screws.

356 Chapter 9 / Computer-Aided Techniques for PCB Layout Design

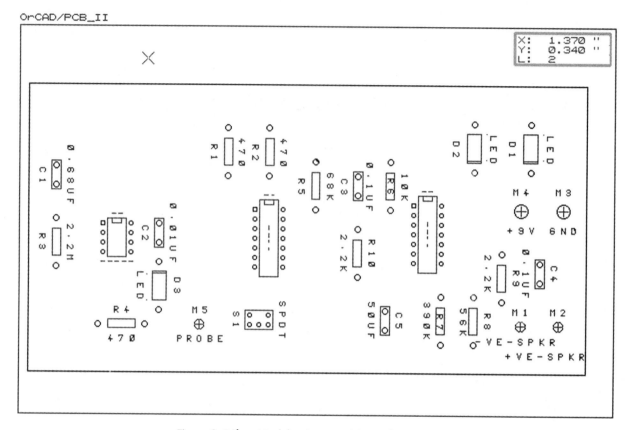

Figure 9–16b. Modules Arranged Properly

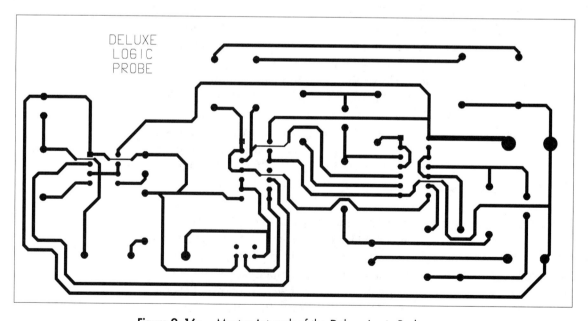

Figure 9–16c. Master Artwork of the Deluxe Logic Probe

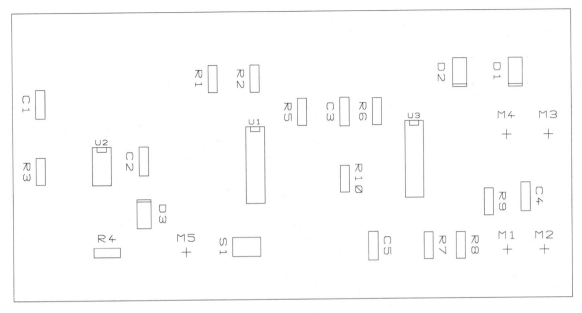

Figure 9–16d. Silkscreen Template of the Deluxe Logic Probe

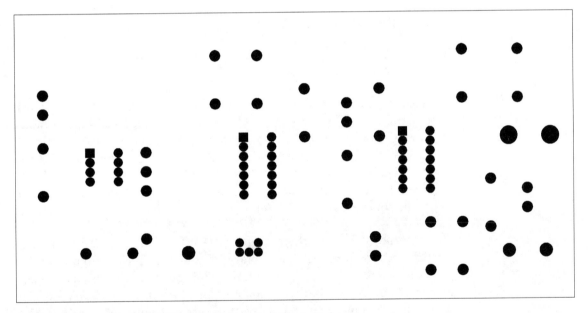

Figure 9–16e. Solder Mask Template of the Deluxe Logic Probe

Route board. Routing of the board can begin. Use the procedure described in Section 9–1.10. Remember that the MHOLEs for SPEAKER, PROBE, +9V, and GND must be edited to provide appropriate netnames before routing. Figure 9–16c shows the complete artwork for Layer 1 of the routed board. Figure 9–16d shows Silk Screen Template, Figure 9–16e Solder Mask, Figure 9–16f Drill Template, Figure 9–16g Module Report, and Figure 9–16h the complete board after routing.

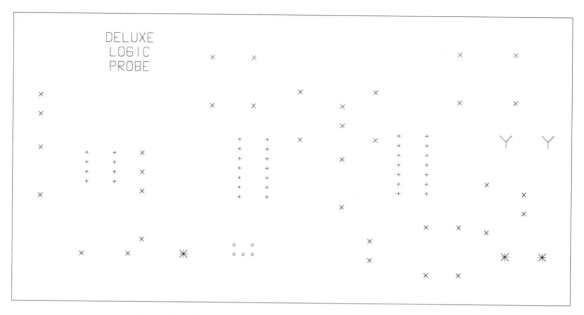

Figure 9–16f. Drill Template of the Deluxe Logic Probe

9–6.2 The Other Projects

Instead of going step by step through the other three projects in this book, I will provide the results of each step for the layout design of the board.

The electronic cricket. Figure 9–17a shows all modules just downloaded directly from the PCB library on a 4″×4″ board. Figure 9–17b shows modules arranged on the board by using optimized placement techniques. Figure 9–17c shows the complete artwork for layers 1 and 2 of the routed board. Figure 9–17d shows Silkscreen Template, Figure 9–17e Solder Mask, Figure 9–17f Drill Template, Figure 9–17g Module Report, and Figure 9–17h the complete board after routing.

The infrared object counter. Figure 9–18a shows all modules just downloaded directly from the PCB library on a 4″×4″ board. Figure 9–18b shows modules arranged on the board by using optimized placement techniques. Figure 9–18c shows the complete artwork for layers 1 and 2 of the routed board. Figure 9–18d shows Silkscreen Template, Figure 9–18e Solder Mask, Figure 9–18f Drill Template, Figure 9–18g Module Report, and Figure 9–18h the complete board after routing.

The mini stereo amplifier. Figure 9–19a shows all modules just downloaded directly from the PCB library on a 6″×3″ board. Figure 9–19b shows modules arranged on the board by using optimized placement techniques. Figure 9–19c shows the complete artwork for layers 1 and 2 of the routed board. Figure 9–19d shows Silkscreen Template, Figure 9–19e Solder Mask, Figure 9–19f Drill Template, Figure 9–19g Module Report, and Figure 9–19h the complete board after routing.

Figure 9-16g. Module Report of the Logic Probe

Module Refer.	Module Value	Module File Name	Orient	Pin	X	Net Y	Name
C2	0.01UF	CK06	(270)	2	1.510	2.170	VSS
				1	1.510	2.370	N00006
C3	0.1UF	CK06	(270)	2	3.660	1.670	N00014
				1	3.660	1.870	N00011
C4	0.1UF	CK06	(90)	2	5.610	2.770	N00019
				1	5.610	2.570	N00017
C1	0.68UF	CK06	(270)	2	0.410	1.570	VSS
				1	0.410	1.770	N00002
R6	10K	RC07	(270)	1	4.010	2.020	N00010
				2	4.010	1.520	N00014
R9	2.2K	RC07	(90)	1	5.210	2.470	N00017
				2	5.210	2.970	N00015
R10	2.2K	RC07	(90)	1	3.660	2.220	N00011
				2	3.660	2.720	N00015
R3	2.2M	RC07	(270)	1	0.410	2.620	N00001
				2	0.410	2.120	N00002
R7	390K	RC07	(90)	1	4.560	2.920	N00018
				2	4.560	3.420	N00019
U1	4011	14DIP300	(90)	1	2.560	2.020	N00007
				2	2.560	2.120	N00007
				3	2.560	2.220	N00008
				4	2.560	2.320	N00007
				5	2.560	2.420	PROBE
				6	2.560	2.520	PROBE
				7	2.560	2.620	VSS
				8	2.860	2.620	N00003
				9	2.860	2.520	N00003
				10	2.860	2.420	N00004
				11	2.860	2.320	
				12	2.860	2.220	VSS
				13	2.860	2.120	VSS
				14	2.860	2.020	+9V
U3	4011	14DIP300	(90)	1	4.260	1.970	N00010
				2	4.260	2.070	N00010
				3	4.260	2.170	N00011
				4	4.260	2.270	N00010
				5	4.260	2.370	N00008
				6	4.260	2.470	N00012
				7	4.260	2.570	VSS
				8	4.560	2.570	N00007
				9	4.560	2.470	N00018
				10	4.560	2.370	N00016

Figure 9–16g. Continued

Module Refer.	Module Value	Module File Name	Orient	Pin	X	Net Y	Name
				11	4.560	2.270	N00017
				12	4.560	2.170	N00016
				13	4.560	2.070	N00016
				14	4.560	1.970	+9V
R1	470	RC07	(270)	1	2.260	1.670	N00007
				2	2.260	1.170	N00013
R2	470	RC07	(270)	1	2.710	1.670	N00008
				2	2.710	1.170	N00009
R4	470	RC07	(0)	1	0.860	3.220	N00002
				2	1.360	3.220	N00005
C5	50UF	CK06	(90)	2	3.960	3.270	N00020
				1	3.960	3.070	N00015
R8	56K	RC07	(90)	1	4.910	2.920	N00016
				2	4.910	3.420	N00019
R5	68K	RC07	(270)	1	3.210	2.020	N00012
				2	3.210	1.520	N00014
D1	LED	DO7	(270)	2	5.510	1.620	VSS
				1	5.510	1.120	N00013
D2	LED	DO7	(270)	2	4.910	1.620	VSS
				1	4.910	1.120	N00009
D3	LED	DO7	(90)	2	1.510	2.570	VSS
				1	1.510	3.070	N00005
U2	LM555	8DIP300	(90)	1	0.910	2.170	VSS
				2	0.910	2.270	N00004
				3	0.910	2.370	N00002
				4	0.910	2.470	N00001
				5	1.210	2.470	N00006
				6	1.210	2.370	N00002
				7	1.210	2.270	N00002
				8	1.210	2.170	+9V
S1	SPDT	RCKRSPDT	(270)	XXXX	2.510	3.120	
				XXXX	2.710	3.120	
				1	2.510	3.220	VSS
				2	2.610	3.220	N00003
				3	2.710	3.220	PROBE
M1	+VE_SPKR	MHOLE2	(0)	XXXX	5.410	3.220	N00020
M2	-VE_SPKR	MHOLE2	(0)	XXXX	5.810	3.220	VSS
M3	GND	MHOLE1	(0)	XXXX	5.860	2.020	VSS
M4	+9V	MHOLE1	(0)	XXXX	5.410	2.020	+9V
M5	PROBE	MHOLE2	(0)	XXXX	1.960	3.220	PROBE

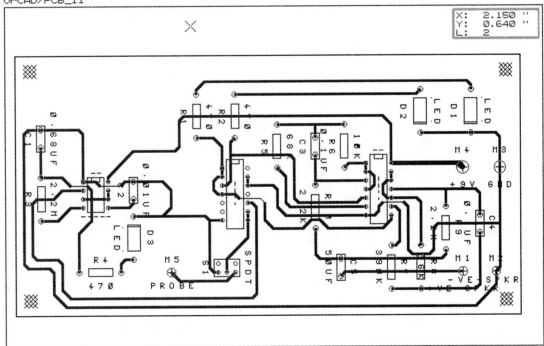

Figure 9-16h. Complete Board after Routing

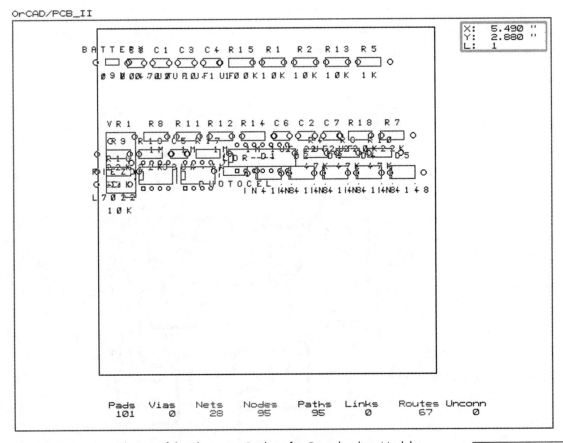

Figure 9-17a. Board View of the Electronic Cricket after Downloading Modules

362 Chapter 9 / Computer-Aided Techniques for PCB Layout Design

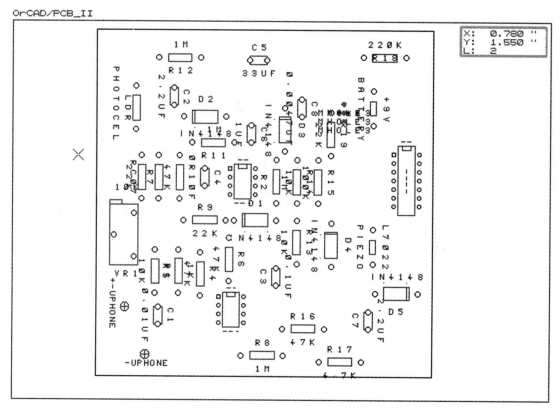

Figure 9-17b. Modules Arranged Properly

Figure 9-17c. Layer 1: Master Artwork of the Electronic Cricket

9-6. The Projects 363

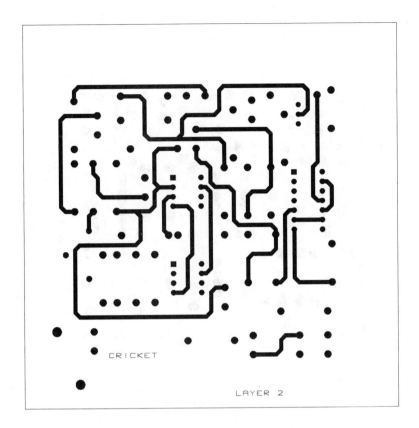

Figure 9-17c. Layer 1: Continued

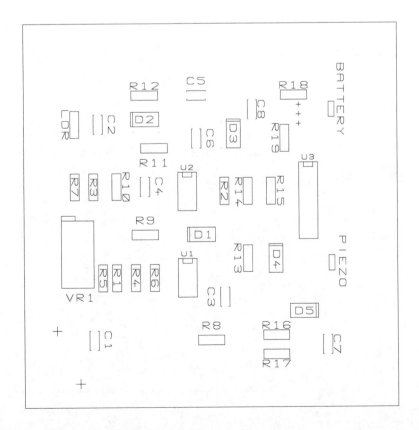

Figure 9-17d. Silkscreen Template of the Electronic Cricket

364 Chapter 9 / Computer-Aided Techniques for PCB Layout Design

Figure 9–17e. Solder Mask Template of the Electronic Cricket

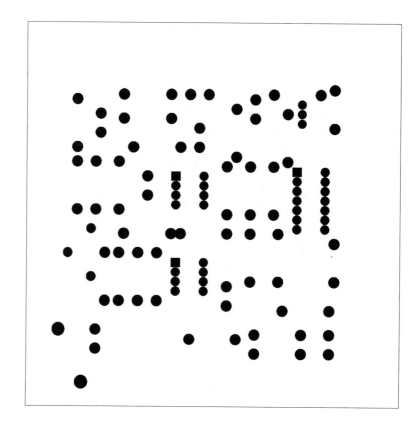

Figure 9–17f. Drill Template of the Electronic Cricket

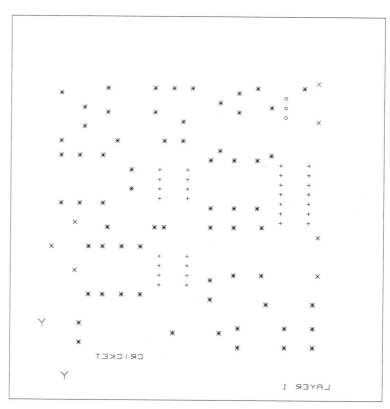

Figure 9–17g. Module Report of the Electronic Cricket

Module Refer.	Module Value	Module File Name	Orient.	Pin	X	Y	Net Name
BATTERY	+9V	RC05	(270)	1	4.300	1.200	SW+
				2	4.300	0.800	-_UPHONE
C8	0.0047UF	CK05	(90)	1	3.450	0.900	N00022
				2	3.450	1.100	N00023
C1	0.01UF	CK05	(90)	1	1.750	3.300	+_UPHONE
				2	1.750	3.500	N00003
C3	0.1UF	CK05	(270)	1	3.150	3.050	N00005
				2	3.150	2.850	N00012
C4	0.1UF	CK05	(90)	1	2.300	1.700	N00020
				2	2.300	1.900	-_UPHONE
R15	100K	RC07	(90)	1	3.650	1.600	N00007
				2	3.650	2.100	N00009
R1	10K	RC07	(90)	1	1.850	2.500	N00001
				2	1.850	3.000	+_UPHONE
R2	10K	RC07	(270)	1	3.150	2.100	N00001
				2	3.150	1.600	N00002
R13	10K	RC07	(90)	1	3.400	2.300	N00025
				2	3.400	2.800	N00008
R5	1K	RC07	(270)	1	2.000	3.000	N00003
				2	2.000	2.500	N00004
VR1	10K	RJ26X	(180)	1	1.700	2.750	N00001
				2	1.450	2.500	N00001
				3	1.700	2.250	N00016
R8	1M	RC07	(0)	1	2.750	3.400	N00004
				2	3.250	3.400	N00005
R11	1M	RC07	(180)	1	2.650	1.400	N00012
				2	2.150	1.400	N00017
R12	1M	RC07	(180)	1	2.550	0.850	N00011
				2	2.050	0.850	N00002
R14	1M	RC07	(90)	1	3.400	1.600	N00006
				2	3.400	2.100	N00007
C6	1UF	CK05	(90)	1	2.850	1.200	N00007
				2	2.850	1.400	N00008
C2	2.2UF	CK05	(90)	1	1.800	1.050	N00001
				2	1.800	1.250	-_UPHONE
C7	2.2UF	CK05	(270)	1	4.250	3.550	N00014
				2	4.250	3.350	N00015
R18	220K	RC07	(180)	1	4.150	0.850	N00021
				2	3.650	0.850	N00022
R7	22K	RC07	(90)	1	1.550	1.550	N00019
				2	1.550	2.050	N00001
R9	22K	RC07	(0)	1	2.050	2.300	N00019
				2	2.550	2.300	N00020
R19	22K	RC07	(90)	1	3.800	1.050	N00022
				2	3.800	1.550	N00024
C5	33UF	CK05	(0)	1	2.750	0.850	N00011
				2	2.950	0.850	-_UPHONE
R17	4.7K	RC07	(0)	1	3.450	3.550	N00014
				2	3.950	3.550	N00018
U3	4069	14DIP300	(90)	1	3.900	1.650	N00006
				2	3.900	1.750	N00009
				3	3.900	1.850	N00009
				4	3.900	1.950	N00008
				5	3.900	2.050	N00013

Figure 9-17g. Continued

Module Refer.	Module Value	Module File Name	Orient	Pin	X	Y	Net Name
				6	3.900	2.150	N00018
				7	3.900	2.250	-_UPHONE
				8	4.200	2.250	N00015
				9	4.200	2.150	N00018
				10	4.200	2.050	N00024
				11	4.200	1.950	N00021
				12	4.200	1.850	N00023
				13	4.200	1.750	N00024
				14	4.200	1.650	+9V
R4	47K	RC07	(90)	1	2.200	2.500	N00001
				2	2.200	3.000	N00004
R6	47K	RC07	(90)	1	2.400	2.500	N00001
				2	2.400	3.000	N00010
R10	47K	RC07	(90)	1	2.000	1.550	N00016
				2	2.000	2.050	N00012
R16	47K	RC07	(0)	1	3.450	3.350	N00013
				2	3.950	3.350	N00014
U1	741	8DIP300	(90)	1	2.600	2.600	
				2	2.600	2.700	N00004
				3	2.600	2.800	N00010
				4	2.600	2.900	-_UPHONE
				5	2.900	2.900	
				6	2.900	2.800	N00005
				7	2.900	2.700	+9V
				8	2.900	2.600	
U2	741	8DIP300	(90)	1	2.600	1.700	
				2	2.600	1.800	N00012
				3	2.600	1.900	N00020
				4	2.600	2.000	-_UPHONE
				5	2.900	2.000	
				6	2.900	1.900	N00017
				7	2.900	1.800	19V
				8	2.900	1.700	
LDR	PHOTOCEL	RC07	(90)	1	1.550	0.900	_UPHONE
				2	1.550	1.400	N00019
D1	IN4148	DO7	(0)	2	2.650	2.300	N00020
				1	3.150	2.300	N00025
D2	IN4148	DO7	(0)	2	2.050	1.100	N00017
				1	2.550	1.100	N00011
D3	IN4148	DO7	(90)	2	3.250	1.000	N00011
				1	3.250	1.500	N00006
D4	IN4148	DO7	(90)	2	3.700	2.300	N00008
				1	3.700	2.800	N00013
D5	IN4148	DO7	(180)	2	4.250	3.100	N00015
				1	3.750	3.100	N00021
PIEZO	L7022	RC05	(270)	1	4.300	2.800	-_UPHONE
				2	4.300	2.400	N00023
R***	RC07	RC07	(90)	1	1.750	1.550	-_UPHONE
				2	1.750	2.050	N00016
M***	MHOLE3	mhole3	(0)	XXXX	3.950	1.050	N00002
M***	MHOLE3	mhole3	(0)	XXXX	3.950	0.950	SW+
M***	MHOLE3	mhole3	(0)	XXXX	3.950	1.150	
M***	MHOLE2	mhole2	(0)	XXXX	1.350	3.300	+_UPHONE
M***	MHOLE2	mhole2	(0)	XXXX	1.600	3.850	-_UPHONE

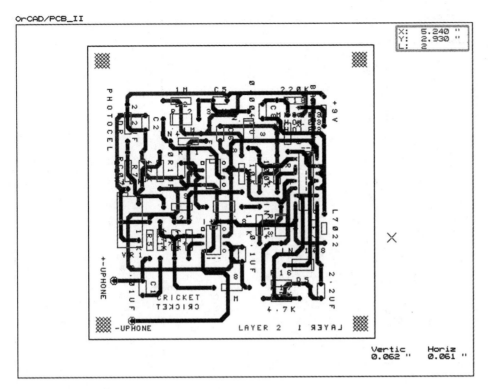

Figure 9–17h. Complete Board after Routing

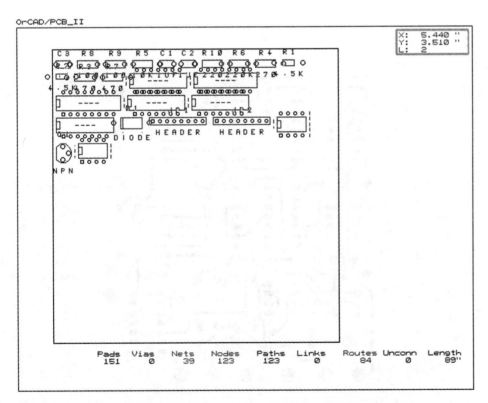

Figure 9–18a. Board View of the Infrared Object Counter after Downloading Modules

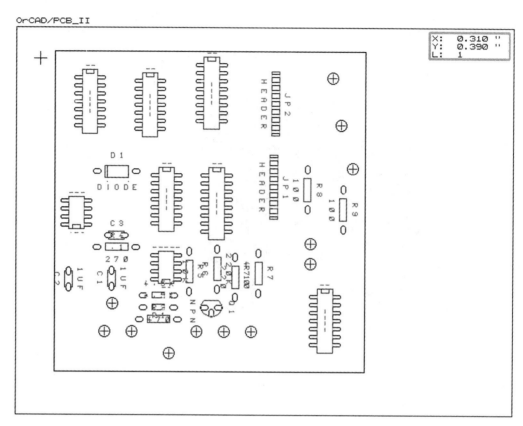

Figure 9–18b. Modules Arranged Properly

Figure 9–18c. Layer 1: Master Artwork of the Infrared Object Counter

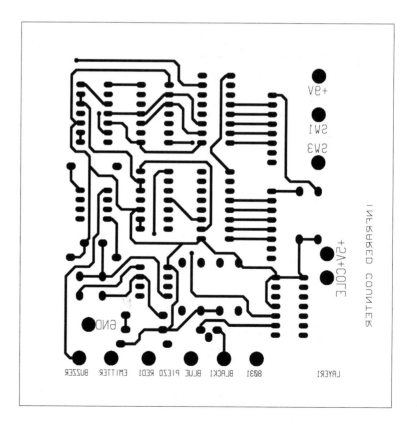

9-6. The Projects

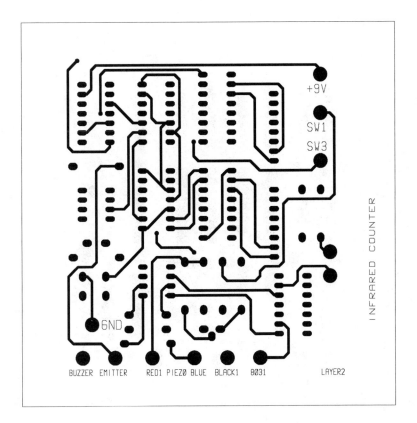

Figure 9–18c. Layer 2: Continued

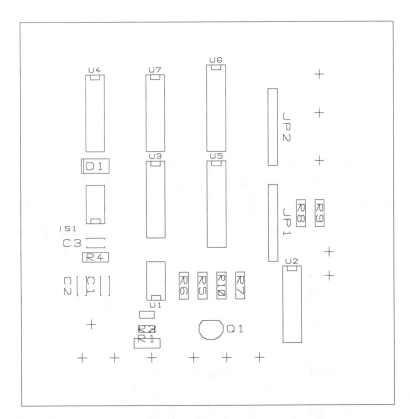

Figure 9–18d. Silkscreen Template of the Infrared Object Counter

370 Chapter 9 / Computer-Aided Techniques for PCB Layout Design

Figure 9–18e. Solder Mask Template of the Infrared Object Counter

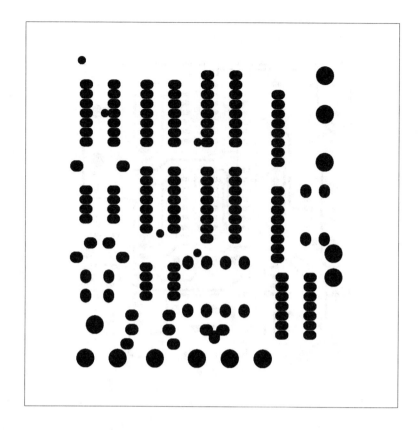

Figure 9–18f. Drill Template of the Infrared Object Counter

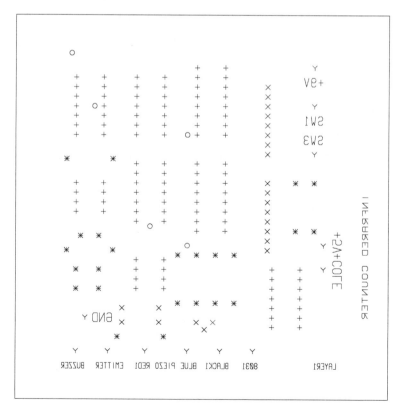

Figure 9-18g. Module Report of the Infrared Object Counter

Module Refer.	Module Value	Module File Name	Orient.	Pin	X	Y	Net Name
C3	.1	CK05	(0)	1	1.200	2.600	+5V
				2	1.400	2.600	GND
R8	100	RC07	(90)	1	3.500	2.050	N00002
				2	3.500	2.550	+5V
R9	100	RC07	(90)	1	3.700	2.050	N00021
				2	3.700	2.550	+5V
R5	10K	RC07	(90)	1	2.250	2.800	BLUE
				2	2.250	3.300	N00028
C1	1UF	CK05	(270)	1	1.400	3.150	N00008
				2	1.400	2.950	GND
C2	1UF	CK05	(270)	1	1.150	3.150	N00016
				2	1.150	2.950	GND
R10	220	RC07	(90)	1	2.650	2.800	SW3
				2	2.650	3.300	+5V
R6	220K	RC07	(270)	1	2.450	3.300	+5V
				2	2.450	2.800	BLUE
R4	270	RC07	(0)	1	1.050	2.750	+5V
				2	1.550	2.750	N00032
R1	4.5K	RC06	(180)	2	1.650	3.500	N00001
				1	2.050	3.500	+5V
R2	4.5K	RC06	(180)	2	1.650	3.350	N00001
				1	2.050	3.350	N00008
R3	470	RC07	(0)	1	1.600	3.650	+5V
				2	2.100	3.650	+ANOD
R7	470	RC07	(90)	1	2.850	2.800	+5V
				2	2.850	3.300	RED1
U7	7420	14DIP300	(90)	1	1.800	0.950	N00012
				2	1.800	1.050	N00004
				3	1.800	1.150	
				4	1.800	1.250	N00019
				5	1.800	1.350	N00020
				6	1.800	1.450	N00018
				7	1.800	1.550	GND
				8	2.100	1.550	
				9	2.100	1.450	N00012
				10	2.100	1.350	N00004
				11	2.100	1.250	
				12	2.100	1.150	N00020
				13	2.100	1.050	N00019
				14	2.100	0.950	+5V
U5	7447	16DIP300	(90)	1	2.450	1.850	N00006
				2	2.450	1.950	N00010
				3	2.450	2.050	
				4	2.450	2.150	
				5	2.450	2.250	
				6	2.450	2.350	N00012
				7	2.450	2.450	N00004
				8	2.450	2.550	+5V
				9	2.750	2.550	N00014
				10	2.750	2.450	N00013
				11	2.750	2.350	N00011
				12	2.750	2.250	N00007
				13	2.750	2.150	N00005
				14	2.750	2.050	N00017
				15	2.750	1.950	N00015

Figure 9-18g. Continued

Module Refer.	Module Value	Module File Name	Orient.	Pin	X	Y	Net Name
				16	2.750	1.850	+5V
U6	7447	16DIP300	(90)	1	2.450	0.850	N00023
				2	2.450	0.950	N00025
				3	2.450	1.050	
				4	2.450	1.150	
				5	2.450	1.250	
				6	2.450	1.350	N00019
				7	2.450	1.450	N00020
				8	2.450	1.550	GND
				9	2.750	1.550	N00029
				10	2.750	1.450	N00027
				11	2.750	1.350	N00026
				12	2.750	1.250	N00024
				13	2.750	1.150	N00022
				14	2.750	1.050	N00031
				15	2.750	0.950	N00030
				16	2.750	0.850	+5V
U2	7474	14DIP300	(90)	1	3.250	2.950	+5V
				2	3.250	3.050	COLE
				3	3.250	3.150	N00009
				4	3.250	3.250	+5V
				5	3.250	3.350	N00003
				6	3.250	3.450	BLACK1
				7	3.250	3.550	GND
				8	3.550	3.550	
				9	3.550	3.450	
				10	3.550	3.350	
				11	3.550	3.250	
				12	3.550	3.150	
				13	3.550	3.050	
				14	3.550	2.950	+5V
U3	7490	14DIP300	(90)	1	1.800	1.850	N00004
				2	1.800	1.950	SW1
				3	1.800	2.050	SW1
				4	1.800	2.150	
				5	1.800	2.250	+5V
				6	1.800	2.350	GND
				7	1.800	2.450	GND
				8	2.100	2.450	N00010
				9	2.100	2.350	N00006
				10	2.100	2.250	GND
				11	2.100	2.150	N00012
				12	2.100	2.050	N00004
				13	2.100	1.950	
				14	2.100	1.850	N00003
U4	7490	14DIP300	(90)	1	1.150	0.950	N00020
				2	1.150	1.050	SW1
				3	1.150	1.150	SW1
				4	1.150	1.250	
				5	1.150	1.350	+5V
				6	1.150	1.450	GND
				7	1.150	1.550	GND
				8	1.450	1.550	N00025
				9	1.450	1.450	N00023

9–6. The Projects

Figure 9–18g. Continued

Module Refer.	Module Value	Module File Name	Orient.	Pin	X	Y	Net Name
				10	1.450	1.350	GND
				11	1.450	1.250	N00019
				12	1.450	1.150	N00020
				13	1.450	1.050	
				14	1.450	0.950	N00012
D1	DIODE	DO7	(0)	2	1.050	1.800	N00033
				1	1.550	1.800	+9V
JP1	HEADER	8SIP100	(90)	1	3.200	2.050	N00002
				2	3.200	2.150	N00005
				3	3.200	2.250	N00007
				4	3.200	2.350	N00011
				5	3.200	2.450	N00013
				6	3.200	2.550	N00014
				7	3.200	2.650	N00015
				8	3.200	2.750	N00017
JP2	HEADER	8SIP100	(90)	1	3.200	1.050	N00021
				2	3.200	1.150	N00022
				3	3.200	1.250	N00024
				4	3.200	1.350	N00026
				5	3.200	1.450	N00027
				6	3.200	1.550	N00029
				7	3.200	1.650	N00030
				8	3.200	1.750	N00031
U1	LM555	8DIP300	(270)	1	2.100	3.150	GND
				2	2.100	3.050	N00008
				3	2.100	2.950	N00009
				4	2.100	2.850	+5V
				5	1.800	2.850	N00016
				6	1.800	2.950	N00008
				7	1.800	3.050	N00001
				8	1.800	3.150	+5V
Q1	NPN	TO92	(90)	1	2.450	3.500	BLACK1
				2	2.535	3.575	N00028
				3	2.625	3.500	RED1
IS1	OPT	8DIP300	(270)	1	1.450	2.350	N00032
				2	1.450	2.250	N00018
				3	1.450	2.150	
				4	1.450	2.050	RED
				5	1.150	2.050	N00033
				6	1.150	2.150	
				7	1.150	2.250	
				8	1.150	2.350	
M***	MHOLE1	MHOLE1	(0)	XXXX	3.800	2.700	+5V
M***	MHOLE1	MHOLE1	(0)	XXXX	3.800	2.950	+COLE
M***	MHOLE1	MHOLE1	(0)	XXXX	1.250	3.450	GND
M***	MHOLE1	MHOLE1	(0)	XXXX	3.050	3.800	BLACK1
M***	MHOLE1	MHOLE1	(0)	XXXX	3.700	0.850	SW1
M***	MHOLE1	MHOLE1	(0)	XXXX	2.700	3.800	RED1
M***	MHOLE1	MHOLE1	(0)	XXXX	1.900	3.800	BLUE
M***	MHOLE1	MHOLE1	(0)	XXXX	1.150	3.800	RED
M***	MHOLE1	MHOLE1	(0)	XXXX	2.350	3.800	+ANOD
M***	MHOLE1	MHOLE1	(0)	XXXX	3.700	1.250	SW3
M***	MHOLE1	MHOLE1	(0)	XXXX	1.500	3.800	+9V
M***	MHOLE1	MHOLE1	(0)	XXXX	3.700	1.750	

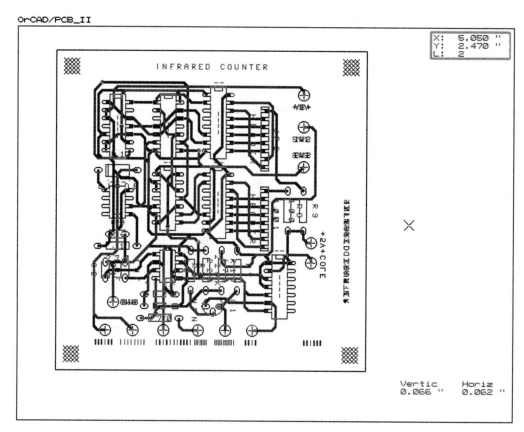

Figure 9–18h Complete Board after Routing

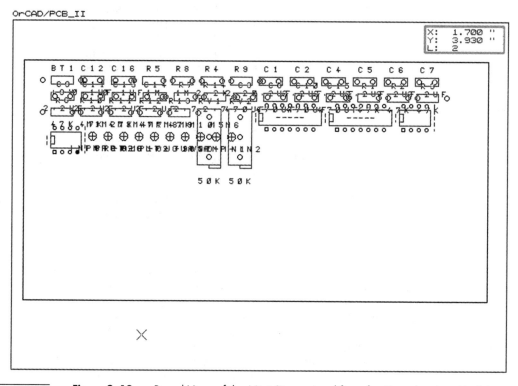

Figure 9–19a. Board View of the Mini Stereo Amplifier after Downloading Modules

9–6. The Projects

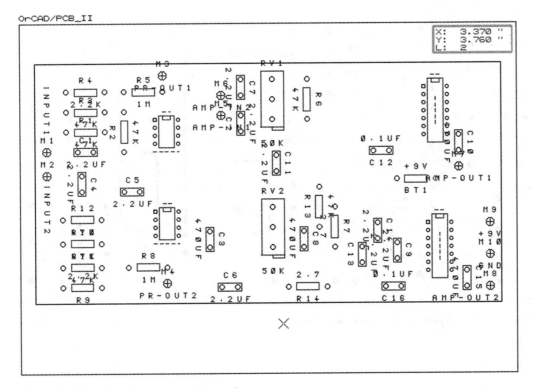

Figure 9–19b. Modules Arranged Properly

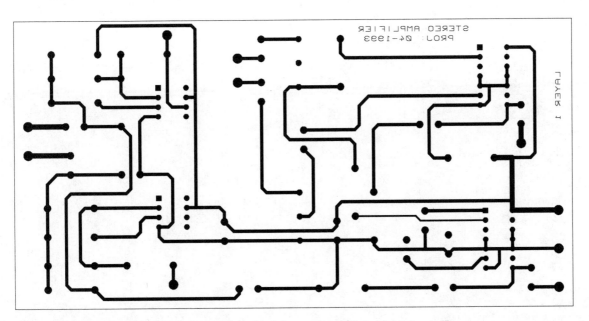

Figure 9–19c. Layer 1: Master Artwork of the Mini Stereo Amplifier

Chapter 9 / Computer-Aided Techniques for PCB Layout Design

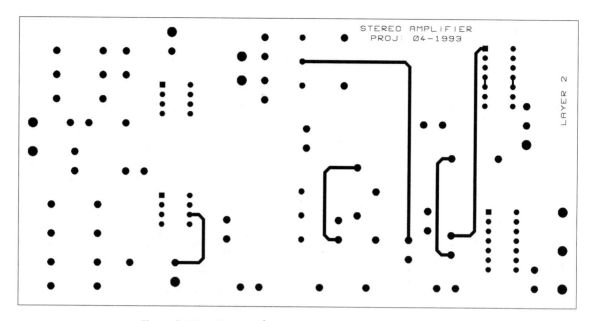

Figure 9–19c. Continued Layer 2

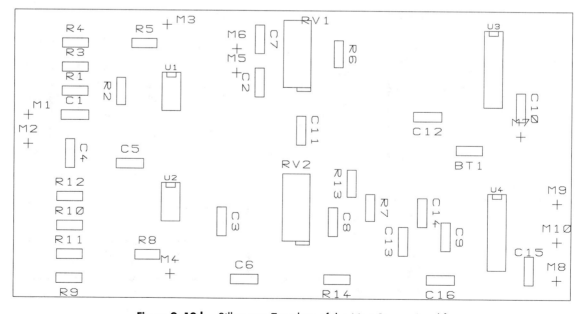

Figure 9–19d. Silkscreen Template of the Mini Stereo Amplifier

9–6. The Projects

Figure 9–19e. Solder Mask Template of the Mini Stereo Amplifier

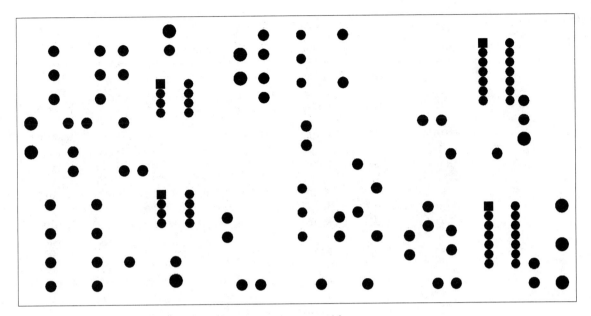

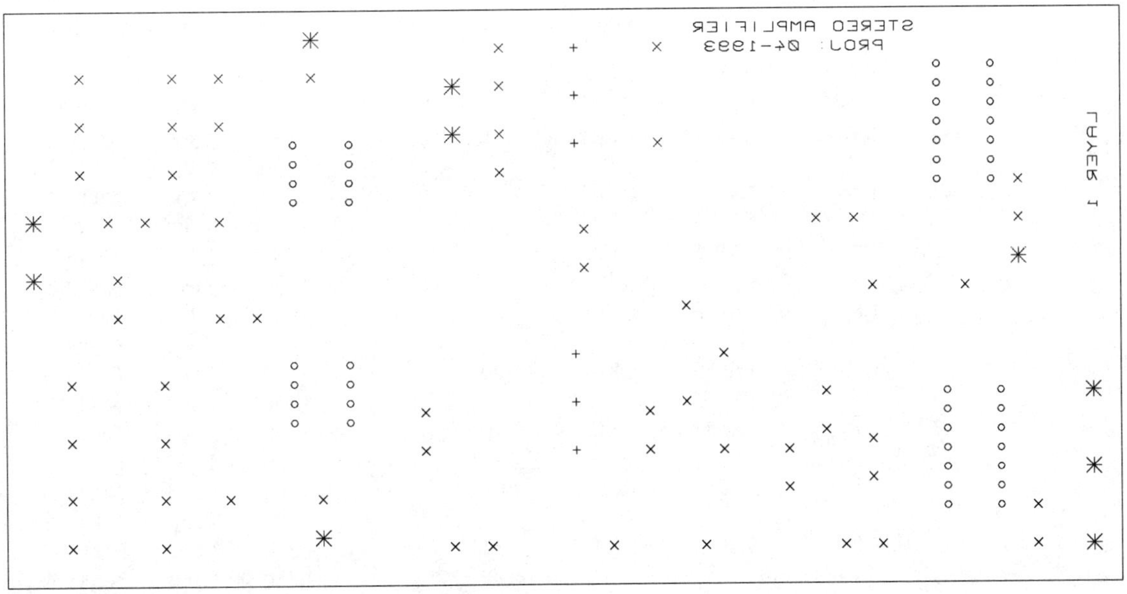

Figure 9–19f. Drill Template of the Mini Stereo Amplifier

Figure 9-19g. Module Report of the Mini Stereo Amplifier

Module Refer.	Module Value	Module File Name	Orient.	Pin	X	Y	Net Name
BT1	+9V	RC07	(180)	1	5.350	1.950	N00002
				2	4.850	1.950	VSS
C12	0.1UF	CK06	(180)	2	4.550	1.600	N00010
				1	4.750	1.600	N00007
C16	0.1UF	CK06	(180)	2	4.700	3.300	N00020
				1	4.900	3.300	N00017
R5	1M	RC07	(0)	1	1.350	0.850	N00005
				2	1.850	0.850	PR_OUT_1
R8	1M	RC07	(0)	1	1.400	3.050	N00015
				2	1.900	3.050	PR_OUT_2
R4	2.2K	RC07	(0)	1	0.600	0.850	N00001
				2	1.100	0.850	N00002
R9	2.2K	RC07	(180)	1	1.050	3.300	N00001
				2	0.550	3.300	N00012
C1	2.2UF	CK06	(0)	2	0.950	1.600	N00001
				1	0.750	1.600	PR_IN_1
C2	2.2UF	CK06	(270)	2	2.850	1.150	IN_1
				1	2.850	1.350	N00003
C4	2.2UF	CK06	(90)	2	0.800	2.100	N00012
				1	0.800	1.900	PR_IN_2
C5	2.2UF	CK06	(0)	2	1.550	2.100	VSS
				1	1.350	2.100	N00012
C6	2.2UF	CK06	(0)	2	2.800	3.300	VSS
				1	2.600	3.300	N00001
C7	2.2UF	CK06	(90)	2	2.850	0.900	IN_2
				1	2.850	0.700	N00013
C9	2.2UF	CK06	(90)	2	4.850	2.950	VSS
				1	4.850	2.750	N00011
C11	2.2UF	CK06	(90)	2	3.300	1.850	N00008
				1	3.300	1.650	N00009
C13	2.2UF	CK06	(270)	2	4.400	2.800	N00018
				1	4.400	3.000	N00019
C14	2.2UF	CK06	(90)	2	4.600	2.700	VSS
				1	4.600	2.500	N00021
R7	2.7	RC07	(90)	1	4.050	2.300	N00010
				2	4.050	2.800	VSS
R14	2.7	RC07	(180)	1	3.950	3.300	N00020
				2	3.450	3.300	VSS
C3	470UF	CK06	(90)	2	2.450	2.800	VSS
				1	2.450	2.600	N00002
C8	470UF	CK06	(90)	2	3.650	2.800	VSS
				1	3.650	2.600	N00002
C10	470UF	CK06	(90)	2	5.650	1.600	OUT_1
				1	5.650	1.400	N00007
C15	470UF	CK06	(90)	2	5.750	3.300	OUT_2
				1	5.750	3.100	N00017
R1	47K	RC07	(0)	1	0.600	1.350	N00001
				2	1.100	1.350	N00004
R2	47K	RC07	(270)	1	1.350	1.600	N00001
				2	1.350	1.100	N00005
R3	47K	RC07	(0)	1	0.600	1.100	N00001

9-6. The Projects

Figure 9-19g. Continued

Module Refer.	Module Value	Module File Name	Orient.	Pin	X	Y	Net Name
				2	1.100	1.100	N00005
R6	47K	RC07	(90)	1	3.700	0.700	N00006
				2	3.700	1.200	VSS
R10	47K	RC07	(0)	1	0.550	2.750	N00012
				2	1.050	2.750	N00014
R11	47K	RC07	(0)	1	0.550	3.050	N00012
				2	1.050	3.050	N00015
R12	47K	RC07	(0)	1	0.550	2.450	N00012
				2	1.050	2.450	N00015
R13	47K	RC07	(270)	1	3.850	2.550	N00016
				2	3.850	2.050	VSS
RV1	50K	RJ24X	(0)	1	3.250	0.700	N00013
				2	3.250	0.950	N00018
				3	3.250	1.200	VSS
RV2	50K	RJ24X	(0)	1	3.250	2.300	N00003
				2	3.250	2.550	N00008
				3	3.250	2.800	VSS
U3	LM384	14DIP300	(90)	1	5.200	0.800	N00011
				2	5.200	0.900	N00006
				3	5.200	1.000	
				4	5.200	1.100	VSS
				5	5.200	1.200	VSS
				6	5.200	1.300	N00009
				7	5.200	1.400	VSS
				8	5.500	1.400	N00007
				9	5.500	1.300	
				10	5.500	1.200	VSS
				11	5.500	1.100	VSS
				12	5.500	1.000	VSS
				13	5.500	0.900	
				14	5.500	0.800	N00002
U4	LM384	14DIP300	(90)	1	5.250	2.500	N00021
				2	5.250	2.600	N00016
				3	5.250	2.700	
				4	5.250	2.800	VSS
				5	5.250	2.900	VSS
				6	5.250	3.000	N00019
				7	5.250	3.100	VSS
				8	5.550	3.100	N00017
				9	5.550	3.000	
				10	5.550	2.900	VSS
				11	5.550	2.800	VSS
				12	5.550	2.700	VSS
				13	5.550	2.600	
				14	5.550	2.500	N00002
U1	LM741	8DIP300	(90)	1	1.750	1.200	
				2	1.750	1.300	N00005
				3	1.750	1.400	N00004
				4	1.750	1.500	VSS
				5	2.050	1.500	
				6	2.050	1.400	PR_OUT_1

Figure 9-19g. Continued

Module Refer.	Module Value	Module File Name	Orient.	Pin	X	Y	Net Name
				7	2.050	1.300	N00002
				8	2.050	1.200	
U2	LM741	8DIP300	(90)	1	1.750	2.350	
				2	1.750	2.450	N00015
				3	1.750	2.550	N00014
				4	1.750	2.650	VSS
				5	2.050	2.650	
				6	2.050	2.550	PR_OUT_2
				7	2.050	2.450	N00002
				8	2.050	2.350	
M1	INPUT1	mhole2	(0)	XXXX	0.350	1.600	PR_IN_1
M2	INPUT2	mhole2	(0)	XXXX	0.350	1.900	PR_IN_2
M3	PR_OUT1	mhole2	(0)	XXXX	1.850	0.650	PR_OUT_1
M4	PR_OUT2	mhole2	(0)	XXXX	1.900	3.250	PR_OUT_2
M7	AMP_OUT1	mhole2	(0)	XXXX	5.650	1.800	OUT_1
M8	AMP_OUT2	mhole2	(0)	XXXX	6.050	3.300	OUT_2
M9	+9V	mhole2	(0)	XXXX	6.050	2.500	N00002
M10	GND	mhole2	(0)	XXXX	6.050	2.900	VSS
M5	AMP_IN1	mhole2	(0)	XXXX	2.600	1.150	IN_1
M6	AMP_IN2	mhole2	(0)	XXXX	2.600	0.900	IN_2

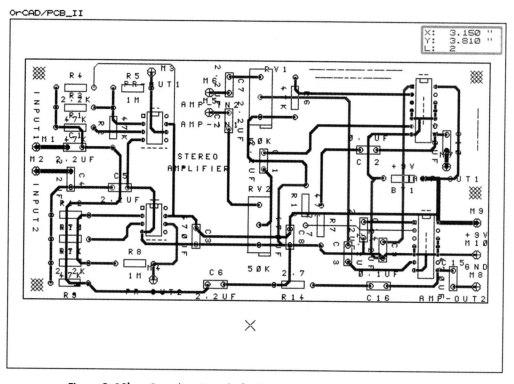

Figure 9-19h. Complete Board after Routing

QUESTIONS

1. Where in the PC Board Layout Tools should you go to define the number of layers for your board?

2. If you want your board to follow certain rules (for example, track-to-track, via-to-via isolation, minimum track width, pad drill size), where would you define them?

3. In addition to the general rules, if you want certain nets to have special characteristics, where would you define them?

4. If you have started routing a board and notice that you need a larger board, can you change the board size? If no, what should you do? If yes, how would you do it?

5. During the unloading of a module from PCB library by using netlist, a prompt appears on the screen: *File Not Found*. What should you check first to find the problem?

6. Write three factors that you should normally consider in order to arrange the module on a board.

7. Write the two different commands needed to move a module from one part of the board to another part.

8. Why is it sometimes necessary to define a special zone or zones on a board?

9. What kind of a zone would you define if a small output transformer needed to be mounted on the board?

10. Which type of board is comparatively difficult to route, single or double sided?

11. Name the two types of vectors that are used to position the modules on a board.

12. Write the command that you could use to find the width of a track, netname of a net, or reference designator of a module.

13. What is the purpose of putting mirrored text on Layer 1 and regular text on Layer 2?

14. What are the five fundamental variables that affect routing?

15. What is solder mask, and why is it used?

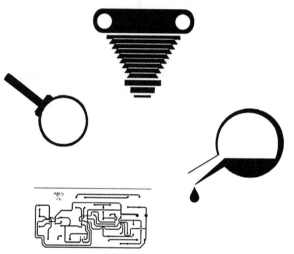
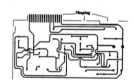

PCB Fabrication

10

THE PRINTED CIRCUIT BOARD FABRICATION PROCESS

10-1. PRINTED CIRCUIT BOARD FABRICATION IN A SCHOOL LABORATORY

Fabricating commercial-grade printed circuit board in large quantities involves about a hundred operating steps. The sequence of these steps and their process chemistry are generally different from one manufacturer to another due to their proprietary nature. These sequences also depend on the wide variety of equipment, material, advantages, manufacturing volume, and preferences of individual manufacturers. Many of these steps are generally not followed in a school laboratory process—for example, board drilling using a CNC machine or the electroless copper deposition process for Plated Through-Hole (PTH). These processes are important to know but quite expensive to perform in a laboratory setting. They are not cost effective for fabricating a small number of boards. Therefore, only a short description of each of those processes not generally performed in fabricating a PC board in a school laboratory is included in this chapter. The concern that first arises in a school laboratory is the *safety* of an individual working in the PCB laboratory. Therefore, safety is described first.

10-2. COMMON SAFETY PRACTICES IN A SCHOOL LABORATORY

Safety is everyone's responsibility. The person who is doing the job must take all safety precautions. You must take all possible safety measures to safeguard yourself from any physical damage, because it is your life. However, educating you about various safety practices is the responsibility of your supervisor. Your supervisor must inform you about all the possible dangers in connection with your particular assignment. He or she

Objectives

After completing this chapter, you should be able to

1. Identify the common safety practices in a school laboratory
2. Describe the safety precautions required for chemicals, equipment, electricity, disposal, and storage
3. Successfully take the safety test
4. Describe the major steps involved in fabricating PCBs in industry and in the laboratory
5. Fabricate PC boards using a laboratory-based chemical process
6. Identify defects on a PC board, repair them, and verify their continuity or discontinuity

must provide you this safety education for both legal and moral reasons. He must obtain the Material Safety Data sheet from the manufacturer of etchants, photodeveloping chemicals, tinning chemicals, and any other chemical used in the PCB fabrication process. The manufacturer must provide this information, if it is requested. It is highly recommended that the person in charge create the following types of labels by using the data provided in the Material Safety Data sheet and post them where the chemicals are used. He or she should also consult OSHA guidelines for laboratory standards and chemical hygiene.

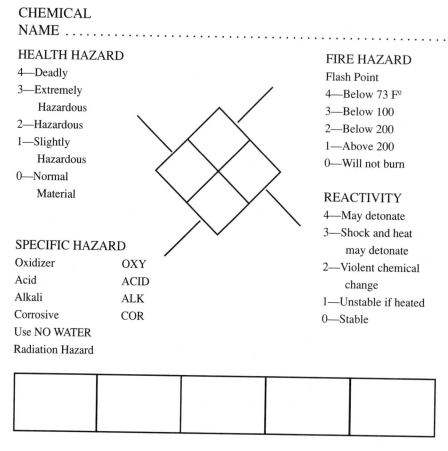

In a school laboratory environment, the course instructor must explain to you about the potential danger from chemicals, electricity, moving machinery, and hand tools, and demonstrate how various safe practices can save you from accidents. Once the education is given to you, it becomes your responsibility to practice it and keep yourself safe. Your supervisor can caution you from time to time, but he or she can hardly do anything if you do not practice safety. In a school PCB laboratory you must not work alone. You must have a partner who should be watching you from a safe distance while you are working. In the case of an accident, both of you will not be affected, and your partner can call for help. While you are working on equipment, it is your responsibility to warn other people around you about the possibility of an accident and any safety precautions they need to take. Always look after your safety and the safety of the people around you. However, do not lose your confidence because of these dangers. You have to develop good habits to keep yourself safe. Developing the knowledge base for safe practices is helpful for when you will be working in an industrial setting.

10-2.1 Chemicals

Photo developing. Various chemicals associated with developing artwork negatives, developing photosensitized PC boards, and photoresist remover are potential sources of slow poison. They may not burn your skin like acids and alkali, but they can cause serious damage to you in the long run. Some photo developing chemicals emit toxic fumes and must not be inhaled. As a safety measure, use rubber chemical safety gloves while handling these chemicals. If you use a photo chemical that emits toxic fumes, wear a mask that is made for toxic fumes and work under a fumehood. If a fumehood is not available, a room with forced ventilation will suffice.

Etchant. There are various kinds of etchant used for PCB etching. Some of them are less dangerous than others. Some emit poisonous fumes; some don't. You must be aware of the danger level. The danger level and the exact chemical composition of any chemical can be obtained from the manufacturer. Many chemicals are sold by their trade name; therefore, the chemical name must be obtained. A sticker describing the danger level of the etchant and its chemical composition must be posted on the etcher by the course instructor. No matter how mild the etchant is, it can cause serious damage to your eyes. Since the etchant could contain acids or alkali of various concentration levels, you must guard your eyes by wearing safety glasses. Also, protect your hands by wearing rubber gloves, and your apparel by wearing a protective apron. If etchant used in the laboratory emits toxic fumes, the etcher must be kept and operated under a fumehood or in a room with forced ventilation.

Tinning solution. Various chemicals used for tin plating could be a potential source of slow poison. Some immersion tin salts are suspected to be carcinogenic. They can cause skin, eyes, and mucous membrane irritation. These salts are acidic and, therefore, need alkaline for neutralization. As a safety measure, use rubber chemical safety gloves and an apron while handling these chemicals. Store them in a dry area and keep them away from strong oxidizers. If you use them in a poorly ventilated room, wear a mask that is made for toxic fumes.

10–2.2 Equipment

UV exposure. Although most commercial PC board exposure systems are closed and safe, you should be aware of the danger involving ultraviolet (UV) light. UV light can cause cataracts and other serious damage to your eyes if you stare at it. It can also cause skin cancer. Your school laboratory must buy a safe UV exposure system. A simpler one is a closable box with UV light bulbs. The box must remain closed when the UV lamps are on. You must not look at them when they are on.

Etchers. Most desk-top etchers in school laboratories are the sprayer type. Therefore, while working with these kinds of etchers, you must wear protective eyeglasses, chemically safe rubber gloves, and an apron to protect your eyes, skin, and clothes. All spray etchers must have a limit switch that will stop the etcher operation as soon as the top cover is opened. That limit switch goes inoperative very quickly, so don't depend on it. If you want to open the top cover to inspect the condition of your board, you must stop the etcher. If you are using the etcher, then you must advise people coming near it to wear eye protection. Any malfunction of the etcher that you notice should be reported to whoever is in charge.

Moving machinery. In a school PCB laboratory the most common moving machine is a high-speed drill press or a computer numerical control (CNC) drill machine. To drill a PC board, carbide drill bits are generally used. These types of drill bits are very brittle, and when they break while drilling, the broken parts become like splinters. To safeguard your eyes from these splinters, you must wear protective eyeglasses at all times. The other common danger regarding moving machinery is that the loose parts of your clothing may get caught. Things that can cause serious danger are neckties, long sleeves, bracelets, and necklaces. Any loose clothing that has a chance of coming close to the moving parts of the drill press can cause a problem. Remember to clamp the piece you are drilling to the base of the drill machine. This may prevent the drill bit from catching the piece and causing a situation where a hand or finger could be cut.

Hand tools. For all sharp and pointed tools such as razor blades, punches, and chisels, you need a hard-cover pouch. You should never carry these tools in your pocket unless they are covered. You must immediately deburr all new edges that develop after cutting, drilling, punching, or shearing metal. You must not cut anything with a sharp tool while resting the object on your palm. Take time to clamp the object on the work table. Soldering irons and hot molten solder are also dangerous.

You must wear protective eyeglasses while soldering. You must not flick solder from the soldering iron, or it can rebound into your eyes or to those of people around you. Always use a safe stand for a soldering iron. If you leave a soldering iron on a wooden table, it will burn it or may even cause a fire.

Shear tools. In the PCB laboratory, you may need to cut the PC board or a piece of metal with a shear tool. Protect your finger from the cutting edge. Generally, use both of your hands to operate the shear tool.

10-2.3 Electricity

There are many kinds of electrical hazard that can take place in the PCB laboratory. If you are not careful, even a soldering iron can be the reason for a severe electrical shock. Remember that 110 volt AC is enough to kill you. If the rubber insulation of the power cord of the soldering iron is defective (the rubber insulation has melted even in one place due to mishandling), you must put several turns of insulated tape around the defective point. Never neglect it, because these small things can be the cause of a serious electrical shock. You must know the proper emergency procedure to treat shock victims. Ask your instructor to provide you with more information about the causes and prevention of electric shock, along with emergency treatment procedures.

10-2.4 Disposal and Storage

Etchant. Disposing of etchant is a problem. You must not dispose etchant into a drain. Not only is etchant an environmental hazard; if you have a copper sewer, it can eat through the pipe. The proper way of disposing of etchant is to take it to the original vendor for recycling or to a hazardous chemical disposal agency. Your university will have a procedure for handling this disposal. Contact your safety department personnel. PCB fabrication industries follow government standards of treating etchant before disposing of it.

There are etchants available that are water soluble. These are a little easier to deal with than ones that are not water soluble. But in both cases they needed to be treated before disposal. In a school laboratory, water-soluble etchant is more desirable than the other type.

Photochemicals. Photochemicals should also not be disposed of into a drain. If they are improperly disposed of, they can cause an environmental problem. You must ask the original vendor of your photochemical about the disposal procedure and confirm this through the OSHA regulations booklet.

10.3. TYPICAL FORMAT OF A SAFETY TEST

Name of the University
Name of the Department
Course Number, Course Title, Semester, Year
SAFETY TEST

T F 1. Excel liquid developer and stripper can cause skin irritation after repeated contact.
T F 2. Large quantities of Excel board developing chemicals can be disposed of into the sewer drain.
T F 3. Large quantities of ammonium persulphate chemical waste from etcher can be disposed of into the sewer drain.
T F 4. Forced ventilation is required in the PCB photographic developing area for the Excel process.
T F 5. Eye protection should be worn when handling Excel developer and etchant.
T F 6. Eye protection must be worn when handling acids and alkalis.
T F 7. Always pour acid into water when diluting the acid.
T F 8. Never eat food in rooms where chemicals are used or mixed.
T F 9. Eye protection is not required in the etcher room when the etcher is in operation.
T F 10. Eye protection is not required when soldering components to the PCBs.
T F 11. Permission from the laboratory instructor is required before the spray etcher can be used.
T F 12. The laboratory instructor need not be in the PCB laboratory when the etcher is in operation and photodeveloping chemicals are being used.
T F 13. An apron and an eye protecting glass should be used when handling or mixing chemicals in the dark room.
T F 14. If acid gets into a student's eyes, he or she should rinse them out, immediately, with tap water.
T F 15. The instructor should be notified, immediately, of any accidents or injuries.
T F 16. It is perfectly safe to work alone in the PCB laboratory.
T F 17. The instructor should be notified, immediately, of any etcher malfunction or darkroom accident (spilled acid, etc.).
T F 18. The small carbide drills used to drill PCBs are very brittle, and eye protection is required when they are used.
T F 19. Ultraviolet light is not harmful to human eyes and skin.
 20. Read the following label and identify each box and what it means.

Date _____ Student's Name (print) _____ Student's Signature _____

Safety Test Evaluation
The students are not allowed to perform the fabrication process unless they understand thoroughly and answer correctly every question. Therefore, any questions missed must be explained to the student by the instructor.

Source: Adapted from J. A. Markum and M. P. Silva, *Intermediate Electronic Fabrication.* Technical Education Press, Seal Beach, CA 1976, pp. 11–12.

STUDENT SAFETY STATEMENT

My instructor, Professor _____, has given me a handout and a lecture on safety followed by an 18-question safety test covering the etcher, the UV exposure system, the handling of hazardous materials (etchants, photochemicals, etc.), and eye safety. The areas in which I had questions were explained to my entire satisfaction. I agree that my number one consideration when working in the PCB laboratory is safety. I will never knowingly violate any safety rules. If I have questions on safety, I will ask the instructor before proceeding further.

I have been instructed that no shop machinery such as the high-speed drill press, shear tool, etcher, and UV exposure device will be operated, nor any exposed voltage be used, unless my laboratory instructor has given me permission and is present in the shop area.

_____   _____
Student's Name (print)   Date

_____   _____
Student's Signature   Course Number

Source: Adapted from J. A. Markum and M. P. Silva, *Intermediate Electronic Fabrication.* Technical Education Press, Seal Beach, CA, 1976, pp. 11–12.

10-4. PRINTED CIRCUIT BOARD MATERIALS

Printed circuits are metal conducting patterns, usually produced on dielectric substrates by photochemical process. The metal conducting pattern serves as the conducting path for electrical signals from one component to another. The dielectric substrate can be rigid or flexible. The dielectric substrates generally used are Epoxy glass sheets, melamine, and Teflon. All these substrates must be bonded with copper foils by using epoxy resin. Various thicknesses of copper foil and dielectric substrate are used to meet different requirements.

There are various types of substrate materials used to produce printed circuit boards. These materials are classified into various categories by the NEMA (National Electric Manufacturing Association). The common ones used by most manufacturers of microcomputers, radios, and military equipment are FR-4 and FR-3. FR-3 has higher arc resistance and lower moisture absorption, and can used for applications up to 10 MHz. FR-4, however, has a few more additional properties such as flame retardants and low dielectric losses. Because of its favorable dielectric properties, it can be used for applications up to 40 MHz of frequency.

Principally, copper is the foil material bonded to the dielectric substrate. Copper has many advantages, such as low cost, high conductivity, excellent soldering characteristics, good tensile strength, smooth surface, and availability in various thicknesses. The thickness of such rolled copper foil can be as low as 0.001 in. The thickness of copper foils for the printed circuit board is generally specified by the weight per square foot of copper in ounces. For example, the foil of one ounce per square foot has an approximate thickness of 0.0014 in.

There are various types of adhesives used to bond the conducting foil to the dielectric substrate. Some of these common types of adhesives are vinyl-modified phenolics and modified epoxies. The following factors should be considered for selecting the grade of a printed circuit material for a particular application:

1. Maximum continuous current: determined by the thickness of the copper foil and conductor width
2. Maximum voltage: determined from the dielectric strength of the board material
3. Maximum frequency: determined from the dielectric losses or dissipation factor of the board material
4. Physical strength required by the application: determined by the board grade and bonding resin material
5. Soldering temperature: determined from the bonding resin material
6. Arc, tracking, and insulation resistance: determined from the type of base material
7. Dielectric breakdown voltage: determined from the type of base material

10–5. THE MANUFACTURING PROCESS

10–5.1 Planning and Board Material Selection

Once the artwork pattern for the electronic circuit is designed, two specifications of the PCB are already known. They are board size and the number of layers the board needs. From the individual board size, panel size and number of boards per panel are determined. Common panel sizes are 12″ × 18″, 18″ × 24″, and 20″ × 26″. PCB sizes are not generally derived from the available panel sizes, although doing so would minimize waste and keep down production costs. PCB sizes are generally driven by the product requirements. The type of dielectric substrate needed for the board is determined by the type of electronic circuit and its requirements. The previous section discusses the board type and its selection procedure. Considering those factors, board type and copper foil thickness can be determined for a particular electronic application. Typical board specifications for electronic applications are given below:

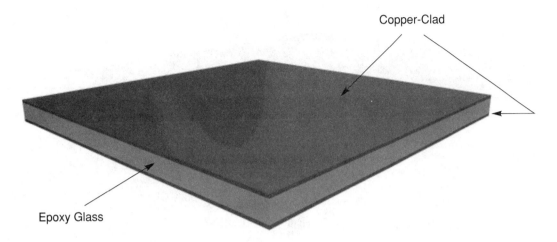

Figure 10-1. Copper-Cladded Epoxy Glass Board

- Board Material, FR4 0.062 1/1. This means NEMA grade FR4–type dielectric material of 0.062 in. thick, clad with 1 oz. of copper on each side. This board will be used for double-sided applications.
- Board Material, FR3 0.031 2/0. This means NEMA grade FR3–type dielectric material of 0.032 in. thick, clad with 2 oz. of copper on one side. This board will be used for single-sided applications. Figure 10–1 shows a copper-cladded epoxy glass board for PCB fabrication.

10-5.2 Drilling

Since the first operation is going to be drilling, using a computerized numerically controlled (CNC) machine, three or four panels are stacked and pinned together by drilling two holes. These holes and steel dowel pins keep the panels together on the drill machine table.

Commercially, drilling of PCBs is performed by a high-speed CNC drill machine. Today an expensive one may have multiple (for example, five) air-bearing spindles. Three or four panels are stacked to be drilled by one spindle. Drill data is generated by the software during the artwork design process. The data can be generated by OrCAD/PCB in various industry-accepted formats. One such format is Excellon. This data in soft form is entered directly in the CNC machine. These drill data are the location of drill hole, their sizes, spindle speed, drill bit feed, and retraction rate. Today a modern CNC drill machine can drill a hole with a diameter as small as 0.010 in. and as large as 0.25 in., automatically change drill bits, detect broken drill bits, and detect missing holes. The challenge of making a small hole on a dielectric substrate is not the size of the bit but keeping that bit straight while it is drilling. Figure 10–2 shows an illustrated view of a board just after drilling.

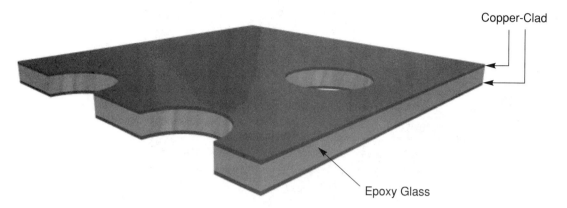

Figure 10–2. Board after Drilling

10–5.3 Deburr

Deburring is removing the rough edge and any copper burrs at the rim of the hole. In a commercial PCB fabrication process, this is an important step after drilling. Although with the improved drilling process, the board can be produced almost burr free, passing the board through some kind of deburring machine will ensure quality later. A deburring machine removes burrs mechanically, using brushes and abrasive wheels.

10–5.4 Electroless Copper Plating

After drilling and deburring, the boards are ready for electroless copper deposition and the copper flash-plating process. In this process, the walls of the drilled holes receive a thin coating of copper. Generally, this coating is 100×10^{-6} to 150×10^{-6} inches (or 100μ-in. to 150μ-in.). This process is used only to provide a primary metalized base for further deposition of copper. The first part of the electroless copper deposition process deposits only 25×10^{-6} inches (or 25μ-in.) of copper surface over the wall of each hole in the board. The second part, which is the flash plating, builds up copper up to 150×10^{-6} inches (or 150μ-in.). The second part of the process also prevents oxide deposition in the walls. The actual chemistry of the process is quite lengthy and not the focus of this book. Formaldehyde is used for the formation of the process chemistry for the electroless copper process. Formaldehyde is a known carcinogen. Recently, a cost-effective direct metalization process has been invented, and the PCB industry is adopting it rapidly. Figure 10–3 shows an illustrated view of an electroless copper-plated board (the holes through the board are plated with copper).

10–5.6 Master Artwork Transfer

Master artwork for a printed circuit board must be transferred onto the copper-clad dielectric panel as a part of the process of PCB fabrication. The

10-5. The Manufacturing Process

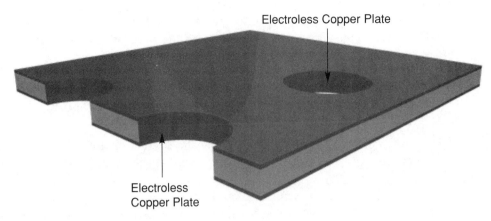

Figure 10-3. Electroless Copper-Plated Board

entire process is also popularly known as imaging. There are a number of different processes available to transfer the artwork onto the copper-cladded panels. Regardless of the method used, the outcome is the same. Generally, any of the following three ways can accomplish this: direct layout, photographic method, and printing method.

Direct layout. In this method, the artwork foil pattern is directly laid onto the copper surface of the board. The artwork pattern is created by using preprinted, etch-resist artwork materials such as trace tape, donut pads, transistor pads, integrated circuit pads, and dual-in-line package (DIP) pads. In this method, the pattern can also be drawn using a template and special etch-resist felt-tip marker. This method may be suitable for prototype and non-repetitive work. Today, with computers on your desktop and CAD packages at your fingertips, very few people fabricate PC board by using this method.

Photographic method. The most common method of transferring artwork onto the copper surface of a board is by a photographic imaging technique. This method is used for both small and large PCB fabrication processes. In this method, the copper surface of the board is first coated with light-sensitive material; then the circuit image is transferred from the original source. Photosensitive material, also called photoresist, is a light-sensitive photopolymer. There are two types of photographic mask, positive and negative, and the two types of photoresist used for this purpose are positive acting and negative acting. It is not true that positive mask is always used with positive photoresists and negative mask with negative photoresists. What type of mask will be used for what type of photoresists depends on the process steps.

Photographic mask. *Positive:* It exactly looks like the original artwork except that it is on photographic film. This means the patterns, such as traces and donuts, are dark and are part of the mask where no pattern is clear. Positive mask needs negative-acting photoresists.

less-expensive

Negative: It is the opposite of the original artwork. This means that the pattern area is translucent and other parts of the mask where there are no traces are dark. Negative mask needs positive-acting photoresists.

Photoresists. *Positive Acting:* When the areas of the positive-acting photoresist polymer are exposed to UV light, they become soluble and are washed away by the developing chemicals.

Negative Acting: When the areas of negative-acting photoresist polymer are exposed to UV light, they become insoluble to the developing chemicals.

As described earlier, creation of a positive or a negative photographic mask depends on the type of photoresist material of the board, and that in turn depends on the PCB fabricating process. Boards with both positive and negative photoresists are available. Both type of masks are used commercially. Circolex process uses negative mask, Excel Circuits uses positive mask, and the dry-film Resists, manufactured by Du Pont and the Dynachem Corporation, can use both types of masks. The process by Excel Circuits is used in this book for the fabrication of PCBs. Dry-film Resists as manufactured by Dynachem Corporation are the aqueous type and are presently being extensively used by the PCB industry. This dry film is a negative-acting photoresist and a commonly used positive photo mask for the imaging process.

The photographic negative or positive mask is created from the original pattern by using a contact printer and photographic film. Later this mask is used to transfer the artwork pattern onto the photosensitized board by using the same contact printer. A process that Excel Circuits uses is one where a positive mask is used with positive-acting photoresists. The positive photo mask is created by directly plotting it from the computer onto a Mylar sheet. The Mylar sheet mask is used later to transfer the artwork onto the photosensitized board by using the contact printer.

A contact printer consists of a vacuum frame where the artwork mask and the photosensitized copper-cladded board can be held face-to-face together, an ultraviolet (UV) light source, and a timer to control the UV light source. This timer controls the exposure time for the photosensitized material of the board. After the board is exposed to UV light in the contact printer, the photosensitive material on the board undergoes a chemical reaction (becomes polymerized) and is not removed in the developing phase. The area of the board where there is no pattern becomes impervious to the chemical solution used by the next pattern-plating process. Figure 10–4 shows an illustrated view of a board after imaging and chemical developing. A positive photomask is used on a negative-acting photoresist to produce this board. The dry-film photoresist is covering the area where the copper will ultimately be removed by etching.

Printing method. Many industrial processes use direct printing to transfer the artwork pattern onto the copper surface of the board. This is a more eco-

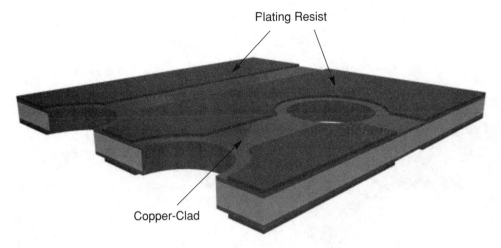

Figure 10-4. Board after Imaging and Chemical Developing

nomical method of transfer because photochemicals are relatively expensive. In this method, a stencil of the master artwork is produced. The stencil is made of fine-mesh metal, nylon, polyester, or silkscreen. Later this stencil is used in conjunction with an automatic printing machine and resist paint or ink in order to print the artwork pattern on the copper surface of the board. Instead of an automatic printer, many PCB manufacturers use a semiautomatic printer. A school laboratory or hobbyist may use a less-expensive wooden screen frame available from a screen printing supplier.

Regardless of the transfer process used, places where there is no pattern on the copper-cladded panel are covered with impervious polymer. Photoresists in the pattern areas that were not exposed to UV light will remain unchanged and will wash away during the developing process. This will expose the pattern areas for the next plating process.

10-5.7 Pattern Plating

Plating the pattern with additional copper followed by a layer of tin-lead alloy is done for two purposes. First, this puts additional thickness (approximately 0.001 in.) of copper on the exposed conductor pattern as well as on the inside surface of all the drilled holes, ensuring that all metal paths and through holes from one side to the other will have good electrical continuity. Second, the tin-lead polymer of 0.0003 in. thickness will act as an etchant resist and provide a protective coating against oxidation. Both copper and tin-lead alloy are plated onto the pattern by an electrochemical process. After the panel is passed through this process, the circuit pattern is covered with additional copper and electroplated with tin-lead alloy. Figure 10-5 is an illustrated view of a board after electroplating with copper, and Figure 10-6 is the view after the tin/lead electrolytic deposition over the copper plating. After this process, the panel is ready for etching.

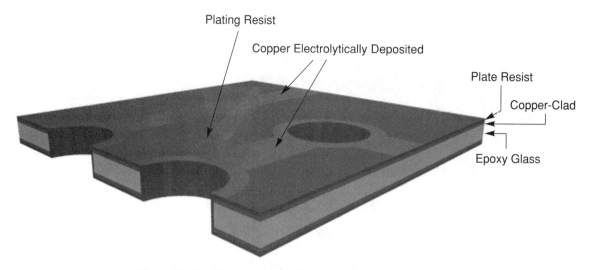

Figure 10–5. Illustrated View of the Board after Electroplating with Copper

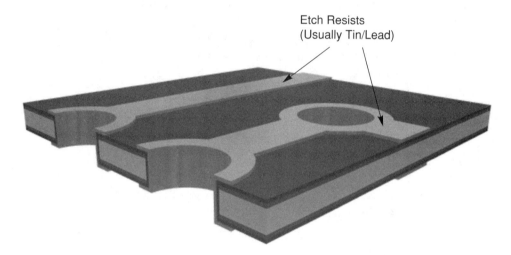

Figure 10–6. Board after Electroplating with Etch Resists (Tin/Lead)

10–5.8 Resist Stripping and Etching

Photoresists that are on the panel are now chemically removed. The types of resist-removing chemical to use depends on the resists used. Dry-film resists made by the Dynachem Corporation can be stripped from the panel by simply immersing them in an alkaline solution. Figure 10–7 is an illustrated view of the board after resist stripping. After the resists strip the panel, it is thoroughly rinsed to remove any film or alkali residue. Now the panel is ready for etching.

The copper conductor pattern is protected from etchant with tin-lead alloy coating whereas the copper layer of the rest of the panel is exposed.

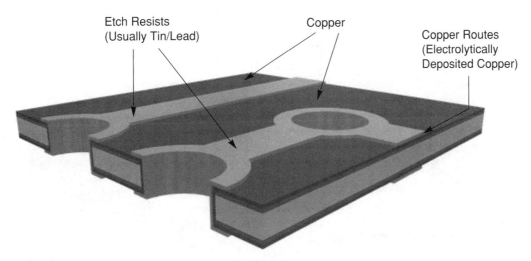

Figure 10–7. Illustrated View of the Board after Photo Resist Stripping

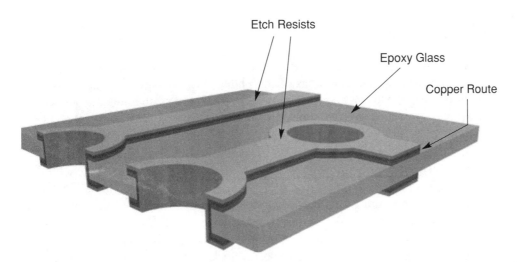

Figure 10–8. Illustrated View of the Board after Etching

Among the many copper etchants, the safest one is modified ammonium persulfate. Since this etchant is an aqueous type, it is extensively used for both laboratory and commercial purposes. It does not attack the tin-lead alloy as ferric chloride does. It also does not release toxic fumes. Ammonium persulfate solution in a conveyorized spray etcher will etch out the exposed copper from the panel, leaving the circuit pattern. Figure 10–8 is an illustrated view of the board after etching. After etching, the tin-lead alloy that exists on the pattern is chemically removed. The PCB panel is now ready to go to the next step. Figure 10–9 is an illustrated view of the board after the removal of etch resists (tin/lead alloy).

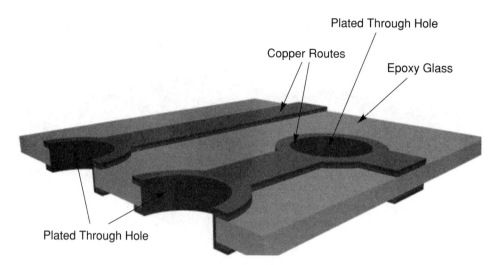

Figure 10-9. Illustrated View of the Board after Removal of Etch Resists

10-5.9 Solder Mask

In this step only the selective area of the pattern will be coated with a nonconductive coating. The areas of the pattern left uncoated are pads, holes for component leads, and test points where the electrical contact is desired. To perform this selective application of solder resists material on the pattern, solder mask is necessary. OrCAD/PCB will generate the mask during the artwork design process. This mask is the pattern for the uncovered areas. The most commonly used solder mask and its application method are epoxy ink and screen printing, respectively. A silk- or metal-screen stencil is prepared by using the computer-produced solder mask pattern. This silkscreen stencil and the printing machine are used to apply the epoxy ink only on the selected places on the pattern. Dry-film solder mask is also used and processed by exposure to UV light. Unexposed dry film becomes unpolymerized and is removed during the developing process. Figure 10–10 is an illustrated view of the board after the application of solder resists.

10-5.10 Solder Coating

In this phase the exposed copper surfaces are covered with tin-lead alloy for two reasons. First, it protects the copper from getting oxidized. This will increase the shelf life of the PC board. Second, it improves the solderability of the board, since electronic components will ultimately be placed and soldered to the routes. One might wonder why the earlier tin-lead coating was stripped and now applied again. The answer to the question is that during the pattern plating process, tin-lead has not been alloyed to the copper, but is adhered to the copper through a metallic bond. For alloying

10–5. The Manufacturing Process

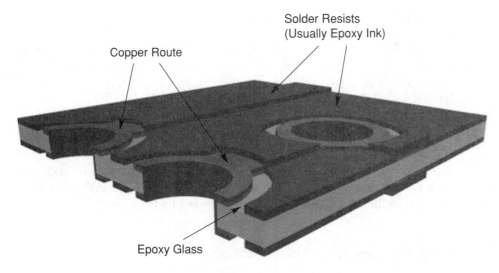

Figure 10–10. Illustrated View of the Board after the Application of Solder Resists

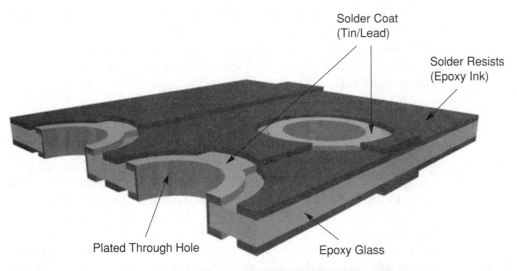

Figure 10–11. Illustrated View of the Board after the Application of Solder

to copper pattern, the tin-lead must be raised to its melting temperature. In the solder coating process the board panels are dipped into molten solder, and the areas not covered by solder mask will have a coat of solder. This process will raise the tin-lead alloy to its melting temperature so that it will form a chemical bond with copper patterns. Any excess solder is removed by a blast of oil or hot air. The process of removing excess solder is called *solder leveling*. By most solder leveling processes, 50µ inches to 1000µ inches of solder thickness on desired areas can be achieved. Figure 10–11 is an illustrated view of the board after the application of solder onto the exposed copper surface (the surface of the copper not covered with solder resists).

10–5.11 Contact Finger Plating

Some PCBs need to have a plated edge connector for durability and better contact. Usually the designer call for the edge connector to be plated with either gold or nickel. If the designer does not call for it, edge connectors are left solder-coated or as bare copper surface.

In this process all parts of the panel are masked except the area to be plated. The solder coating is stripped chemically, if necessary. Finally, nickel or gold is electroplated to the edge connector. Gold plating is preferable over nickel because it has less plugging resistance. The thickness of gold and nickel on the edge connector is 50μ inch and 200μ inch, respectively. Figure 10–12 is an illustrated view of the plated finger of a PC board.

10–5.12 Component Identification

Generally, screen printing is used to print the reference designator for each component. A silk- or metal-screen stencil is prepared from the silkscreen produced by OrCAD/PCB. This stencil and a printing machine are used to perform the operation. Component identification on the PC board is used by the testing and field service personnel to locate a particular component on an assembled board. Figure 10–13 is an illustrated view of a board with component identification.

10–5.13 Scoring and Final Shaping

Finally, each PC board must be sheared out from the panel for the mounting of components. It may also be necessary to cut holes and routes to shape the board. In a commercial environment, boards are shaped or profiled into other than square or rectangular shape by a router machine. A CNC drill router machine is generally used for this purpose. Before shearing the panel into individual boards, scoring is done by a CNC machine to make the shearing easy and less damaging to the board. Figure 10–14 is an illustrated view of the board after final shaping.

10–5.14 Inspection and Electrical Testing

PC boards must be tested against any defect before mounting components. It is less expensive to discard a bare board than a board with components. Testing the board against defects is the PC board manufacturer's responsibility. A high-resolution scanner in conjunction with a CAD tool is used for this purpose. The entire board is scanned by the scanner and checked against the netlist file for any electrical misconnections and flaws. Figure 10–15 displays a number of examples of flaws (defects) on PC boards. There are computer-based inspecting systems that will scan a PC board, detect all the flaws, and pass or reject the board depending on the rules entered in the testing system. The computer will also let the inspector view

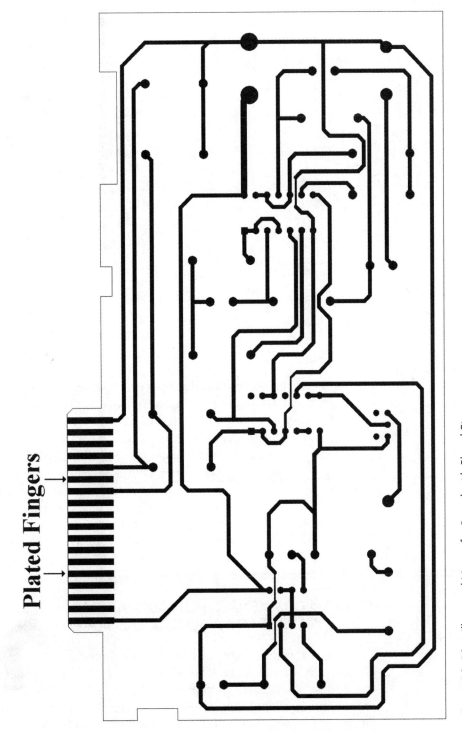

Figure 10–12. Illustrated View of a Board with Plated Fingers

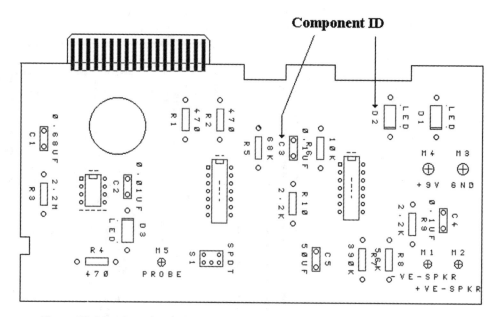

Figure 10-13. Board with Component IDs

the flaws on the screen and let him or her decide on marginal cases. However, there are many other test procedures available for testing PC board. A *bed of nails* is a very common one. Here, visual inspection of the board as a final checkpoint is performed before packaging for shipment. In a laboratory environment, since only a few boards need to be inspected, visual inspection using a magnifier with a scaled reticule is employed. Inspecting a finished PC board under a magnifying lamp is the most inexpensive way to do this in a school laboratory where boards are not mass-produced. Figure 10–16a is a picture of a PC board under a scaled reticule. Figure 10–16b shows typical repairs of a PC board's defects.

10-6. RECOMMENDED PCB FABRICATION PROCESS FOR A SCHOOL LABORATORY

As mentioned earlier, there are many chemical processes by which a printed circuit board can be fabricated. Many chemical companies developed the entire range of chemicals for processing the various stages of PC board fabrication. Plating, photoresists, developing, and etching chemicals are part of the process. If the process chemicals are water soluble and less toxic, they are generally identified as aqua-based. Due to the strict regulation of the Environmental Protection Agency (EPA), aqua-based processes are more desirable in the United States. For a school laboratory, aqua-based process is a must.

One such process developed by Excel Circuits is of aqueous type. There are many advantages to this process, and all these are favorable for a school laboratory environment. The advantages are described below:

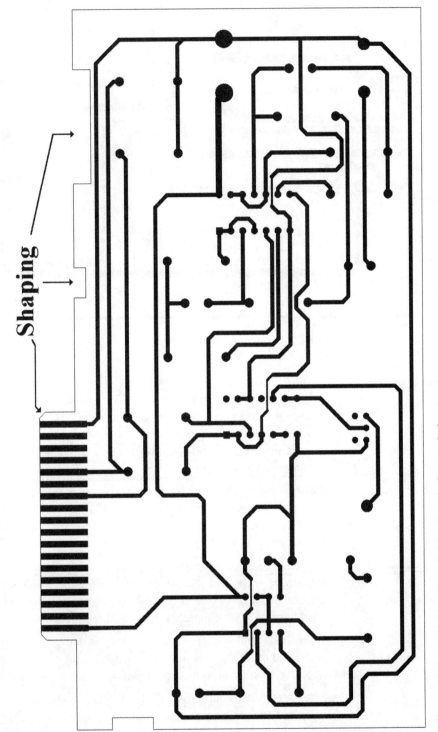

Figure 10–14. Illustrated View of the Board after Final Shaping

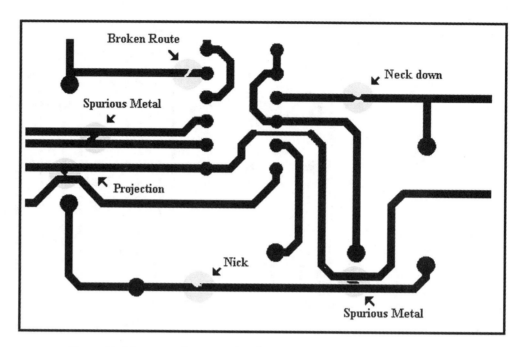

Figure 10-15. Typical PC Board Defects

- It uses a board that has positive photoresists. This means positive artwork is required for tooling the fabrication process. It is easier to produce a positive artwork than to produce a negative artwork in a noncommercial environment.

- Photosensitive chemicals on the board cannot be overexposed. Therefore, they can be exposed to UV light for more than the sufficient amount of time to ensure proper exposure. Exposure timing control is not so critical.

- It is easy to inspect the exposed board because the board area that will remain or withstand the etching turns blue and the area that will etch out turns green. In case of insufficient exposure, the board and the artwork can be placed together again and reexposed.

- Chemicals that are used for developing and etching the boards are water soluble. They do not release toxic fumes, and this reduces the need for forced ventilation or fumehoods for the work area.

- Most processes require the baking of boards right after developing them in order to set the photoresist. The Excel process does not require this.

- Since the developing and etching chemicals are water soluble, they make hardly any stain on the etcher, other equipment, and objects around the work place. This makes it very suitable for a school laboratory environment, where a large dedicated room for just for this purpose is always difficult to obtain.

10-6. Recommended PCB Fabrication Process for a School Laboratory

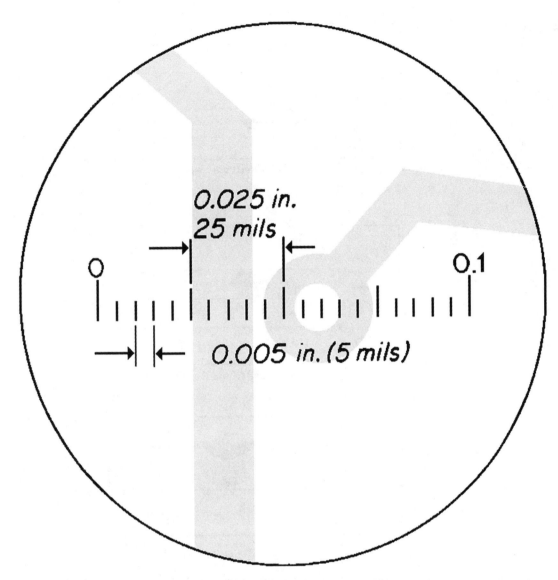

Figure 10-16a. Board under Scaled Reticules
Source: Adapted from *Electronic Techniques*, Robert S. Villanncei, Alexander W. Autgis, and William F. Megow. 3rd Ed.: Englewood Cliffs, NJ Prentice-Hall, 1986, p. 439.

However, there is one disadvantage with this process, and that is the fact that you must buy photosensitized board from Excel Circuits or its authorized dealer. It is not possible to buy photoresists separately and sensitize your own board. Various sizes, thicknesses, and types of board are available from Excel Circuits. Usually, a couple of sizes and thicknesses of boards are sufficient for laboratory applications.

After analyzing the situation, it is clear that Excel is the most suitable chemical process for fabricating PC board in a school laboratory environment. The step-by-step procedure of processing the board by using this process is described below.

Figure 10–16b. Typical Repair of PC Board Defects

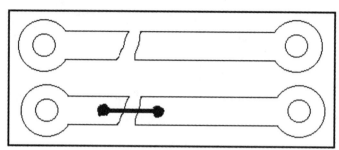
Broken Route Repair

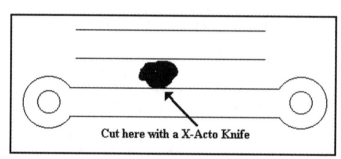
Spurious Metal Repair

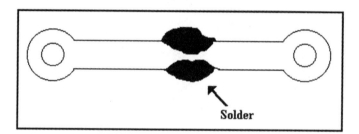
Neck down Repair Using Solder

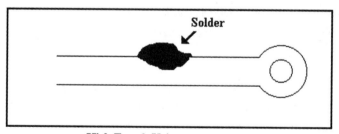
Nick Repair Using Solder

10-7. FABRICATING PCBS BY USING THE EXCEL PROCESS

Before we describe this process, a few points need to be addressed:

- Since the Excel boards are purchased with photoresist material on them, the process of laminating the copper-cladded board with dry-film photo resists will not be required.
- For making prototype PC board in the laboratory, the plated-through-hole (PTH) process for vias is not cost effective. Also, the process is not very safe because of the use of formaldehyde, a known carcinogen. The vias are done by using via-pins, also called eyelets. Therefore, the PTH process will not be required.
- Pattern plating and plating the edge connector fingers with gold are also not necessary for laboratory applications. Therefore, these two processes are also not required.

The main sequences of the laboratory process outlined by Excel Circuits are the following:

- Transfer of artwork to the presensitized board, using a photo imaging technique
- Developing, using photochemicals provided by Excel Circuits
- Etching, using ammonium persulphate solution
- Tin-plating, using Metex Acid Tin Immersion salts

The process described here can be used for both single- and double-sided board. However, processing a single-sided board is little bit different. For a single-sided board fabrication, an Excel board with one side photosensitized is required. Also, for the route side of the board, one artwork with mirror text is required.

Step 1: For laboratory use, Excel boards come with positive-acting photoresists, and a positive mask or artwork is required. OrCAD/PCB software can produce both positive and negative photomasks for the artwork. For a double-sided board, two photomasks of master artwork on Mylar sheets are required, one for Layer 1, which is the route side of the board, and the other for Layer 2, which is the component side of the board. Layer 1 should have mirrored text, and Layer 2 will have regular text. This text will help you determine which way to place the Mylar on the photosensitized board. For a single-sided board, a photomask of Layer 1 master artwork with mirrored text is required.

Generating the photomask of master artwork is a critical tooling process in fabrication, because any flaw will ultimately show up on the board. A high-resolution (at least 600×600 dpi) laser printer should be used for printing the artwork on the Mylar sheet directly. Commercially, PC board manufacturers use high-resolution photoplotters for this purpose. The resolution of these laser photoplotters are generally quite high, and they are specified in mils, such as 0.25 mil to 1 mil.

Step 2: The green side of the board has the photoresist material. For a single-sided board, photoresist is only on one side, and for a double-sided board photoresist will be on both sides. For a single-sided board, place the artwork on the green side of the board in such a way that the mirrored text is seen by you as regular text. Place the artwork and the board such that the board edge and artwork margin are aligned properly.

For a double-sided board, the two layers of the artwork on the Mylar must be aligned first. To help the alignment, three/four aligning targets are usually placed on both layers at the same place on the artwork. They are placed outside the board edge during the artwork design process.

Find a piece of board one inch wide, six inches long, and having the same thickness as the sensitized board. Sandwich this dummy piece of board between Layer 1 and Layer 2 artwork in such a way that it is located outside the border of the artwork and the aligning targets and the two sides of the artwork are aligned accurately by the targets. Figure 10–17 shows where to place the dummy board and how to align the two layers of the artwork. For a double-sided board, the alignment and the direction of these layers are critical. The Layer 1 artwork must be placed on one side of the dummy board in such a way that the mirrored text is seen by you as regular text. The Layer 2 artwork must be placed on the other side in such a way that the text on it is seen by you as regular text. Once the two layers of the artwork are placed and aligned correctly, tape them only on the dummy board side such that the board piece and the Mylar stay together. This arrangement is made so that the photosensitized board can be placed between the two Mylar sheets before placing them all in the UV exposure system. For commercial fabrication, alignment of the board is accomplished by the registration mark on the artwork and on the bare board.

Step 3: The photosensitized board is shipped by the manufacturer individually wrapped by a black polythine paper. These sensitized boards are

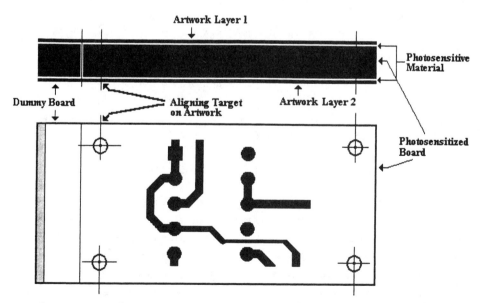

Figure 10–17. Illustrated View of the Artwork Aligned with the Photosensitized Board

not sensitive like photographic film. However, they should not be exposed to the sun or to fluorescent or incandescent light. Red light is generally safe. If the proper red light is not available, you may work under fluorescent or incandescent lights, but do not let the photosensitized board be exposed more than fifteen seconds once it is taken out of its wrapper. Photosensitive material may not expose totally, but degrade instead. Unwrap the boards and place them carefully and quickly between the Mylar sheets you prepared in Step 2. Remember not to stare at the UV lights while they are on. Place the boards in the UV exposure device and close the latch. Set the exposure time to 5 minutes. Do not worry too much about the exposure time because the photosensitive material on Excel boards cannot be overexposed. Switch on the exposure device and wait for the timer to finish. Take the boards out for developing. Before putting the boards in the developing solution, inspect them carefully for proper exposure. A properly exposed board will have blue traces, and the part etched out will be green. If you find that the exposure was not done properly, align the artwork with the board and put it back in the exposure system. Types of light sources and their exposure times are provided below:

Type of Light	Exposure Time
Direct sunlight	5–10 minutes
Direct incandescent light (No. 2 photoflood lamp)	10–15 minutes
Fluorescent light	40–60 minutes
Ultraviolet light	2–5 minutes

Step 4: The board must be developed in the developing solution before it can be etched. The developer used by the Excel process is a caustic solution. Therefore, eye protection must be worn, and care should be taken to avoid contact with the skin. Developing solution is shipped in plastic bottles in concentrated form and must be mixed with water for proper strength. The ratio of water and the developing chemicals depends on the strength of the shipped chemicals. Generally, water and developing chemicals are mixed at a 1:4 ratio by volume.

It is necessary to follow the instructions provided by Excel Circuits for the dilution ratio. The temperature of the developing solution should be about 70°F to 80°F. Water that is warmer than this may remove the resists from the board. Place the board in the developing solution and slowly rock the pan so that the solution is washing over the board. For a double-sided board, turn it every 5 minutes. Use laboratory tongs or rubber gloves to turn the board. It is not safe to dip your bare fingers in the developing solution. The developing time is approximately 5–10 minutes. Develop the board until the background of the board is copper colored and the tracings are still dark green. This step can be performed under fluorescent light. After 10 minutes, rinse the board thoroughly in warm water. Inspect the board carefully to see if the entire unwanted background of the board is copper colored. If not, dry the board with paper towels, realign the artwork with the board again, and expose it in the UV exposure system. At the end of the developing process, the board is ready for etching. Commercially, mechanized immersion or spray developers are used for this purpose. Some developing processes need additional machinery to perform the process.

Step 5: Once the developing is complete, remove the board from the solution and rinse the board thoroughly with water. Now the board is ready for etching. The etchant used by the Excel process is ammonium persulphate. It will cause eye and skin irritation upon contact. Wear eye protection while working around the etcher. According to the Material Safety Data Sheet, airborne dust from the chemical may cause eye, nose, skin, and throat irritation.

Ammonium persulphate crystal etchant is shipped in either 4-lb. or 1-lb. plastic bags. These crystals must be mixed with water at a ratio of 1 lb. of crystals to 1 qt. of water to make etchant solution.

$$\frac{\text{Water}}{\text{Ammonium Persulphate Crystal}} = \frac{1 \text{ Qt}}{1 \text{ Lb}}$$

A desk-top spray etcher is the best way to etch boards in a school laboratory environment. The Kepro Bench Top Spray Etcher Model BTE-202 can be used for this purpose. It can etch two double- or single-sided boards at one time. The boards are hung from the middle bracket with the help of four screw-type clamps. Spray nozzles run parallel to the bracket on both sides of the boards. The etching process will take anywhere from 1 to 5 minutes, depending on the strength and temperature of the solution and the thickness of the copper foil. Temperature of the etching solution should be approximately 140°F. As more and more boards are etched with the solution, its strength decreases and etching time increases. It is necessary to check the boards every minute to monitor their progress. If the boards are etched for too long, it may remove part of the traces that are covered with polymerized photoresists. When the boards are completely etched, take them out and rinse them thoroughly with both warm and cold water. Dry them with paper towels. Care must be taken in handling the board so that the traces are not damaged. Commercially, etching is also done by using a spray etcher, but a conveyor system is used to load the boards and carry them through the etcher spray area and the water-wash area.

Step 6: The polymerized photoresists should be removed to improve the solderability. The safest way to do this is to re-expose them. Put the etched, dried board in the UV exposure device without the artwork and expose it for about 3 minutes. This is called an open exposure. Follow the same procedure described in Step 4 to develop the board after the open exposure. The developer solution will remove the resists, leaving the clean copper traces on the board.

Step 7: After the boards are removed from the etcher, washed, and dried carefully, they need to be tin coated. Tin coating is done to improve the shelf life and solderability, and make the traces and pads stronger. For this process the board still needs to go through two more steps. The tin coating salts are shipped in dry form in plastic bags. Mix them in a glass or plastic container with warm water according to the following ratio: 1 lb. of salt to 1 qt. of water.

$$\frac{\text{Water}}{\text{Tin Coating Salt}} = \frac{1 \text{ Qt}}{1 \text{ Lb}}$$

Immerse the clean board and leave it in the tin coating solution for about 10 minutes. Tin will deposit on the copper surface. If a spotty deposit appears, the board was not clean. Take it out of the solution and clean it lightly with scouring powder and a wet cloth. Do not scrub hard with steel wool because it may damage the traces. Put it back into the solution and leave it in for the same length of time as before. Once the tin coating is complete, take the board out of the solution and wash it in running water. At this point, the board is ready for drilling.

In a commercial setting, drilling and final shaping of the PCB are the integral parts of PC board fabrication. Drilling is done first during PCB processing and shaping is done last, before testing. For our laboratory process, drilling and final shaping are done as the last parts of the processing.

10–8. QUALITY ASSESSMENT OF PCBS

The quality assessment of PCBs involves much high-precision equipment and many inspections. The cost of a printed circuit board is approximately 30 to 40 percent of the cost of the components that go on the board. Even compared with a very large-scale integrated (VLSI) circuit, such as a microprocessor, PCBs are often the largest and costliest unit that go in an electronic product.

A printed circuit board with components mounted and soldered correctly will cost several times more if it has to be rejected due to flaws. These defects may cause enormous customer dissatisfaction and in turn may seriously hurt the business. For example, if a fault is detected during the PC board fabrication process, it may cost the manufacturer only $1/fault. However, if that same fault is detected during final inspection of the boards, it may cost $20/fault. Because of their importance and cost, PCBs used in electronics equipment are inspected carefully for quality. Although PCB manufacturers are supposed to inspect them for defects, to avoid losses, many electronics manufacturing companies perform their own inspection of the boards. Extremely thorough testing costs lots of money. Therefore, such a comprehensive effort is done only for critical applications.

There are numerous attributes of a PCB that can be inspected. Many of them are inspected visually, some of them by a combination of computer-aided tools and visual inspection, and a few of them by using computer-aided tools alone. Board manufacturers incorporate testability for multilayer board inspection. One such means of testability is to incorporate special types of through-holes and patterns on a multilayer board for locating inner-layer registration errors. In another type of testing, a PC board is scanned by a very-high-resolution special scanner, and the scanned file is compared with its netlist information to detect the broken routes and other defects. Each of these defects is then compared against certain user-defined rules. Later, as the computer displays each of these defects, the inspector can make decisions about each one of them.

10-9. PC BOARD FABRICATION FOR THE PROJECTS

To fabricate PC board for the projects, you need the following:

- To pass the Safety Test
- Positive master artwork for both sides on Mylar sheets (in at least 600 × 600 dpi resolution)
- Photosensitized Excel board (three double- and one single-sided board for the four projects)
- Excel developing, etching, and tinning chemicals
- UV exposure system
- Desktop spray etcher system
- Two nonmetallic (for example, plastic) trays for developing process
- Safety gears

Once you have all the above items, you may begin the process and follow the procedure described in Section 10–7 to make the boards. If this is your first time, you must work under supervision by your laboratory instructor. Note the following important points to help ensure the quality of your boards:

- **Resolution of artwork on Mylar,** because any particles opaque to UV light on the artwork will come right onto your etched board.
- **Alignment of artwork with the board** before putting the boards in the exposure system. Artwork must be aligned carefully for a double-layer board.
- **Etching time** of the board, because it depends on the strength of the solution. If you are not careful, the solution may etch out some of the thin routes on the board.
- **After etching, do not scrub the board** with steel wool or hush scrubber, because it may delaminate the routes and pads.
- **Before tinning,** you must remove the polymerized photoresists by exposing the board without the artwork in the UV exposure system and developing the board again in the developing solution.

Figures 10–18a and 10–18b show the two sides, Layer 1 and Layer 2, of the PC board for the Mini Amplifier project after etching and tin coating.

10-10. DEFECT REPAIR

Inspect the fabricated boards under a magnifying lamp and a loupe with scaled reticule to look for defects. If you find one, try to repair it. Broken routes can be repaired by using jumper wires. Shorted routes and pads may

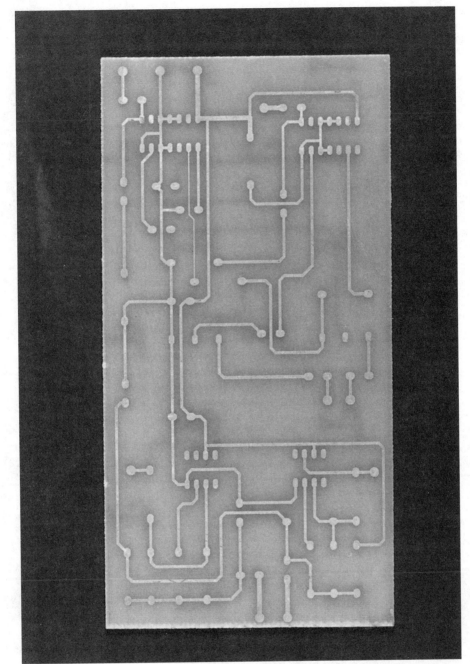

Figure 10–18a. Layer 1 of the PC Board of the Mini Amplifier Project after Etching and Tin Coating

Figure 10–18b. Layer 2 of the PC Board for the Mini Amplifier Project after Etching and Tin Coating

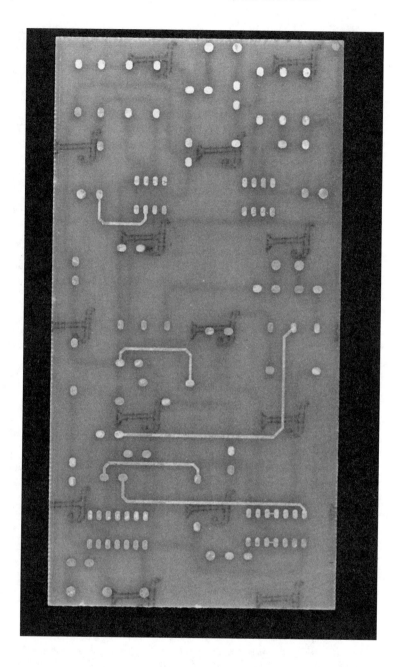

be repaired by carefully removing the part of the copper route at the shorted point, using an X-Acto knife or a razor blade. If you have any other defect-inspecting tools available in your lab, use them to find the flaws on the boards. Do not just assume that the board you fabricated is flawless. It is important to detect the flaws, if any, at this time, because after putting the component on the board, it may be difficult to see and repair the flaws.

The PC board fabrication is not yet complete, because the board must be drilled for mounting components and eyelets must be placed for vias and other holes through which component leads will pass. This process will be described in the next chapter.

10-11. TOOLS, MACHINERY, AND CHEMICAL REQUIREMENTS FOR A LABORATORY-BASED PC BOARD FABRICATION FACILITY

The following tools, chemical, and machinery can provide you with an inexpensive PC board fabrication facility at your laboratory. The types and sources of materials have proven to work well. You may choose to use another type or source.

10-11.1 Safety Requirements and Chemical Disposal

Contact the person in charge of safety in your organization. He or she should be able to tell you the procedure of disposing of chemical wastes. If you use the chemicals suggested, you should not have much problem disposing of them. Remember not to dispose of them into a drain; this may be illegal without proper permission in your area. Also, remember that if the chemical can eat away copper from your board, it can also eat your sewer pipe. Request the Material Safety Data sheet from the manufacturer or supplier of etching, photo-developing, and tin-coating chemicals. Analyze the data and establish a method for disposing of the chemical waste. Also discuss with safety personnel any other safety issue you might have in your laboratory and its surroundings.

10-11.2 Tools and Machinery

(For addresses and telephone numbers, see Appendix C.)

- Desktop Etcher ⇨ Kepro Circuit, Model BTE-202
- High-Speed Drill Press ⇨ Dumore Corporation Series 37
- CNC Drill and Router Machine ⇨ Kepro Circuit, CNC-1218 RCC6
- Laser Printer, Resolution at Least 600 × 600 ⇨ HP Laser IV
- High-Speed Carbide Drill Bits ⇨ Precision Carbide Tools Co. Inc.
- UV Exposure System ⇨ Kepro Circuit, Model BTX-200A
- Eyelets and Via Pins, Replacements for *Plated-Through-Hole* (PTH) ⇨ International Eyelets
- Magnifying Lamp and Other PC Board Visual Inspection Tools ⇨ Edmund Scientific

10-11.3 Chemicals and Photosensitized Blank PC Board

(For addresses and telephone numbers, see Appendix C.)

- Etchant ⇨ Ammonium Persulphate Crystal, PCB Kits
- Photodeveloping Chemical ⇨ PCB Kits

- Tin Plating Salt ⇨ PCB Kits
- Photosensitized Blank PC Board ⇨ PCB Kits

Tools and machinery for the fabrication of PC board should be put in a laboratory where there are running water, a wash basin, and fluorescent or incandescent lighting. You do not need a darkroom. If you have a good ventilation system in your building and if you use the chemicals mentioned, you will not need any forced ventilation for the fabrication laboratory.

QUESTIONS

1. Why is safety everyone's responsibility? Explain.
2. What are the four areas of safety that one needs to learn before using the PCB fabrication laboratory?
3. When you are using a high-speed drill press, what must you wear?
4. Why should you never dispose of laboratory chemical waste into a sewer?
5. How would you determine the type of board substrate material required for your project?
6. What is the most common type of board substrate material used by industrial PC board fabricators?
7. What are the thirteen major steps of the PC board fabrication process?
8. Why is plated through-hole not done while fabricating a board in a laboratory environment?
9. Instead of plated through-holes, what else can be used for a double-sided board?
10. Why is the transfer of artwork to a photosensitized board critical?
11. Can a photosensitized Excel board be overexposed under UV light?
12. Write the chemical name of the etchant in the Excel process.
13. How do PC board manufacturers assess and ensure quality of PCBs?
14. What are the two different ways you can remove polymerized photoresists from the Excel board?

Assembling and Soldering

11

ASSEMBLING AND SOLDERING OF ELECTRONIC PARTS ON PRINTED CIRCUIT BOARDS

11-1. PROJECT ASSEMBLY TOOLS

There are several different types of hand tools and power tools that are necessary for assembling the components on the PC board and packaging the project. The following are some common tools for assembling parts:

Hand tools

- Screwdrivers
- Nutdrivers
- Wrenches
- Pliers
- Wire strippers
- Soldering irons
- Double fine smooth cut files
- "Third hands"
- IC extractors
- X-Acto knives

Power tools

- High-speed drill press and carbide and sheet-metal bits
- Sheet-metal hole cutters

Common project assembly parts

- Enclosures (metal or other material)
- Sheet-metal screws
- Machine screws and associated hexagonal nuts

Objectives

After completing this chapter, you should be able to

1. Describe the hand and desktop tools required for assembling laboratory projects
2. Acquire the parts and tools required for assembling a school laboratory project
3. Drill and put eyelets on a PC board
4. Give final shape to a PC board by using hand tools
5. Prepare the selected enclosure for mounting off-board parts and the PC board
6. Select soldering material and solder parts on the PC board
7. Assemble a laboratory project and check its operation

- Fiber shoulder washers
- Threaded standoffs with associated machine screws
- Cable clamps
- Crimp terminals
- Terminal posts
- Eyelets
- Solder-type terminals
- Eyelet-type terminals

11-2. PCB DRILLING

PC board drilling is not the same as drilling through wood or sheet metal. Lots of care must be taken in drilling it. One criterion for a drill press is that it must be high speed. Drill bits should be wide-shank and carbide-tip. Depending on the project requirement, only a few sizes of bits may be necessary. Like wires, drill-bit sizes are specified by number. A larger number indicates a bit of smaller diameter. For most holes on a PC board for component leads, bit sizes required range from No. 50 to No. 80. Most electronic parts such as transistors, ¼ W resistors, and IC chips require bit sizes ranging from No. 55 to No. 72. You also need to consider the type and size of eyelets and pins you will be using. The chuck of a drill press should have the capability of accepting small drills, from No. 50 to No. 80. Table 11-1 shows various component sizes, their lead diameter, and drill-bit number. This table will help you select different sizes you need for your project. Some of the sources for these drill bits are provided in Appendix D.

There is another direct way of finding the drill hole size: by knowing the diameter of the lead wire and by using the following table.

Diameter of the Lead	Required Diameter of the Drill	Drill Size No.
0.005"–0.013"	0.016"	78
0.014"–0.018"	0.025"	72
0.020"–0.025"	0.031"	68
0.028"–0.032"	0.040"	60
0.036"–0.045"	0.052"	55

Before you start drilling a PC board, get the drilling template, NC drill file, and ASCII printout from the TOL file produced by OrCAD/PCB to help you determine the location and sizes of holes. Get all the different sizes of drill bits that you need. Follow a few simple rules while drilling a processed PC board:

- Always drill through from the copper side of the board because it is difficult to locate the center of the pads from the other side of the

Table 11-1. Drill-Bit Sizes for PC Board Electronic Components

Component or Device	Lead Diameter in Inches	Drill Size No.	Decimal Equivalent of Drill Bit in Inches
⅛ W Resistor	0.016	75	0.0210
¼ W Resistor	0.019	72	0.0250
½ W Resistor	0.027	66	0.0330
1 W Resistor	0.041	64	0.0360
2 W Resistor	0.045	55	0.0520
Disc Capacitor	0.030	64	0.0360
TO-5	0.019	72	0.0250
TO-18	0.019	72	0.0250
DO-14	0.022	70	0.0280
8 or 14 Pin DIP	0.023	69	0.0292

board and it will minimize the possibility of pulling the copper foil from the base material.

- The sizes of holes through which component leads pass are not critical, but they are still important. Holes that are too large may cause the solder to fracture due to vibration. Holes that are too small may pull out the copper foil while forcing the component leads through the holes. Forcing the component leads through the holes may also damage the components.

- Get a ½″ thick piece of wood that is much larger than your PC board and clamp the board to it. If you like, clamp it with the base of the drill press. This will ensure no lateral movement of the PC board during drilling. The lateral movement of the drilled piece is the main cause of breakage of carbide drill bits.

- While drilling via holes, check the available sizes of via pins and eyelets. The diameter of a via eyelet depends on the thickness of the board. The outside diameter of an eyelet barrel is the approximate size of a hole and the inside diameter is the approximate size of a component lead that will ultimately pass through it. Several different types of via eyelet are used for fabricating the projects in this book. They are as follows:

Board Thickness	Barrel Outside Dia. ∓0.002″	Flange Length ∓0.005″	Flange Dia. ∓0.005″	Barrel Inside Dia. ∓0.002″
0.031″	0.040″	0.062″	0.060″	0.030″
0.031″	0.059″	0.062″	0.090″	0.045″
0.031″	0.030″	0.073″	0.046″	0.021″
0.062″	0.030″	0.094″	0.046″	0.021″
0.062″	0.039″	0.093″	0.060″	0.030″
0.062″	0.058″	0.093″	0.090″	0.046″
0.062″	0.048″	0.095″	0.067″	0.037″
0.062″	0.059″	0.110″	0.090″	0.048″

After you drill the board, light deburring is necessary, using a very fine-grade silicon carbide sandpaper. After deburring, put the appropriate size eyelets in each hole by using eyelet setting tools. These tools are available from the eyelets' manufacturer. The setting tools come in a pair. One is called ANVIL (AR), with a retractable spindle, and the other one is called FORM Tool (FR), with a fixed spindle.

In a commercial manufacturing process, boards are drilled as a first step toward PC board fabrication process. A CNC machine is generally used for this purpose. OrCAD software will generate a drilling pattern in various industry-accepted formats to feed CNC drill router machines. Excellon is a common industry-accepted CNC machine for PC board drilling.

Drilling the boards for the projects. If you are going to drill the boards manually using a high-speed drill press, you need a list of drill sizes and their X-Y coordinate locations. OrCAD/PCB can generate these files. To find the drill sizes and their location, you should be able to read and interpret the output of three different files and a Drill Template Legend provided by OrCAD. The files are with NCD (ASCII format) and TOL (ASCII and Layout format) extensions. The template legend is provided below. During execution of the **Item to Plot** command located under **QUIT Plot,** the file with TOL extension is generated.

Drill Template Legend
1 = O
2 = +
3 = X
4 = *
5 = Y
6 = Z
7 = T
8 = H

You need to print the TOL file by using two different tools. Print the TOL file by using M2EDIT and the Print PCB tool; also using M2EDIT, print the NC Drill File with NCD extension. If you look at the hard copy of these files and the template legend, you should be able to find the drill sizes and their location. Once all this information is available and you have the appropriate drill bits, you may begin drilling the board by using a high-speed drill press. Figures 11–1a through 11–1c provide the three files for the Deluxe Logic Probe project.

11–3. FINAL SHAPING OF THE PC BOARD

The PC board needs to be cut into its final shape for several reasons. If the board has an edge connector, it needs to be shaped so that it fits in the female edge connector slot. The PC board needs to be cut into a specific size so that it fits in the enclosure box. Other types of shaping may also be required.

11-3. Final Shaping of the PC Board

```
                    X-coordinate of the hole
 Drill size in inches          Y-coordinate of the hole on
                    ↘    ⇓    ↙  the board
              ASCII(0.026,2.51,3.12)
              ASCII(0.026,2.51,3.22)
              ASCII(0.026,2.61,3.22)
              ASCII(0.026,2.71,3.22)
              ASCII(0.026,2.71,3.12)
              ASCII(0.035,0.91,2.17)
              ASCII(0.035,0.91,2.27)
              ASCII(0.035,0.91,2.37)
              ASCII(0.035,0.91,2.47)
              ASCII(0.035,1.21,2.47)
              ASCII(0.035,1.21,2.37)
              ASCII(0.035,1.21,2.27)
              ASCII(0.035,1.21,2.17)
              ASCII(0.035,2.56,2.12)
              ASCII(0.035,2.56,2.02)
              ASCII(0.035,2.56,2.22)
              ASCII(0.035,2.56,2.32)
              ASCII(0.035,2.56,2.42)
              ASCII(0.035,2.56,2.52)
              ASCII(0.035,2.56,2.62)
              ASCII(0.035,2.86,2.62)
              ASCII(0.035,2.86,2.52)
              ASCII(0.035,2.86,2.42)
              ASCII(0.035,2.86,2.32)
              ASCII(0.035,2.86,2.22)
              ASCII(0.035,2.86,2.12)
              ASCII(0.035,2.86,2.02)
              ASCII(0.035,4.26,1.97)
              ASCII(0.035,4.26,2.07)
              ASCII(0.035,4.26,2.17)
              ASCII(0.035,4.26,2.27)
              ASCII(0.035,4.26,2.37)
              ASCII(0.035,4.26,2.47)
              ASCII(0.035,4.26,2.57)
              ASCII(0.035,4.56,2.57)
              ASCII(0.035,4.56,2.47)
              ASCII(0.035,4.56,2.37)
              ASCII(0.035,4.56,2.27)
              ASCII(0.035,4.56,2.17)
              ASCII(0.035,4.56,2.07)
              ASCII(0.035,4.56,1.97)
              ASCII(0.045,0.41,1.57)
              ASCII(0.045,0.41,1.77)
              ASCII(0.045,0.41,2.12)
              ASCII(0.045,0.41,2.62)
              ASCII(0.045,0.86,3.22)
              ASCII(0.045,1.36,3.22)
              ASCII(0.045,1.51,3.07)
              ASCII(0.045,1.51,2.57)
              ASCII(0.045,1.51,2.37)
              ASCII(0.045,1.51,2.17)
              ASCII(0.045,2.26,1.67)
```

Figure 11-1a. NCD File in ASCII Format for the Deluxe Logic Probe

Figure 11-1a. Continued

Drill size in inches — X-coordinate of the hole — Y-coordinate of the hole on the board

ASCII(0.045,2.71,1.67)
ASCII(0.045,2.71,1.17)
ASCII(0.045,2.26,1.17)
ASCII(0.045,3.21,1.52)
ASCII(0.045,3.66,1.67)
ASCII(0.045,3.66,1.87)
ASCII(0.045,3.66,2.22)
ASCII(0.045,4.01,2.02)
ASCII(0.045,4.01,1.52)
ASCII(0.045,4.91,1.62)
ASCII(0.045,4.91,1.12)
ASCII(0.045,5.51,1.12)
ASCII(0.045,5.51,1.62)
ASCII(0.045,5.21,2.47)
ASCII(0.045,5.61,2.57)
ASCII(0.045,5.61,2.77)
ASCII(0.045,5.21,2.97)
ASCII(0.045,4.91,2.92)
ASCII(0.045,4.56,2.92)
ASCII(0.045,4.56,3.42)
ASCII(0.045,4.91,3.42)
ASCII(0.045,3.96,3.27)
ASCII(0.045,3.96,3.07)
ASCII(0.045,3.66,2.72)
ASCII(0.045,3.21,2.02)
ASCII(0.08,1.96,3.22)
ASCII(0.08,5.41,3.22)
ASCII(0.08,5.81,3.22)
ASCII(0.13,5.41,2.02)
ASCII(0.13,5.86,2.02)

Figure 11-1b. Deluxe Logic Probe TOL File

Tool Symbol number representing drill size	Diameter Drill hole size in inches
Symbol 1	0.026"
Symbol 2	0.035"
Symbol 3	0.045"
Symbol 4	0.080"
Symbol 5	0.130"

Figure 11-1c. Deluxe Logic Probe TC File

Drill size and number	Number of times each drill is used
T1C0.026	5 holes drilled
T2C0.035	36 holes drilled
T3C0.045	36 holes drilled
T4C0.08	3 holes drilled
T5C0.13	2 holes drilled

The total number of holes drilled : 82

Clamp the PC board between two pieces of wood, keeping out the portion that needs to be cut. Place the wood very close to the cutting edge to avoid twisting and in turn fracturing. Use a hacksaw, coping saw, or jigsaw as a cutting tool. A guillotine cutter can also be used. It is difficult to make a straight-line cut on a PC board with a guillotine cutter. Therefore, avoid it if possible.

Commercially, cutting and final shaping of the PC board is done by using a CNC router machine. For the projects in this book, special shaping of the board is not required.

11-4. DRILLING DESIRED HOLES AND OPENINGS ON THE ENCLOSURE

Before you start this step, get the enclosure drawing for your project. Get all the off-board parts that need to be mounted on the enclosure or cabinet. Get a set of drill bits appropriate for sheet metal drilling, a jigsaw, a hand-operated sheet-metal nibbler, a fine round file, and a flat file. Use appropriate types of drill bits and other tools to make openings in the enclosure. (If the enclosure is made of sheet metal, use drill bits made for sheet-metal drilling.) Hold the work piece steady at its position while drilling. A little horizontal movement of the work piece may cause it to get caught in the bit and cause you physical injury. If possible, clamp the work piece in a drill press. If some of the holes you need to make on the enclosure are too big for a drill bit, use a sheet-metal punch. Holes that are between ½ in. and 3 in. in diameter can best be made by a hand-operated punch; however, holes whose diameter are less than ½ in. are best done by using a drill machine. A hand punch requires a guide hole of a different diameter than you need to make using a drill machine.

11-5. COMPONENT ASSEMBLY

At this point the board should be ready for mounting and for attaching components. Most electronic parts are attached to the board by soldering. Components are positioned at the appropriate place on the component side, and their leads are passed through holes to the route side. Since these laboratory boards do not have plated through holes, it may be necessary to solder the component on both sides or use eyelets. If routes are on both sides of a through hole, you need to use an eyelet or solder on both sides in order to make contact. The excess leads can be clipped off with end-cutting pliers. In places where plated through holes cannot be made, the number of via points must be restricted while you are designing the artwork with the OrCAD/PCB tools. The software can be used to design the artwork of a board without any via points. However, for complicated boards a few vias become absolutely necessary. In such cases, instead of plated through holes,

Funnel Flange

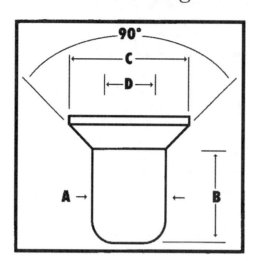

Flat Flange

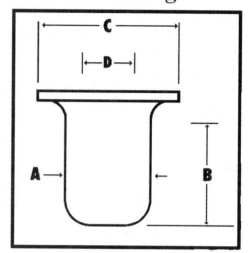

Rolled Flange

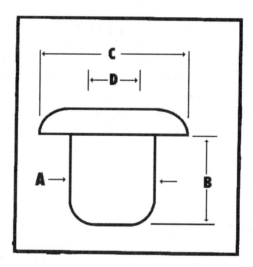

Figure 11-2a. Different Types of Eyelets
Source: Extracted from *Eyelets Ordering Guide* of International Eyelets Inc., Vista, CA, 1993.

a via is created by using an eyelet or by soldering both sides of a double- or single-ended gold or aluminum terminal. Figures 11–2a and 11-2b show various kinds of eyelets and via pins used in place of plated through vias for a PC board. The use of via pins and eyelets will make the board neater and easily solderable. Special tools are needed to place eyelets on the board. See Appendix C for the sources of these products.

The most important aspect of assembling and soldering the components on the printed circuit board and putting it in the enclosure is your patience.

11-5. Component Assembly

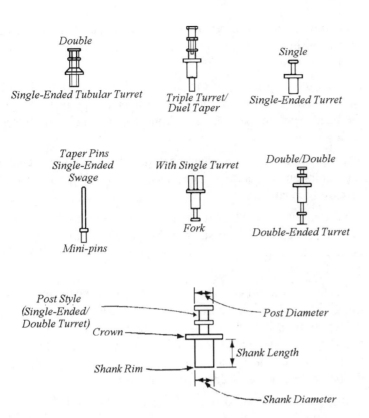

Figure 11-2b. Different Types of Via Pins and Terminal Posts
Source: Extracted from Robert Villanucci, Alexander Autgis, and William Megow, *Electronics Techniques*, Prentice-Hall, Inc., 1986, p. 259.

If you follow a few basic rules, you can save time and money and reduce your frustration.

- Get the drawings for the project: schematic; artwork; silkscreen; assembly; enclosure, if one has been designed specifically for this project; and the developmental plan.
- Get the parts for the project: processed and drilled PC board, enclosure box, and components.
- Get the tools you need: soldering iron, wire cutter, desoldering pump, "third hand," and rosin-core solder.
- Prepare leads for each component before insertion. Generally, resistors, diodes, capacitors, transistors, and SCR leads need to be prepared before inserting them in their respective places on the board.
- Integrated circuits should not be directly soldered on the board. You should use DIP (dual in-line package) sockets for them. For a non-plated through-hole you are going to need a high-profile socket so that you can solder the socket on both sides of the board. If you use via eyelets, you can use low-profile ones.

Board assembly can be best performed by following the sequence described below:

- First, mount and solder via pins and via eyelets. You may need to use special insertion tools to insert via pins and via eyelets into the PC board.
- Second, mount and solder the parts that are closer to the board—parts such as resistors. Do not mount wire-wound type resistors flush with the board surface, because heat from it may delaminate the pads and routes.
- Third, mount and solder other small and discrete components, such as transistors and capacitors. Keep them about 1/8″ away from the surface of the board.
- Fourth, mount and solder IC sockets. Try not to use IC sockets whose bottoms touch the board surface. You need the kind whose pins are designed to keep the bottom of the socket about 1/16″ from the surface of the board.
- Fifth, mount the off-board components on the enclosure—the potentiometer, speaker, power transistor that needs a large heat sink, fuse holder, switch, binding post, etc. Connect the off-board parts to the board by using insulated wire and solder-type terminal posts that you have already connected to the board. Avoid directly soldering wires to the board; always use solder-type terminal posts.
- Sixth, solder the enclosure side of the off-board components and wires. Wires that connect off-board parts to the PC board should not be too long, but just long enough so that you can easily open the box for testing, adjusting, and servicing.

Commercially, components are placed on the PC board by automatic machines and solder is applied on the desired places of the board either by floating the board on the molten solder or by a precisely controlled solder wave.

Figures 11–3a and 11–3b show the route side and the component side of the Deluxe Logic Probe project board after drilling and stuffing with eyelets.

Figures 11–4a and 11–4b, 11–5a and 11–5b, and 11–6a and 11–6b show the route side and the component side of the boards for the other three projects.

11–6. SOLDERING IRON, SOLDER, AND THEIR APPLICATION TECHNIQUES

Soldering irons are generally rated in watts. For electronic application, 20- to 60-watt irons are used. Never use a high-watt soldering iron, because excessive heat will delaminate the pads and routes. The temperature at the iron tip may range from 675°F to 950°F. Exact tip temperature depends on the iron's length and diameter. The tip of a soldering iron is interchangeable and is available in various sizes and shapes. For PC board soldering, you need only a 20- to 35-watt temperature-controlled soldering iron having a

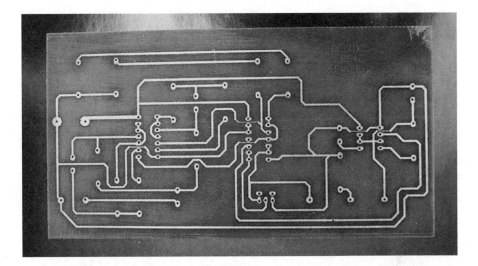

Figure 11-3a. Route Side (Layer 1) of the Logic Probe Project after Drilling and Stuffing with Eyelets

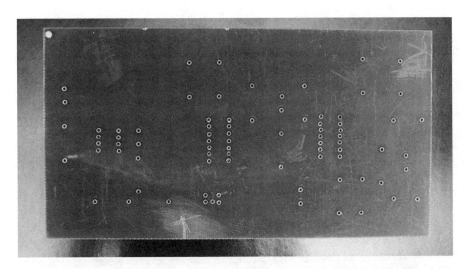

Figure 11-3b. Component Side (Layer 2) of the Logic Probe Project after Drilling and Stuffing with Eyelets

Figure 11-4a. Route Side (Layer 1) of the Electronic Cricket Project after Drilling and Stuffing with Eyelets

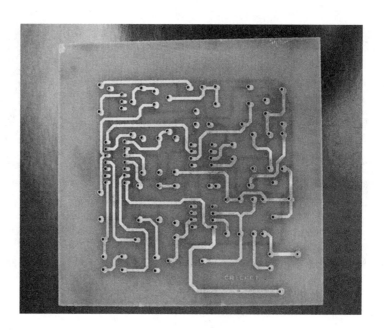

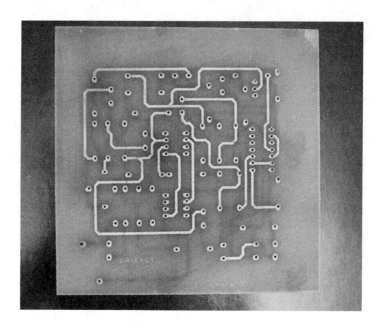

Figure 11-4b. Component Side (Layer 2) of the Electronic Cricket Project after Drilling and Stuffing with Eyelets

11-6. Soldering Iron, Solder, and Their Application Techniques

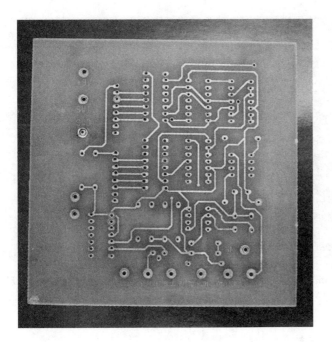

Figure 11-5a. Route Side (Layer 1) of the Infrared Object Counter Project after Drilling and Stuffing with Eyelets

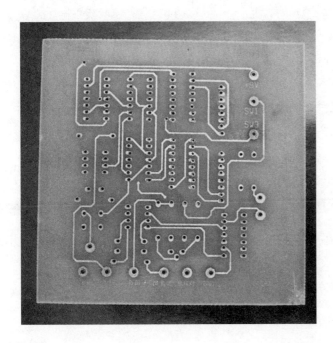

Figure 11-5b. Componet Side (Layer 2) of the Infrared Object Counter Project after Drilling and Stuffing with Eyelets

432 Chaper 11 / Assembling and Soldering of Electronic Parts on PCBs

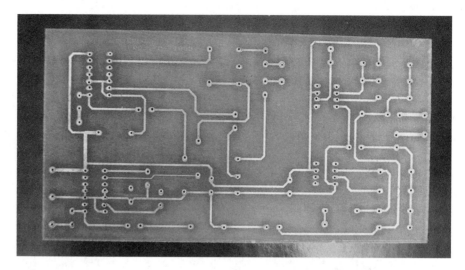

Figure 11-6a. Route Side (Layer 1) of the Mini Stereo Amplifier Project after Drilling and Stuffing with Eyelets

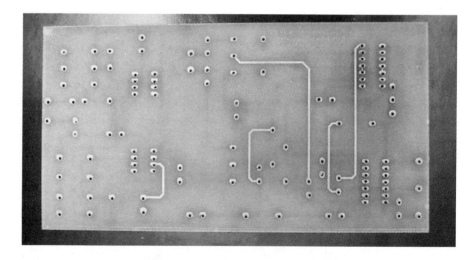

Figure 11-6b. Component Side (Layer 2) of the Mini Stereo Amplifier Project after Drilling and Stuffing with Eyelets

conical chisel-shaped tip. The tip temperature control will ensure consistent soldering.

It is not difficult to learn how to do a good soldering job. The key point about soldering is to not melt the solder on the iron tip; rather, heat the pads or routes to help the solder flow. Another key is to apply a minimum amount of solder to a joint, just enough to ensure proper alloying. Excess solder should be removed by using a solder sucker.

There are several different types of soldering flux available. Rosin flux–activated solder will give you the best results. It is noncorrosive at room temperature and attacks the oxide film on the metal surface near the melting temperature of 60/40 tin-lead solder. Use 60/40 rosin flux–core wire solder. The 60/40 number indicates the tin to lead ratio. The percentage of rosin flux is about 1% to 4% per unit volume of tin/lead alloy.

Commercially, PC board soldering is performed by a wave-soldering machine. Modern PC board assembly outfits use many other efficient soldering techniques as well.

11–6.1 Hand Soldering of PCB

Before you start soldering a PC board, secure it in a circuit board holder (also called a "third hand") and tilt it almost in a horizontal position. Rosin liquid flux should be applied by an artist's brush to all joints to be soldered. The flux works as an oxide-removing agent because it attracts the oxide film near the melting point of 60/40 tin-lead solder. One problem that may occur is that prolonged heating may upset the foil bond and the foil may get separated from the base material.

As you begin soldering with a clean soldering iron, first put a small amount of solder on the tip. This will enable the rapid transfer of heat from the tip to the joint you are soldering. Place the surface area of the entire tip in contact with the pad and the component. After approximately 1 or 2 seconds, for most joints, apply rosin flux–core wire solder to the terminal pad and component lead simultaneously, on the opposite side of where the iron tip is touching. Stop applying any more solder as soon as it starts flowing. Apply just a little solder to ensure proper alloying. If you have applied excessive solder, remove it by using a solder sucker or a desoldering bulb. Since solder follows heat flow, the entire connection, including the pad, will have a nice contour of solder. Remove the soldering iron and let the joint cool down. Do not move the PC board or the lead until the solder has completely solidified. Clean the tip of your iron by wiping it with a wet sponge or cloth. At this point you may solder another joint. If the copper routes are not covered with tin coating, you may want to coat all the routes with solder to inhibit oxidation. When soldering the terminals of off-board parts, especially large ones, you may need to apply more heat by holding the tip of the iron on the lead for about 4 to 5 seconds. If you practice and follow this soldering procedure, soldering your entire PC board will not take long.

11-6.2 Industrial Assembly and Soldering Techniques

In industrial settings, the soldering of PC boards is generally accomplished in one of the following ways: dip and wave. Wave soldering is very elaborate and thus widely acceptable in industrial environments. The large soldering capacity coupled with controllable process variables produces consistently high-quality results, making wave soldering a preferred process for the industry.

Soldering a PC board involves two major steps. They are application of flux and application of solder.

Application of flux. Liquid flux must be applied before the solder. The sole purpose of applying flux is to remove the oxide film and thereby produce a good soldering joint. Flux may be applied in four different ways: wiping, dipping, spraying, and sponging. Of these four methods, sponging is the most effective way to apply flux.

Application of solder. In dip soldering, after the application of flux, 60/40 tin/lead solder is applied by floating the board on the surface of the molten solder. In wave soldering, eutectic solder (tin/lead ration of 63/37) is applied by forming a wave of solder. Precise control of process variables such as solder temperature, board preheat temperature, fluxing, width of solder wave, and inclination of the board will result in accurate, consistent soldering of PC boards.

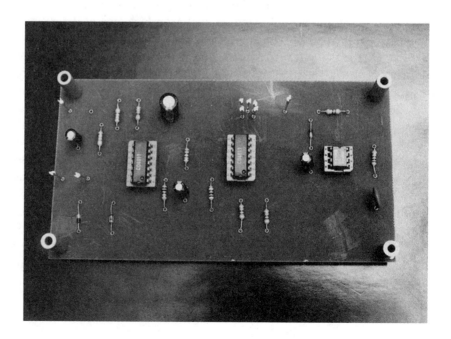

Figure 11-7. Photograph of the Assembled PC Board for the Logic Probe Project

11-7. ASSEMBLED CIRCUIT BOARDS OF THE PROJECTS

Figures 11–7 through 11–10 show the photographs of the assembled boards of the projects.

11-8. PACKAGING OF THE PROJECTS

The assembled PC board is not the end of the task. The PC board and the off-board parts need to be packaged properly into an appropriate cabinet. Packaging is done not only for aesthetic beauty, but also for convenience of operation, protection from the environment, and for safety reasons. If the assembled PC board is for a new product, the packaging must be designed to fulfill the product requirements and to improve the marketability. Designing packaging for a product requires lots of considerations. A few of these considerations are the following:

- Type of materials
- Size and shape
- Cost
- Convenience of operation
- Ease of fabrication
- Appearance
- Safety of use

Figure 11-8. Photograph of the Assembled PC Board for the Electronic Cricket Project

Figure 11-9. Photograph of the Assembled PC Board for the Infrared Object Counter Project

Figure 11-10. Photograph of the Assembled PC Board for the Mini Stereo Amplifier Project

Figure 11-11. Photograph of the Completed Logic Probe Project after Packaging

For our purpose of packaging an electronic project, it is more convenient to buy an enclosure than to design one. There are numerous companies that sell ready-made enclosures made of metal, plastic, fiberglass, or wood in many sizes and shapes.

After purchasing an enclosure for your project, you need to drill holes, cut slots, and make openings to mount the PC board and the off-board parts. Finally, functional labeling of each external component needs to be done by using dry transfer labels. If you are looking for durability in these labels, you should use engraving or metalized adhesive plates. The labeling will enable proper operation and use of the electronic circuit.

The PC board must be supported inside the enclosure by using insulated, threaded standoffs. You may also use a snap-in–type nylon circuit board support for this purpose. Supporting the board in the enclosure, either by standoffs or nylon supports, requires at least four holes, usually on corners, a little bit away from the board edge. If your PC board is interfacing with other circuits by an edge connector, you should use grooved runners made of plastic and designed for use with edge connectors.

Figures 11–11 through 11–14 show the photographs of the four completed projects after packaging.

11-9. FINAL TESTING OF THE PROJECTS

Although the projects are complete, they have little value if they do not work. Don't feel defeated; almost all projects need testing, adjusting, and troubleshooting. The keys to successful and quick testing are methodical

438 Chaper 11 / Assembling and Soldering of Electronic Parts on PCBs

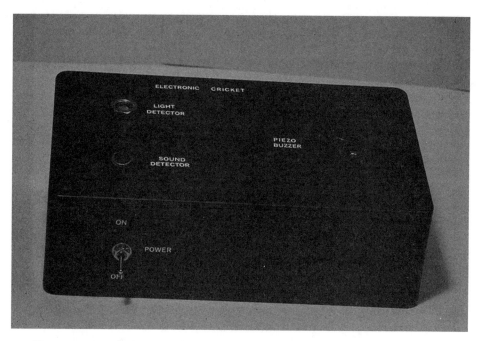

Figure 11-12. Photograph of the Completed Electronic Cricket Project after Packaging

Figure 11-13. Photograph of the Completed Infrared Object Counter Project after Packaging

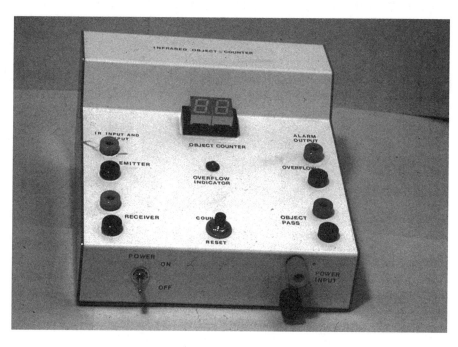

11-9. Final Testing of the Products

Figure 11-14. Photograph of the Completed Mini Stereo Amplifier Project after Packaging

approach, systematic methods, and patience. Before you start testing your project, arrange the following test equipment. Most of it you will find in your electronics laboratory: digital multimeter, regulated variable power supply, function generator, oscilloscope, logic probe, and plenty of alligator-to-alligator leads and other appropriate probes and leads for meters and oscilloscopes. The following steps will lead you to a successful conclusion to your project.

11-9.1 Visual Inspection

Get the circuit diagram, the silkscreen, and the artwork, and check for missing components or connections. Also check for wrong connections or loose connections. If everything looks all right, you need to first check the overall performance of the circuit; if it fails, then check it by isolating it into logical blocks.

11-9.2 Operational Check

First, isolate the power supply and check to see if the desired voltage is delivered by the power supply unit. If no voltage is coming out of the power supply, check the main fuse. If it is all right, get the circuit diagram of your power supply and debug other parts of it systematically. If the power supply is all right, the operation of each part of the circuit needs to be tested. Operational tests are not so complicated if you know the behavior of every part of the circuit. If the circuit operates but not to the desired extent, it may need calibration. If the electrical part of the circuit is not functioning at all, check the signal input and output transducers for component failure.

11-9.3 Calibration for Desired Performance

Check the description of the circuit operation. Your circuit may need calibration of some variable resistance, capacitor, inductor, etc. You need to know the voltage levels at various points of your circuit. Check to see if these voltages are of the desired value for the operation of the circuit. If they are all right and your circuit is still not working in the desired manner, you need to check for the wave form pattern or sequence of logic levels at various points of your circuit. For this last check, apply the appropriate input to the circuit and connect the output to the logic analyzer or oscilloscope. You need to have a thorough understanding of your circuit to be able to calibrate it properly.

11-10. A FEW TIPS ABOUT DETECTING DEFECTIVE PARTS

Many times, defective parts can be detected easily while they are in the circuit. Therefore, keep the power to the circuit on while testing the following parts. (**Important note:** Make sure that the operating voltage is not lethal. Before testing, power your circuit from the laboratory-regulated DC power supply.)

Resistors. Feel the temperature of each resistor with your fingertip. Be careful; they may be hot enough to burn your finger. If one is unusually hot, check the voltage across it. If the voltage is zero or very close to zero, the resistor is bad. Replace the resistor before proceeding further.

Transistors. Feel the temperature of each transistor with your fingertip. If a transistor is unusually hot, check the base and collector voltage with respect to the emitter. Base-to-emitter voltage is generally much smaller than collector-to-emitter voltage. If the transistor is too hot, either the transistor is bad or something unusual is making the transistor hot. Do not get confused about the large power transistors mounted on heat sinks; they are generally much hotter than the smaller ones. No base or collector voltage is an indication of a bad transistor. If base-to-emitter voltage is high, the transistor may also be bad. Take it out of the circuit and test its V_{CE}-I_C characteristics by using a transistor tester.

Diodes. Silicon diodes generally should have 0.7V across them, and germanium diodes should have 0.3V across them. Zero voltage across a diode is an indicator of a shorted diode. Much more than 0.7V across a diode is an indicator of an open diode.

11-10. A Few Tips about Detecting Defective Parts

Zener diodes. Zener diodes are generally placed in the circuit in reverse biased mode. This reverse voltage depends on the Zener diode specification. If your voltage measurement does not give you the voltage the Zener diode is supposed to have, the diode is probably defective.

Inductors. No DC voltage across an inductor is an indication of a shorted unit. Voltage across an inductor is not always an indication of a good inductor. If the inductor is open, you may find voltage across it.

Transformers. Switch your DMM to AC mode. Measure the voltage on both sides. No voltage across it is indicative of a defective unit.

Capacitors. Capacitors are least likely to fail. However, one that is unusually hot and has no voltage across it is probably a leaky or bad capacitor.

Integrated circuits (IC). There is no single test that can quickly detect a bad IC. It depends on its type. As a general rule of thumb, an IC that is too hot is likely to be bad. To be sure, you need to do a functional test. If it is a simple AND, OR, or NAND gate, check its functional table and measure its logic level with a logic probe. If a gate is not behaving according to its functional table, check its power and ground leads. If they are receiving the right voltage, the gate is defective. Since all ICs should be put on the PC board by using sockets, replacing the IC is easy. If the gate is behaving correctly but other gates are supplying an incorrect voltage level to the gate, then you need to troubleshoot the gates in the portion of the circuit that is supplying the incorrect inputs.

For an analog IC, such as an op-amp, measuring the input and the output voltage is enough to know if it is defective. In some situations, you may need to display output and input wave form on an oscilloscope to find the IC's desired performance and to know if it is defective.

CMOS ICs. CMOS ICs are sensitive to static, and you must take adequate precaution when using them. Wear a grounding strap around your wrist and ground it properly. If you suspect that one of them is bad, replace it with a good one. They may become defective during handling because of static voltage from your body.

In all, defective ICs may puzzle you by displaying weird behavior in your circuit, so be careful about them. As you find a defective one, replace it. If you burn the new one in the same place for the second time, you must now find the source of the fault before replacing the IC for the third time. *Twice burned* means that you must find the fault before replacing an IC or any other component.

QUESTIONS

1. What is the major difference between PC board drilling in industry and in the laboratory?

2. Why are drill bits with carbide tips required for PC board drilling?

3. What are the two most common ways that PC boards are soldered in industry?

4. Why is rosin flux applied to a joint before soldering?

5. What type of solder is generally used in industry for soldering PC board?

6. Why is a PC board packaged carefully in an enclosure?

7. Generally, what is the most expensive part of an electronic product?

MICROCOMPUTERS AND DOS COMMANDS

STARTING THE MICROCOMPUTER

Starting a computer system basically includes loading the Disk Operating System (DOS) into computer memory. All microcomputer systems are run by a set of assembly-level codes called Basic Input and Output System (BIOS). This code is permanently stored in the computer's read only memory (ROM). This code is specific to a computer's hardware, and it loads DOS into computer memory. Once DOS is loaded, it takes over most of the system operations. If the microcomputer has a hard disk, then DOS is usually located in the hard disk. During the starting process, DOS is loaded from the hard disk to system memory. However, DOS can also be loaded from one of the floppy drives. One should have the option of loading DOS either from hard disk or from floppy. This means that the system configuration should be such that during booting, BIOS should go to one of the floppies first and then to hard disk for loading DOS into the system memory. If the system configuration is made to load DOS only from hard disk, one may encounter serious problems in the case of a defective hard drive. For most microcomputers, starting or restarting can be done in several different ways. A few common ways are described below.

Common DOS Commands

Disk FORMAT. Formatting a disk prepares it to be acceptable for data storage by DOS. DOS cannot store data on an unformatted disk. Formatting a disk is almost like allocating space for storing filenames and their whereabouts on the disk. (**Caution:** When a floppy or a hard disk is formatted, all data that was stored earlier on the disk will be lost.) It is strongly recommended that you not format the hard disk without a proper knowledge of its contents. Formatting a hard disk is much more complicated and time-consuming than formatting a floppy disk. Floppy and hard disks are made for various data storage capacities:

- $5\frac{1}{4}''$ double-density disks are generally formatted by DOS for approximately 360 Kbytes of data storage.
- $5\frac{1}{4}''$ high-density disks are generally formatted by DOS for approximately 1.2 Mbytes of data storage.
- $3\frac{1}{2}''$ double-density disks are generally formatted by DOS for approximately 720 Kbytes of data storage.
- $3\frac{1}{2}''$ high-density disks are generally formatted by DOS for approximately 1.44 Mbytes of data storage.

It should be remembered that double-density disk drives can read from and write into double-density disks of the same size. On the other hand, high-density disk drives can read from and write into both high- and double-density disks.

A disk is formatted by placing it in drive A or B and by typing the following command at the DOS C:\> prompt.

C:\>FORMAT A: [Enter ⏎] → Command for a disk in drive A
C:\>FORMAT B: [Enter ⏎] → Command for a disk in drive B

Once the format command is initiated, the microcomputer system will display various messages depending on the DOS version or type of DOS (such as MS-DOS, IBM-DOS, or Zenith DOS) that exists in the computer. Generally, the following types of messages will be displayed on the screen:

Insert the disk in drive A: and press enter when ready. . . .
checking existing format. . .

Formatting 1.44M (if the disk is $3\frac{1}{2}''$) or
Formatting 1.2M (if the disk is $5\frac{1}{4}''$)

Formatting Completed. . . .

Volume label will provide a label for the formatted disk. At this point, press the **Enter** key if the label ID is not desired. DOS may prompt for formatting another disk, but for the present disk, formatting has been completed.

Change directory. Change Directory is the first command that is necessary to enter into an OrCAD subdirectory. DOS commands are rarely needed once the user enters the ESP design environment. If the OrCAD subdirectory is in the C drive, use the following command to go to the MYORCAD subdirectory. MYORCAD is a subdirectory where all the OrCAD files are stored; it is located within the root directory, drive C in this case.

At C:\> prompt, type :cd\myorcad [Enter ⏎]
or type :cd myorcad [Enter ⏎]
Response: C:\MYORCAD>

At C:\> prompt,   type : cd\myorcad\orcad [Enter ↵]
                   Response: C:\MYORCAD\ORCAD>

At C:\MYORCAD> prompt, type: orcad [Enter ↵]

This last command will invoke the OrCAD ESP_MENU, and the user will be in the ESP design environment.

Change drive.   Moving from one drive to another is necessary, especially when the user stores design files on floppy disks. To change directories, type the following to go from C drive to A drive, from C drive to B drive, and from A drive back to C drive, respectively.

At C:\> prompt, type: a: [Enter ↵]
          Response: A:\>

At C:\ prompt, type: b: [Enter ↵]
          Response: B:\>

At A:\ prompt, type: c:, [Enter ↵]
          Response: C:\>

Make directory.   Make Directory is another DOS command that may be necessary before the installation of the software. Let us suppose that a subdirectory named MYORCAD needs to be created within the root directory, C. To do this, use the following:

At C:\> prompt, type md\myorcad [Enter ↵]
          Response: C:\>

The MYORCAD subdirectory will be created, and the C:\> prompt will be returned. Now if the change directory command is executed, the prompt will be C:\MYORCAD>.

Display the files belonging to a directory.   This command displays the files stored on a disk or within a subdirectory on a disk. The command may be used in one of three ways, depending on the need.

At C:\> prompt, type: DIR; this will display files in C drive, including subdirectories within root C directory.

At C:\MYORCAD> prompt, type: DIR; this will display files in the MYORCAD subdirectory, including any directory within MYORCAD.

If the DIR command is followed by /P, there will be a pause when a full screen of information is displayed. Pressing the ENTER key will display the next screen, and so on. If the DIR command is followed by /W, a wide version of the directory display will be shown on the screen. The display mode will not show the date and size information of the files.

Wildcards.   Wildcards are used in the DOS environment to shorten the command line expression for filenames and their extensions.

Appendix A / Microcomputers and DOS Commands

- ? mark is used to represent any single character.
- * mark is used to represent any group of characters.

For example:

- C:\>DIR*.* [Enter ↲] will display all files and directories in C drive. The star to the left of the period calls for all file names, and the star to the right calls for files with all extensions.
- C:\>DIR *.SCH [Enter ↲] will display only the files with an SCH extension.
- C:\>DIR DESIGN.* [Enter ↲] will display files having filename DESIGN and any extension.
- C:\>DIR D*.SCH [Enter ↲] will display files whose filename starts with D and has an extension of SCH.
- C:\>DIR A: [Enter ↲] will display all files in the disk currently in the A drive.
- C:\>DIR\ORCAD [Enter ↲] will display all files in the directory named ORCAD.

COPY one or more files. The **COPY** command is often used in the DOS environment to copy files from one disk to another or from one filename to another. While you are working in the OrCAD EDA environment, this command can be used in places where DOS commands are recognized. A COPY command is also available in Design Management Tools, but that COPY command is not command-line-oriented like the DOS COPY command. Examples:

- C:\>COPY C:CIRCUIT1SCH A:CIRCUIT1.SCH [Enter ↲] will copy a file with name and extension CIRCUIT1.SCH from C drive to A drive.
- C:\>COPY C:CIRCUIT1SCH A: [Enter ↲] will produce the same result as the earlier command line.
- C:\>COPY C:CIRCUIT1SCH A:CIRCUIT2.SCH [Enter ↲] will copy a file with name and extension CIRCUIT1.SCH from C drive to A drive with name and extension as CIRCUIT2.SCH.
- C:\>COPY C:*.BRD A:*.BRD [Enter ↲] or C:\>COPY C:*.BRD A: [Enter ↲] will copy all files with BRD extension to drive A with same extension.

Unlike mainframe computers, DOS keeps only one version of a file, with a backup file in a directory. Two files with exactly the same name and extension cannot stay in a directory under the DOS environment. The newer file will overwrite the older one.

Starting the Microcomputer

Delete one or more files. The **Delete** command is abbreviated as DEL. This is often used under the DOS environment to delete a file or files from a disk. While you are working in the OrCAD EDA environment, this command can be used in places where DOS commands are recognized. Delete a File and Delete a Design commands are also available with Design Management Tools, but they are not command-line-oriented like the DOS DEL command. Examples:

- C:\>DEL C:\ORCAD\CIRCUIT1.SCH [Enter ↵] will delete a file with filename and extension CIRCUIT1.SCH located in a directory named ORCAD in C drive.
- C:\>DEL C:\ORCAD\DESIGN*.* [Enter ↵] will delete all files in the DESIGN subdirectory in C drive.
- C:\>DEL C:\ORCAD\DESIGN*.BRD [Enter ↵] will delete all files with BRD extension in the DESIGN subdirectory in C drive.
- C:\>DEL A:CIR*.SCH [Enter ↵] will delete all files in A drive that have extension SCH and have CIR as the first three letters of the filename.

Remove a directory. Although this DOS command is not commonly used, it is an important one. To remove a directory, first all the files belonging to the directory must be deleted. Then only the **RD** (Remove Directory) command can be issued to remove the directory. For example, assume that a directory named TEST and all the files belonging to the TEST directory are to be removed. First, within the TEST directory type as follows: C:\TEST>del*.* [Enter ↵]. This is a wildcard to remove all files from the TEST directory. Second, type C:\TEST.cd [Enter ↵]. This will shift the control to root directory. Third, type C:\>rd test [Enter ↵]. This command will remove the directory from the C drive.

B

OrCAD/PCB MODULES BY THEIR PACKAGE NAMES

I have mentioned in several places in this text that a circuit schematic is a scheme of logical connections among electrical symbols. Symbols are representation of electrical parts. In a schematic, the physical size of a part does not matter. Logical characteristics and connections among parts are important. However, for a PCB layout, the physical size of each part and its logical connections both have to be known. The PCB layout tool converts logical network connections of a schematic into a drawing of physical tracks.

When the SDT produces a netlist file, it lists nets, reference designator, part value, and package number for each part. A net is a path connecting two or more pins on a PCB layout. When a netlist file is invoked from PC Board Layout Tools, the software uses the packaging number provided with each part of a schematic (content of Part Field 1) to download the appropriate module from the library. It also downloads nets among parts. These nets are later converted into a drawing of physical tracks.

This appendix provides the complete list of package numbers of modules available in OrCAD/PCB II. Actual shapes of modules are provided in Appendix E. The OrCAD PCB II library has many more modules than what is provided here. If you want, you can print a hard copy of all the modules by using the procedure described later in this appendix. Each of these pictures provides actual physical size, shape, number, and location of pins, and package number for each module. Each module is stored in the PCB module directory by its package number. For example, a package number for a 14-pin Dual-In-Line Package (DIP) is 14DIP300.U***, SW***, R***, etc., are reference designators of all parts defined in the schematic.

How to print a hard copy of PCB modules. **Step 1:** Print a list of package numbers of all the available PCB modules. A directory listing is available in the following subdirectory: C:\>MYORCAD\ORCADESP\PCB\MODULE.

While you are in the MODULE subdirectory, print the listing by using the following command:

C:\>MYORCAD\ORCADESP\PCB\MODULE>DIR >PRN

However, you may save the list in the form of a text file. The following command will allow you to send the list of modules to a file:

C:\>MYORCAD\ORCADESP\PCB\MODULE>DIR >A:MOD_LIST.TXT

Table 1 is the list of package numbers of modules available in OrCAD/PCB II. You will need to refer to this list to download modules onto a board.

Step 2: To print the actual shape of a module, you need to fetch the module onto a board. Look at the list and fetch modules you need for a layout worksheet. Each module needs to be fetched by its package number. Save the board file (with BRD extension), and later you can print it using the Print PCB tool. Modules use lots of memory, so you may be able to print only a limited number of modules per board. If you assign more module memory, the number of modules per board will increase.

Table 1. List of Module Packages Available in OrCAD/PCB

		SEVEN SEGMENT LEDs				
TIL302	TIL303	TIL304	TIL305	TIL306	TIL307	
TIL308	TIL309	TIL311				
		CAPACITORS				
CK05	CK06	CK12	CK13	CK14	CK15	
CK16	CK17	CK62	CK62			
CL23D1	CL23D2	CL23D3	CL23D4	CL23D5	CL25C1	
CL25C2	CL25C3	CL25C4	CL25C5	CL65T1	CL65T2	
CL65T3						
CS13A	CS13B	CS13C	CS13D			
CM05	CM06	CM07	CM08			
CY10	CY15	CY20				
CC0805	CC1206	CC1210	CC1812	CC1825		
TC3216	TC3518	TC3527	TC3528	TC6032	TC7227	
TC7243	TC7257					
		CONNECTORS				
10CON150	20CON200	26CON200	34CON200	40CON200	50CON200	
DB15F	DB15M	DB25F	DB25M	DB37F	DB37M	DB50F
DB50M	DB9M	BDNCF	RBNCF	IBMJ1	IBMJ2	DB9F

Table 1. Continued List of Module Packages Available in OrCAD/PCB

CRYSTALS					
HC18UH	HC18UV	HC25UV	HC33UH	HC33UV	HC6UV
XO51B	MODULAR				

DIODES					
DO13	DO14	DO41	DO7	PROTOTYP	

IC PACKAGES					
4DIP300	6DIP300	8DIP300	14DIP300	16DIP300	18DIP300
20DIP300	22DIP400	24DIP300	24DIP400	24DIP600	28DIP600
32DIP600	40DIP600	48DIP600	52DIP600	64DIP600	68DIP600
MO–00310	MO–00314	MO–00410	MO–00414	MO–00416	MO–01840
MO–01924	MO–01928	MO–02036	MO–02040	MO–02116	MO–02124
MO–02136	MO–02220	MO–02242	MO–02336	MO–02350	

HEADERS					
10HH100	14HH100	16HH100	20HH100	26HH100	34HH100
40HH100	50HH100	10HV100	14HV100	16HV100	20HV100
26HV100	34HV100	40HV100	50HV100	60HV100	8IDI300

SURFACE MOUNT DEVICES					
14IDI300	16IDI300	18IDI300	20IDI300	22IDI400	28IDI600
24IDI600					
WWA	WWB				
JEDA28	JEDA44	JEDA52	JEDA68		
18RJPCC	22RJPCC	28RJPCC	32RJPCC	100SJPCC	124SJPCC
20SJPCC	28SJPCC	44SJPCC	52SJPCC	68SJPCC	84SJPCC
28SJEDC	44SJEDC	52SJEDC	68SJEDC	100SJEDC	124SJEDC
156SJEDC	84SJEDC				
14JSOIC	16JSOIC	18JSOIC	20JSOIC	22JSOIC	24JSOIC
26JSOIC	28JSOIC				
124LCC50	156LCC50	16LCC50	18LCC50	20LCC50	24LCC50
28LCC50	32LCC50	44LCC50			
52LCC50	68LCC50	68LCCC	84LCC50	84LCCC	100LCC50
84MPLCC	100MPLCC	132MPLCC	164MPLCC	196MPLCC	244MPLCC
QUAD100	QUAD44	QUAD48	QUAD52	QUAD64	QUAD70
QUAD80	QUAD54				
PRC05	PRC06	PRC07	PRC08		

Table 1. Continued List of Module Packages Available in OrCAD/PCB

RESISTORS

RC05	RC06	RC07	RC08	RC12	RC20	RC22	RC32	RC42
RL05	RL07	RL20	RL32	RL42				
RN50	RN55	RN60	RN65	RN70	RN75	RN80	RWP20	
MK620	MK632							
RB70	RB71							
RC0805	RC1206	RC1210						

MULTITURN POTENTIOMETERS

RJ24X RJ26W RJ26X

SINGLE LINE PACKAGES

4SIP100 5SIP100 6SIP100 7SIP100 8SIP100 9SIP100 10SIP100
22SIP100 24SIP100 30SIP100

SURFACE MOUNT DEVICES

SO–8 SO–14 SO–16 SOL–14 SOL–16 SOL–24 SOL–28
8SOP150 14SOP150 16SOP150 16SOP300 18SOP300
20SOP300 24SOP300 28SOP300

SWITCHES

RCKRDPDT	RCKRSPDT			
TGLDPDT	TGLDPDTR	TGLDPDTV	TGLSPDT	TGLSPDTR
TGLSPDTV				

TEST POINTS/TESTJACKS/MOUNTING HOLES

TSTPT1 TSTPT2 TSTPT3
TJACK200 TJACK400
MHOLE1 MHOLE2 MHOLE3

SURFACE MOUNT DEVICES

MELF MLL34 MLL41
SOD80
DST4
ST4

TRANSISTOR

TO107 TO12 TO17 TO18 TO218 TO218H TO220
TO220H TO3 TO39 TO46 TO5 TO52 TO72 TO92

SURFACE MOUNT DEVICES

OPA512
SOT23 SOT89 S14DIP

C

POTENTIAL SOURCES FOR PC BOARD FABRICATION NEEDS

PC board design and fabrication is a vast area. There are only a few companies that deal with every aspect of PC board design and fabrication. Ours is a world of specialization, and companies specialize in certain areas. Some of these areas of specialization are the following:

1. Software tools for schematic and artwork design, such as schematic and artwork design tools, simulation tools, and thermal analysis tools

2. Process chemicals for fabricating PC boards, such as etchants, developer, electroless copper for plated through-holes, and photoresists

3. Tools and machinery for PC board processing, such as CNCs, and drill bits and shaping tools

4. PC board fabrication and testing, such as fabricating single-, double-, and multilayer board with fine tracks (1 to 5 mil width); testing boards with state-of-the-art scanners; and detecting registration errors of the inner layer of multilayer boards

5. PC board assembling, such as populating boards with components by using automatic machines and testing the functionality of boards

6. Imaging machinery, such as high-resolution laser photoplotters (½ mil resolution) and high-resolution artwork printing services

It is not possible to list here every company that deals with the hardware, software, and machinery related to PCB design and fabrication. I am providing only the sources that I feel may be useful to you. You may find additional sources in PCB design and fabrication-related magazines. The telephone numbers and addresses shown here were those that were available to me at the time of writing this book. The companies are grouped according to their area of specialization. It may happen that some of the companies listed below have added a few more areas or that some may have narrowed their area of specialization. This means that you may need to launch your own search process for finding additional sources.

Potential Sources for PC Board Fabrication Needs

Sources	PCB Fabrication	PC Boards Sensitized	Imaging Service	Process Chemicals	Eyelets & Drill Bits	Tools & Machinery
PCB Kits, 750 Mariday, Lake Orion, MI 48362, (313) 693-0328		✓		✓		
Kepro Circuit Systems, Inc., 630 Axminister Drive, Fenton, Missouri 63026-2992, (800) 325-3878		✓		✓		✓
Shipley Company, Inc., 3945 Freedom Circle, Suite 370, Santa Clara, CA 95054, (800) 546-0974				✓		✓
Precision Carbide Tools Co. Inc., 14567 Big Basin Way, Suite 1-B, Saratoga, CA 95070, (408) 452-0180						✓
Accutronics, Inc., 225 North First Street, Cary, Il 60013, (708) 639-2102	✓					
General Technology Corp., 6816 Washington NE, Albuquerque, NM 87109, (505) 345-5591	✓					
Q-Image Corp., 3375 Scott Blvd., Suite 128, Santa Clara, CA 95054, (800) USA-PLOT			✓			
Express Photo, 1108 West Evelyn Ave., Sunnyvale, CA 94086, (408) 735-7137			✓			
CCS Technologies Corp., 226 North Sherman Ave., Suite A, Corona, CA 91720						✓
International Eyelets Inc., 1240 Keystone Way, Vista, CA 92083, (800) 333-9353					✓	✓
Edmund Scientific Co., 101 E. Gloucester Pike, Barrington, NJ 08007, (606) 573-6270						✓
Dumore Corporation, 1300 17th Street, Racine, WI 53403						✓
Excel Circuit Co., Inc., 32096 Howard, Madison Heights, MI 48071, (313) 588-5100				✓		

A QUICK USER GUIDE TO OrCAD/SDT 386+ AND ORCAD/PCB 386+

Changes in the hardware and software power of the desktop personal computer are so great nowadays that it is sometimes difficult to keep up. Recently, OrCAD has come up with new versions of SDT and PCB. They are called OrCAD/SDT 386 Plus (+) and OrCAD/PCB 386 Plus (+). SDT 386+ has a few minor changes, but PCB 386+ has changed considerably. I am going to show you some of these changes and the procedures to use them. You may not be able to use the full power of PCB 386+ tools from this explanation, but at least it will get you started. The example I am using for this explanation is a full adder circuit.

SCHEMATIC CAPTURE AND NETLIST GENERATION

Schematic capture and netlist generation using OrCAD/SDT 386+ are almost the same as before, with a few changes. The changes are an added tool called **Check Design Integrity,** a new configuration for Key Fields, and a new netlist format (FEDIF.EXE instead of PCBII.CCF). Figure D1 shows the new main menu, called ESP386_MENU, and Figure D2 shows the SDT menu screen, here called SDT386_MENU. If you look carefully, you will see that there is one additional tool in this menu located under **Processors,** and that is **Check Design Integrity.** Design integrity is a three-step process. It does all the three following operations in one shot.

- **Cleanup Schematic:** This part of the tool scans the schematic and removes duplicate or overlapping wires, buses, and junction boxes. The process provides warning messages for other duplicate parts. It will also remove error marks from the schematic.
- **Crossref Check:** This part of the tool scans the schematic and reports to you about identical reference designators, parts with type mismatch, etc. To find this problem, the tool will sort the parts by their

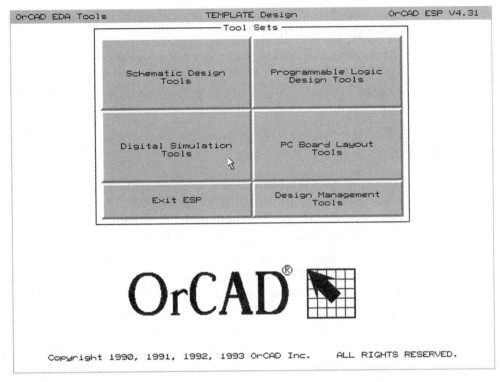

Figure D1. ESP Menu for OrCAD 386 Plus (ESP386_MENU) (Courtesy of OrCAD Inc.)

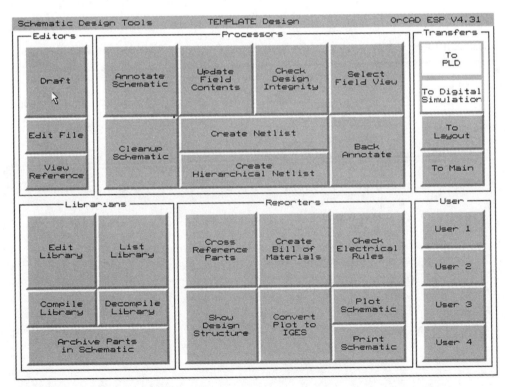

Figure D2. SDT Menu for OrCAD 386 Plus (SDT386_MENU) (Courtesy of OrCAD Inc.)

part values and then by their reference designators. This tool is an addition to OrCAD/SDT 386+.

- **Check Electrical Rules:** This part of the tool scans the circuit and checks the electrical rules defined in the Electrical Rules Matrix. Under default condition, it reports all unconnected wires, pins, module ports, etc.

Procedure

1. Capture a desired schematic. For an example, I have captured a full adder (Design Name: FADDER; Schematic File: FADDER.SCH) circuit. Figure D3a shows the full adder schematic, and Figure D3b shows the half adder schematic referenced by both the sheet symbols.

2. Since the schematic has a complex hierarchical file structure, you need to convert it from complex to simple. To do this, you need to use the **Complex to Simple** tool located under the Design Management tool set menu. Go to Design Management Tools and convert the schematic to simple hierarchical. **Note:** In the original full adder schematic, both the sheet symbols referenced the same half adder schematic. Before generating the netlist for the full adder, it

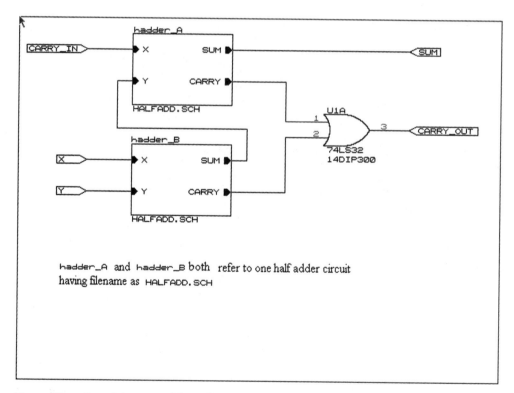

Figure D3a. Root Schematic of the Full Adder Circuit

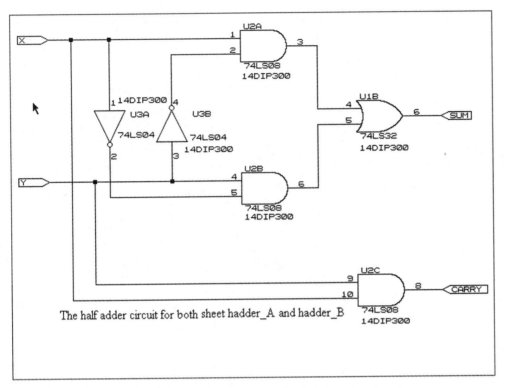

Figure D3b. Schematic of the Half Adder Referenced by Both Hadder_A and Hadder_B

needs to be changed because PCB needs a unique part reference for each part in the entire schematic. This requirement is for the physical existence of the parts on the board layout. This means that each sheet symbol must refer to two different half adder schematics. Both of them must have parts with unique part reference designators.

3. After Complex to Simple conversion, you must work with the simple schematic instead of the original one. In this case, the name of the simple schematic is SFADDER.SCH. Annotate the schematic by using the Annotate tool. Figures D3c, D3d, and D3e show the root of the full adder schematic and the two half adder circuits with unique part reference. Figure D3c is the root schematic, and Figure D3d and Figure D3e are the two schematics of the half adder.

4. Archieve the schematic parts into a single library.

5. Configure Key Fields located under Schematic Design Tools Configuration and make the changes shown in Figure D4a. OrCAD 386+ requires you to keep the package number in Part Field 8, instead of Part Field 1 for PCB II.

458 Appendix D / Quick User Guide to OrCAD/SDT 386+ and OrCAD/PCB 386+

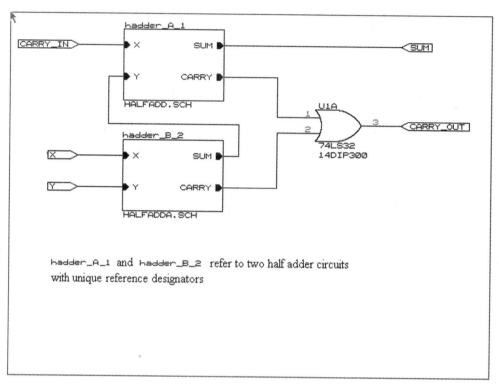

Figure D3c. Root Schematic of the Full Adder after Being Processed by the Complex to Simple Tool

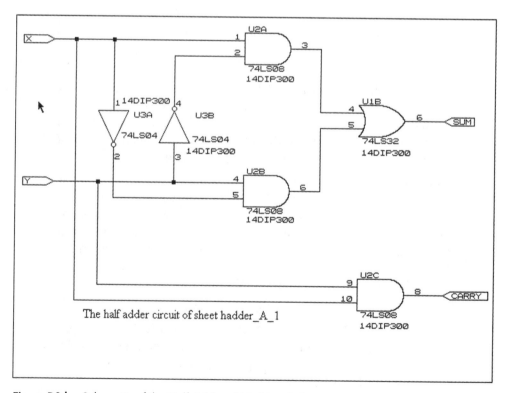

Figure D3d. Schematic of the Half Adder of Hadder_A_1

Schematic Capture and Netlist Generation

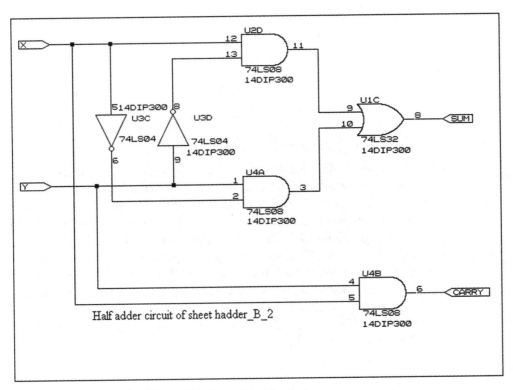

Figure D3e. Schematic of the Half Adder of Hadder_B_2

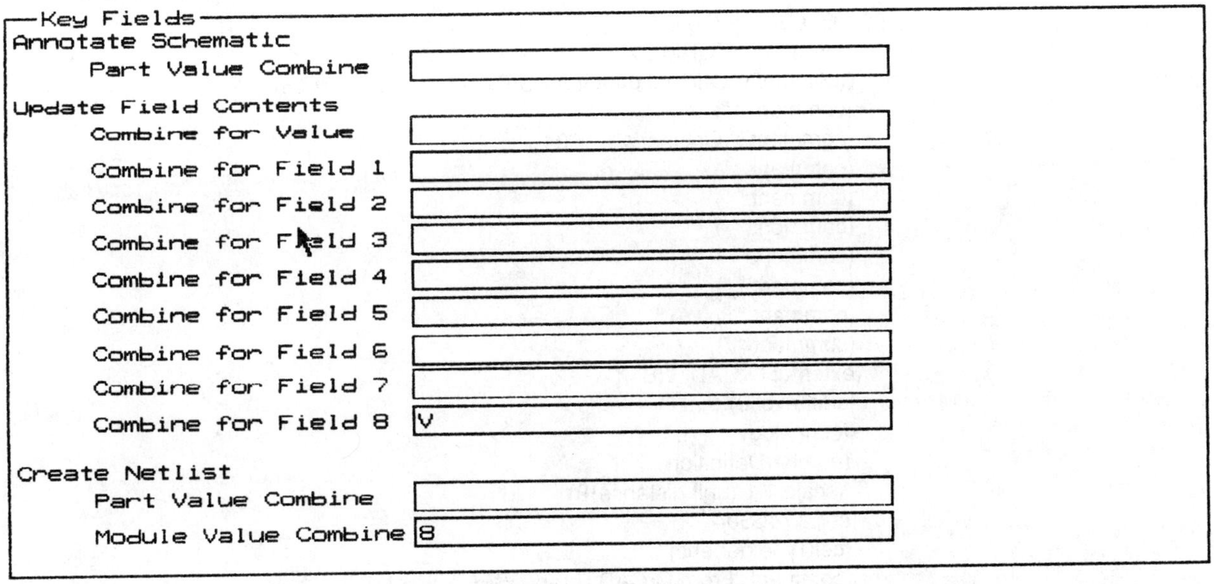

Figure D4a. Key Fields Configuration Screen (Courtesy of OrCAD Inc.)

6. Create an Update file with an STF extension. The update file for the full adder circuit is given below:

        '74LS04'   '14DIP300'

        '74LS08'   '14DIP300'

        '74LS32'   '14DIP300'

7. Execute the **Update Field Contents** tool.
8. Configure INET and ILINK. When you are configuring IFORM, instead of using PCBII.CCF as the netlist formatter, select FEDIF.EXE. FEDIF.EXE is the formatter for the OrCAD/PCB 386+ tool. This formatter will generate netlist files of different format. This netlist file is quite large. Figure D4b shows the netlist file of the full adder circuit for PCB 386+ tools. If the software does not generate the correct netlist or gives you an error, check the version and date of the FEDIF.EXE file. The file is located under the ORCAD\ORCADESP\SDT\NETFORMS subdirectory. According to OrCAD, the date code of this file should be 09-03-93. If every step goes successfully, you now need to go to PCB Tools for layout design.

Figure D4b. Netlist File for the Full Adder Circuit

```
(edif &SFADDER
 (edifVersion 2 0 0)
 (edifLevel 0)
 (keywordMap (keywordLevel 0))
 (status
  (written
   (timeStamp 0 0 0 0 0 0)
   (program "IFORM.EXE")
   (comment "Original data from OrCAD/SDT schematic"))
  (comment " ")
  (comment "  October 13, 1994")
  (comment " ")
  (comment " ")
  (comment " ")
  (comment " ")
  (comment " ")
  (comment " ")
  (comment " "))
 (external OrCAD_LIB
  (edifLevel 0)
  (technology
   (numberDefinition
    (scale 1 1 (unit distance))))
  (cell &74LS04
   (cellType generic)
   (comment "From OrCAD library .\SFADDER.LIB")
   (view NetlistView
    (viewType netlist)
```

Figure D4b. Continued

```
  (interface
   (port &1 (direction INPUT))
   (port &3 (direction INPUT))
   (port &5 (direction INPUT))
   (port &9 (direction INPUT))
   (port &11 (direction INPUT))
   (port &13 (direction INPUT))
   (port &2 (direction OUTPUT))
   (port &4 (direction OUTPUT))
   (port &6 (direction OUTPUT))
   (port &8 (direction OUTPUT))
   (port &10 (direction OUTPUT))
   (port &12 (direction OUTPUT))
   (port &14 (direction INPUT))
   (port &7 (direction INPUT)))))
 (cell &74LS08
  (cellType generic)
  (comment "From OrCAD library .\SFADDER.LIB")
  (view NetlistView
   (viewType netlist)
   (interface
    (port &1 (direction INPUT))
    (port &4 (direction INPUT))
    (port &9 (direction INPUT))
    (port &12 (direction INPUT))
    (port &2 (direction INPUT))
    (port &5 (direction INPUT))
    (port &10 (direction INPUT))
    (port &13 (direction INPUT))
    (port &3 (direction OUTPUT))
    (port &6 (direction OUTPUT))
    (port &8 (direction OUTPUT))
    (port &11 (direction OUTPUT))
    (port &14 (direction INPUT))
    (port &7 (direction INPUT)))))
 (cell &74LS32
  (cellType generic)
  (comment "From OrCAD library .\SFADDER.LIB")
  (view NetlistView
   (viewType netlist)
   (interface
    (port &1 (direction INPUT))
    (port &4 (direction INPUT))
    (port &9 (direction INPUT))
    (port &12 (direction INPUT))
    (port &2 (direction INPUT))
    (port &5 (direction INPUT))
    (port &10 (direction INPUT))
    (port &13 (direction INPUT))
    (port &3 (direction OUTPUT))
    (port &6 (direction OUTPUT))
    (port &8 (direction OUTPUT))
    (port &11 (direction OUTPUT))
    (port &14 (direction INPUT))
    (port &7 (direction INPUT))))))
```

Figure D4b. Continued

```
(library MAIN_LIB
 (edifLevel 0)
 (technology
  (numberDefinition
   (scale 1 1 (unit distance))))
 (cell &SFADDER
  (cellType generic)
  (view NetlistView
   (viewType netlist)
   (interface
    (port &CARRY_IN (direction INPUT))
    (port &SUM (direction OUTPUT))
    (port &CARRY_OUT (direction OUTPUT))
    (port &X (direction INPUT))
    (port &Y (direction INPUT)))
   (contents
    (instance &U3
     (viewRef NetlistView
      (cellRef &74LS04
       (libraryRef OrCAD_LIB)))
     (property PartValue (string "74LS04"))
     (property ModuleValue (string "14DIP300"))
     (property TimeStampValue (string "0735A9FE"))
     (property Field1Value (string " "))
     (property Field2Value (string " "))
     (property Field3Value (string " "))
     (property Field4Value (string " "))
     (property Field5Value (string " "))
     (property Field6Value (string " "))
     (property Field7Value (string " "))
     (property Field8Value (string "14DIP300")))
    (instance &U2
     (viewRef NetlistView
      (cellRef &74LS08
       (libraryRef OrCAD_LIB)))
     (property PartValue (string "74LS08"))
     (property ModuleValue (string "14DIP300"))
     (property TimeStampValue (string "0735AA02"))
     (property Field1Value (string " "))
     (property Field2Value (string " "))
     (property Field3Value (string " "))
     (property Field4Value (string " "))
     (property Field5Value (string " "))
     (property Field6Value (string " "))
     (property Field7Value (string " "))
     (property Field8Value (string "14DIP300")))
    (instance &U4
     (viewRef NetlistView
      (cellRef &74LS08
       (libraryRef OrCAD_LIB)))
     (property PartValue (string "74LS08"))
     (property ModuleValue (string "14DIP300"))
     (property TimeStampValue (string "0735A9FA"))
     (property Field1Value (string " "))
     (property Field2Value (string " "))
```

Figure D4b. Continued

```
    (property Field3Value (string " "))
    (property Field4Value (string " "))
    (property Field5Value (string " "))
    (property Field6Value (string " "))
    (property Field7Value (string " "))
    (property Field8Value (string "14DIP300")))
(instance &U1
 (viewRef NetlistView
  (cellRef &74LS32
   (libraryRef OrCAD_LIB)))
    (property PartValue (string "74LS32"))
    (property ModuleValue (string "14DIP300"))
    (property TimeStampValue (string "0735AA06"))
    (property Field1Value (string " "))
    (property Field2Value (string " "))
    (property Field3Value (string " "))
    (property Field4Value (string " "))
    (property Field5Value (string " "))
    (property Field6Value (string " "))
    (property Field7Value (string " "))
    (property Field8Value (string "14DIP300")))
(net N00001
 (joined
  (portRef &3 (instanceRef &U2))
  (portRef &4 (instanceRef &U1))))
(net N00002
 (joined
  (portRef &2 (instanceRef &U2))
  (portRef &4 (instanceRef &U3))))
(net N00003
 (joined
  (portRef &5 (instanceRef &U1))
  (portRef &6 (instanceRef &U2))))
(net N00004
 (joined
  (portRef &2 (instanceRef &U3))
  (portRef &5 (instanceRef &U2))))
(net N00005
 (joined
  (portRef &11 (instanceRef &U2))
  (portRef &9 (instanceRef &U1))))
(net N00006
 (joined
  (portRef &13 (instanceRef &U2))
  (portRef &8 (instanceRef &U3))))
(net N00007
 (joined
  (portRef &10 (instanceRef &U1))
  (portRef &3 (instanceRef &U4))))
(net N00008
 (joined
  (portRef &6 (instanceRef &U3))
  (portRef &2 (instanceRef &U4))))
(net N00009
 (joined
```

Figure D4b. Continued

```
            (portRef &1 (instanceRef &U1))
            (portRef &8 (instanceRef &U2))))
          (net N00010
           (joined
            (portRef &2 (instanceRef &U1))
            (portRef &6 (instanceRef &U4))))
          (net N00011
           (joined
            (portRef &9 (instanceRef &U2))
            (portRef &3 (instanceRef &U3))
            (portRef &4 (instanceRef &U2))
            (portRef &8 (instanceRef &U1))))
          (net &CARRY_IN
           (joined
            (portRef &CARRY_IN)
            (portRef &10 (instanceRef &U2))
            (portRef &1 (instanceRef &U3))
            (portRef &1 (instanceRef &U2))))
          (net &SUM
           (joined
            (portRef &SUM)
            (portRef &6 (instanceRef &U1))))
          (net &VCC
           (joined
            (portRef &14 (instanceRef &U1))
            (portRef &14 (instanceRef &U3))
            (portRef &14 (instanceRef &U2))
            (portRef &14 (instanceRef &U4))))
          (net &CARRY_OUT
           (joined
            (portRef &CARRY_OUT)
            (portRef &3 (instanceRef &U1))))
          (net &GND
           (joined
            (portRef &7 (instanceRef &U1))
            (portRef &7 (instanceRef &U3))
            (portRef &7 (instanceRef &U2))
            (portRef &7 (instanceRef &U4))))
          (net &X
           (joined
            (portRef &X)
            (portRef &5 (instanceRef &U4))
            (portRef &5 (instanceRef &U3))
            (portRef &12 (instanceRef &U2))))
          (net &Y
           (joined
            (portRef &Y)
            (portRef &4 (instanceRef &U4))
            (portRef &9 (instanceRef &U3))
            (portRef &1 (instanceRef &U4))))))))
        (design &SFADDER
         (cellRef &SFADDER
          (libraryRef MAIN_LIB))))
 eywordMap (keywordLevel 0))
```

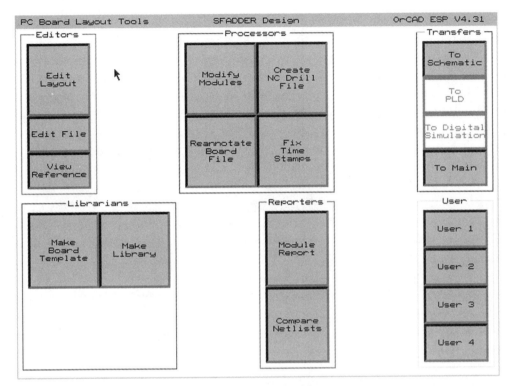

Figure D5. PCB Tool Set Menu for OrCAD 386 Plus (PCB386_MENU) (Courtesy of OrCAD Inc.)

PC BOARD LAYOUT DESIGN

Setting of Environment Variables

1. Invoke the PC Board Layout tool set, either by using the Transfer icon located at SDT386_MENU or by invoking it from ESP386_MENU. The tool set menu called PCB386_MENU (shown in Figure D5) will be displayed on the monitor screen. Check the top of this screen to verify that the appropriate design environment is selected. In this case the design environment is SFADDER.

2. Click on the **Edit Layout** icon located in PCB386_MENU, and select **Configure Layout Tools** from the Transit Menu. Unlike PCB II, in PCB386+ you need to load the module library. The loading procedure is the same as the library loading procedure in SDT. PCB software can download modules only from the libraries in the configured list box. To know which libraries to load, you may have to analyze their contents. The name of the library will give you an idea about the content.

3. In the configuration screen, select the appropriate drivers for your display (monitor) and printer unit. Keep the rest of the settings on their default values, and exit the configuration by left-clicking the mouse on the OK icon. Clicking on the OK icon will save your changes.

Figure D6. Layout Worksheet for OrCAD/PCB386 Plus (PCB386_MAIN) (Courtesy of OrCAD Inc.)

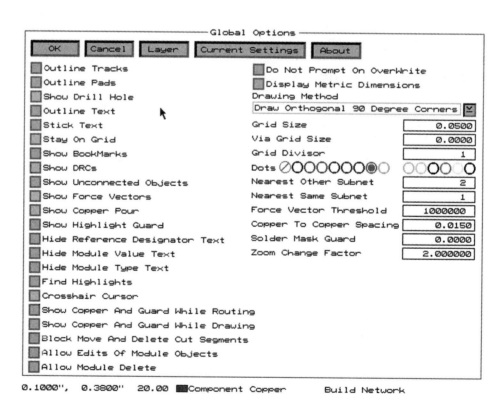

Figure D7. Global Options Screen for OrCAD 386 Plus (Courtesy of OrCAD Inc.)

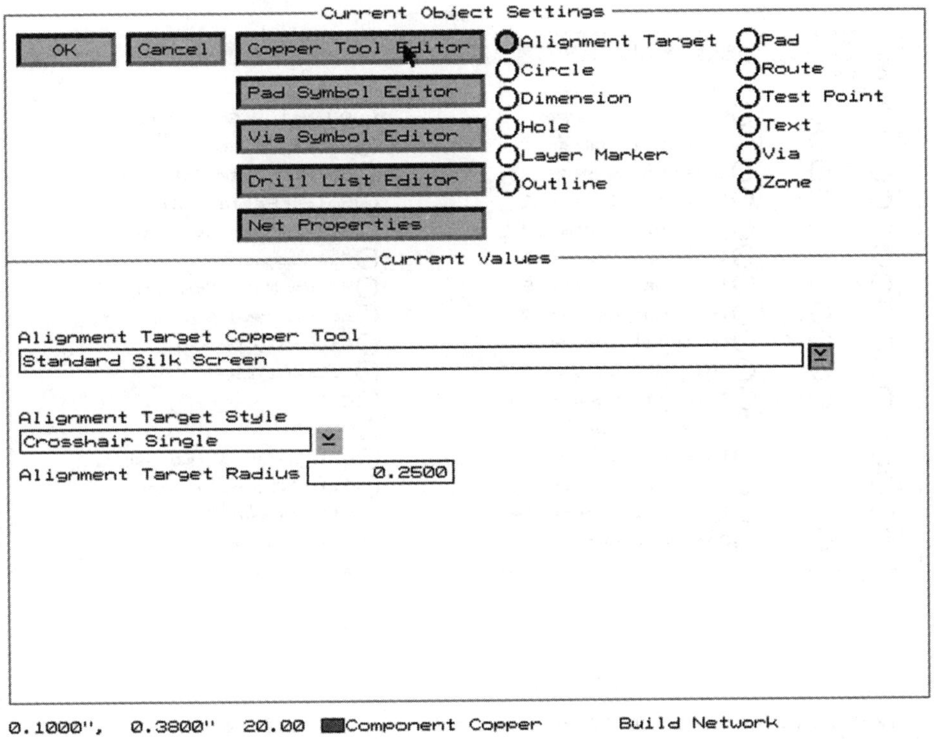

Figure D8. Current Object Settings Screen for OrCAD386 Plus (Courtesy of OrCAD Inc.)

4. Click on **Execute** from the **Edit Layout** Transit Menu. The software will display the PCB386_MAIN menu shown in Figure D6. This is the PC board layout editor worksheet.

5. Select **SET,** and the **Global Options** screen shown in Figure D7 will appear. Set the following Global Options for the full adder circuit. Activate the button by left-clicking the mouse on it. An active button is usually green under default conditions.

   - Show Drill Holes
   - Stay on Grid
   - Show Design Rule Check (DRCs)
   - Cross Hair Cursor
   - Copper to Copper Spacing (from default value of 0.015 to 0.025)
   - Show Copper Pour

6. Click on the **Current Settings** icon, and the screen for the **Current Object Settings,** shown in Figure D8, will appear on the monitor. In this screen you can set many variables concerning pads, vias, drill holes, copper tool editor, and net property editor. Click on each of

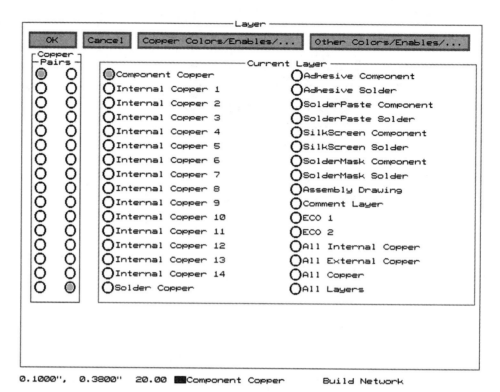

Figure D9. Layer Setting Screen for OrCAD 386 Plus (Courtesy of OrCAD Inc.)

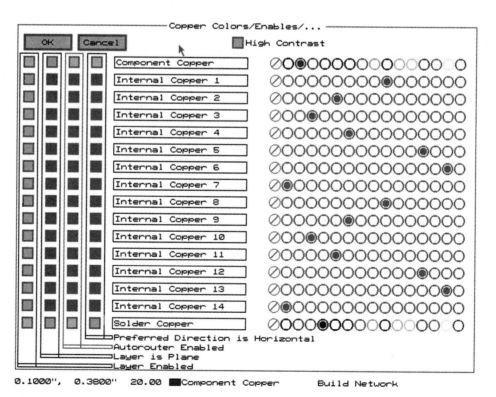

Figure D10. Color and Layer Enable Screen for OrCAD 386 Plus (Courtesy of OrCAD Inc.)

the following icon buttons and verify the setting. For this full adder circuit, keep all the default values.

- Copper Tool Editor
- Pad Symbol Editor
- Via Symbol Editor
- Drill List Editor
- Net Property Editor

You can change some of the current settings, if you feel comfortable about what you are doing.

7. Left-click the mouse on the **Layer** icon, and the software will display the **Layer** screen shown in Figure D9. In this screen you will be able to set some of the variables regarding the layers of the board.

8. Left-click the mouse on the **Copper Color/Enable . . .** icon, and the **Copper Color Enable** screen shown in Figure D10 will appear on your monitor. You can specify the number of layers for your board in this screen. Default setting is double layer, **Component,** and **Solder Copper.** If you need to route a board on a single layer, deactivate the **Component Copper** layer enable button. Figures D11 through D14 show configuration screens for **Pad Symbol Editor, Via Symbol Editor, Drill List Editor,** and **Net Property Editor.** You may need them for another project.

Draw Board Edge and Download Module

1. From the PCB386_MAIN menu, select **PLACE** and then select **Outline** from the **PLACE** submenu shown in Figure D15. Use **Begin** and draw the board edge of desired size. For a full adder circuit, draw a 3″ × 4″ board. Use the arrow keys located on your keyboard if the mouse is too sensitive. If you zoom in, this step may be even easier.

2. From PCB386_MAIN, select **GO TO FUNCTION** and then select **Netlist Loader** from **GO TO FUNCTION** submenu shown in Figure D16. Use **Block** to define the area where the modules will be downloaded. When you select **End of Block,** the software will display the **Load Netlist File?** screen shown in Figure D17.

3. Select the desired netlist file for which you want to download modules. Select the file from the left list box. For the full adder it is SFADDER.NET. If the desired netlist file is not in the list box, most likely you are in the wrong design environment. Select the right design environment from the Design Management Tools screen. If you click on the OK icon after selecting the right netlist file, the modules will be downloaded onto the layout worksheet within the specified area.

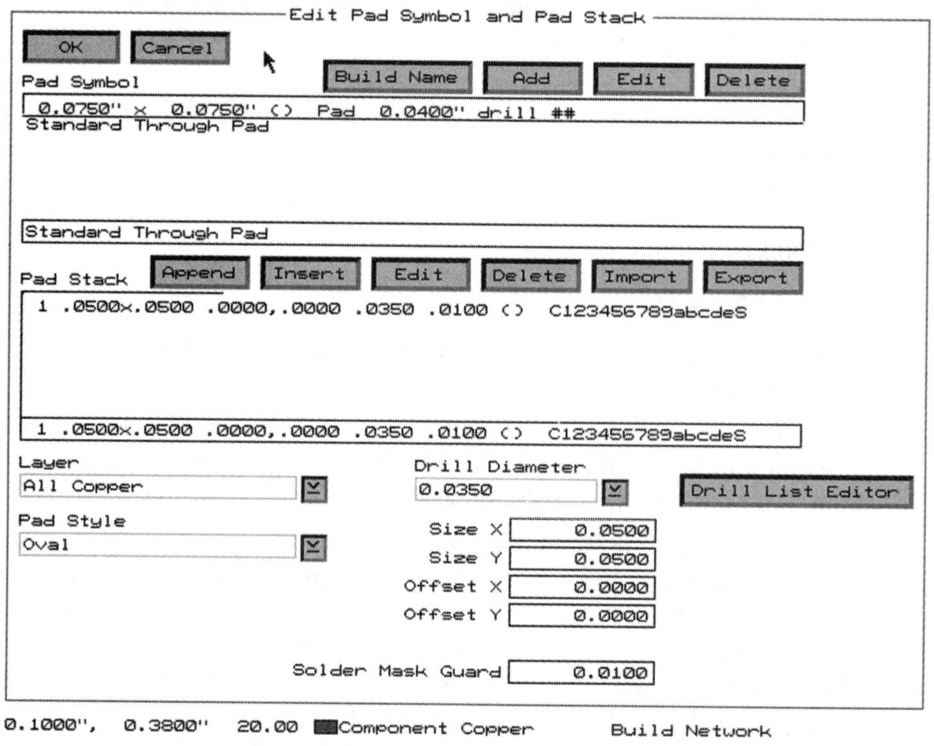

Appendix D / Quick User Guide to OrCAD/SDT 386+ and OrCAD/PCB 386+

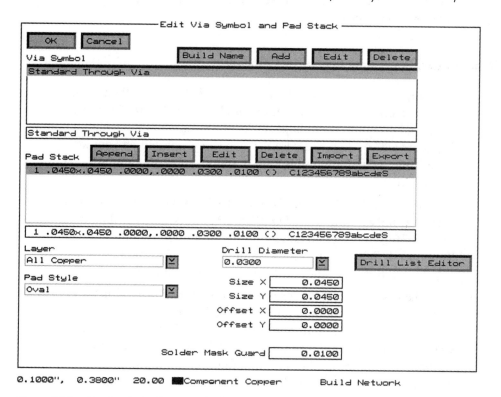

Figure D11. Pad Symbol Editor Screen for OrCAD 386 Plus (Courtesy of OrCAD Inc.)

Figure D12. Via Symbol Editor Screen for OrCAD 386 Plus (Courtesy of OrCAD Inc.)

PC Board Layout Design   471

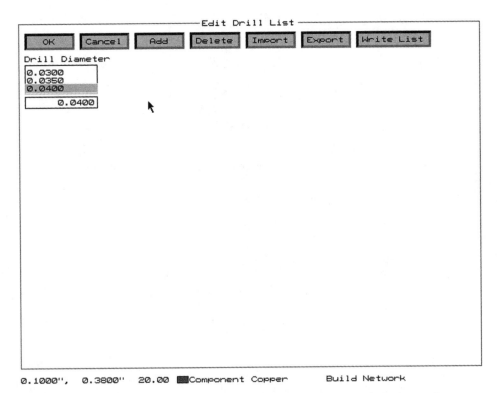

Figure D13. Drill List Editor Screen for OrCAD 386 Plus (Courtesy of OrCAD Inc.)

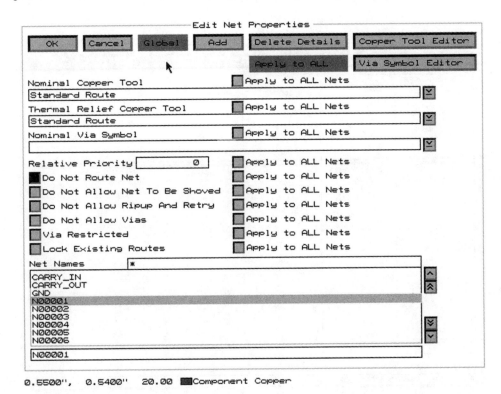

Figure D14. Net Property Editor Screen for OrCAD 386 Plus (Courtesy of OrCAD Inc.)

Figure D15. Submenu of PLACE Command (Courtesy of OrCAD Inc.)

Figure D16. Submenu of GO TO FUNCTION Command (Courtesy of OrCAD Inc.)

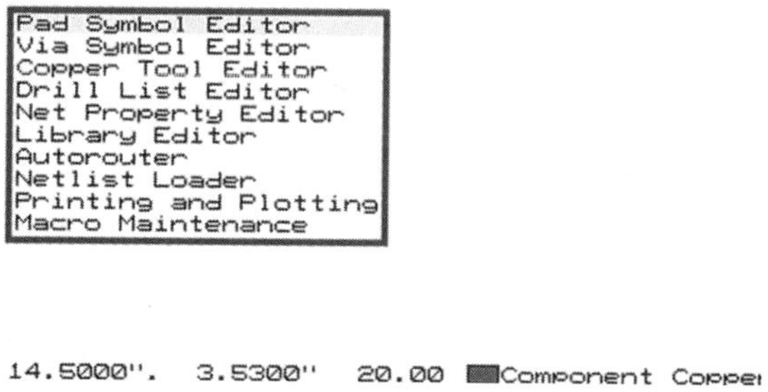

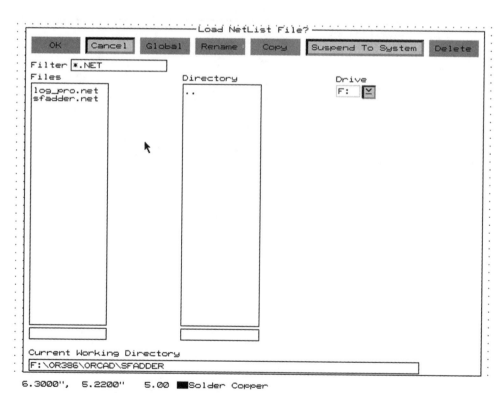

Figure D17. Netlist Loader Screen for OrCAD 386 Plus (Courtesy of OrCAD Inc.)

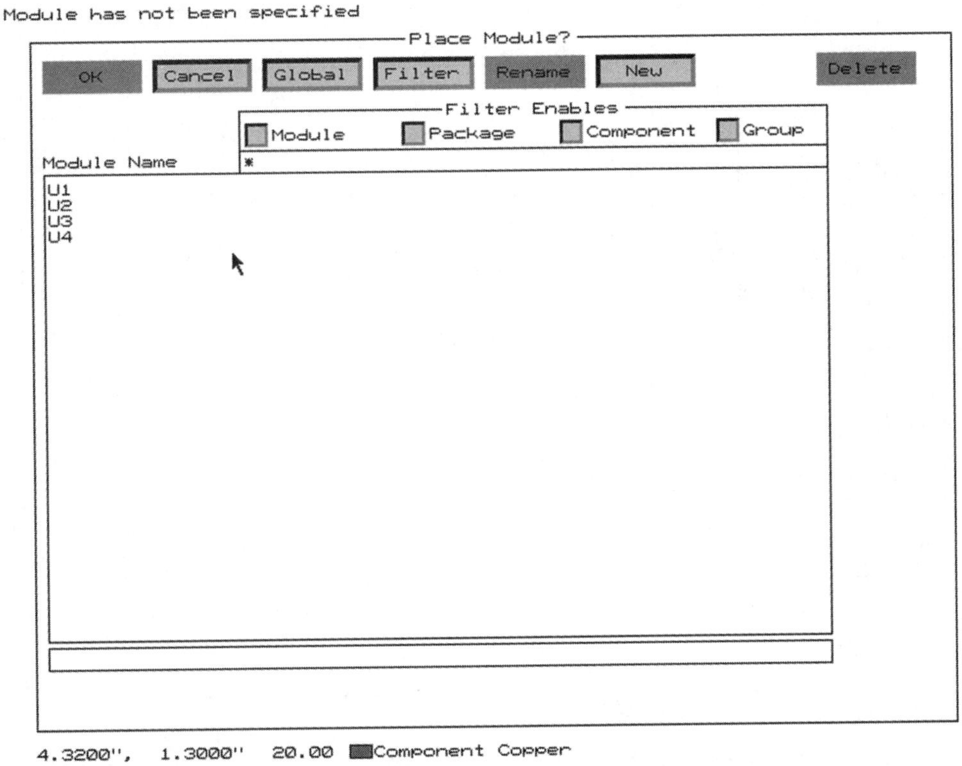

Figure D18. PLACE Module? Screen for OrCAD 386 Plus (Courtesy of OrCAD Inc.)

4. To move a module on the board, select **PLACE** from the PCB386_MAIN menu and then select **Module** from the submenu. The **PLACE Module?** screen shown in Figure D18 will appear with a list of modules for your schematic. The modules are listed by their reference designators. Select any reference designator from the list and click on the OK icon. The module will be at the mouse pointer, so you can move, rotate, or do another available operation before placing the module at the desired position on the board. In this way you can get modules one by one and place them at the desired locations.

5. If you want to load a new module from the module library, click on the **New** icon located in the **PLACE Module?** screen. Figure D19 shows The New Module screen you need to use to load new modules. Select the library where the module is located and select the module from the list located below the library list. (You can also move a module onto the layout worksheet by using the BLOCK command.)

Place Copper Fill Zone

Make sure that in the **Global Options** screen, the **Show Copper Pour** button is active. Select **PLACE** from the PCB386_MAIN menu and then select the **Fill Zone** from **PLACE** submenu. Draw the zone outline. Click the left mouse key, and the zone will be filled with copper.

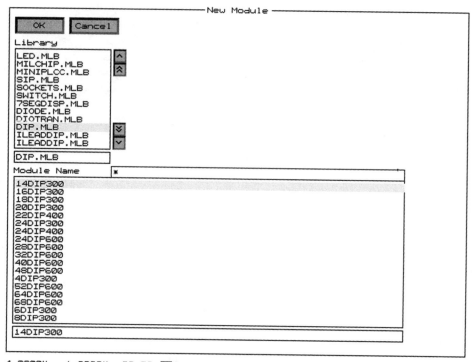

Figure D19. PLACE New Module Screen for OrCAD 386 Plus (Courtesy of OrCAD Inc.)

Figure D20. Autorouter Submenu for OrCAD 386 Plus (Courtesy of OrCAD Inc.)

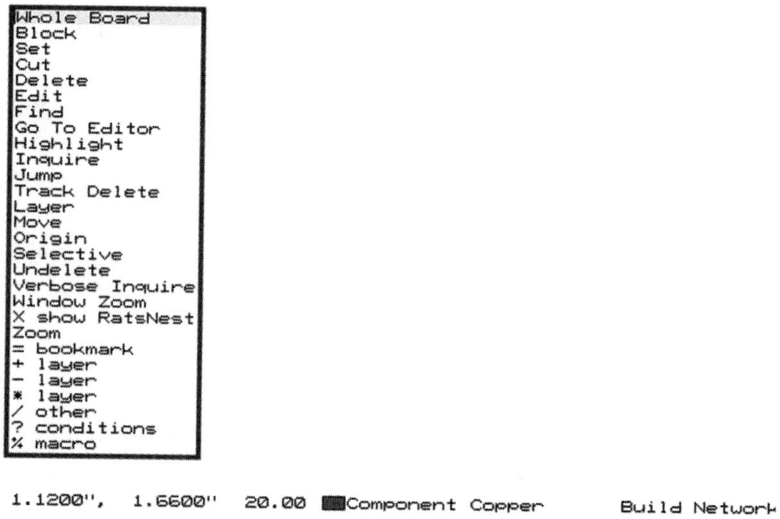

Route Board

1. Click on **SET** from the PCB386_MAIN menu and make the **Show Force Vector** button active. Depending on the force vector, move the modules to their appropriate positions on the board.

2. Click on **GO TO FUNCTION** from the PCB386_MAIN menu and select **Autorouter** from the submenu. Select **Whole Board** from the Autorouter submenu shown in Figure D20, if you want to route the entire board. Select **Autoroute Whole Board** from the next submenu. The autorouter will route the board and report the number of unconnected, connected, and partial routes. Status of a board may be viewed by clicking the mouse on the **Condition?** option from the PCB386_MAIN menu.

3. Each unconnected route will be shown by a highlighted straight line. Select **QUIT** and then **Erase All Routes** from the **QUIT** submenu. Rearrange the modules and route the board again.

4. If you want to route some of the nets manually, select **ROUTE** from PCB386_MAIN and use **Begin** to route one pad to another. To go from one layer to another, use Layer+ and Layer−. Figure D21 shows the board file of the full adder after arranging the modules on the board, and Figure D22 shows the board file after routing.

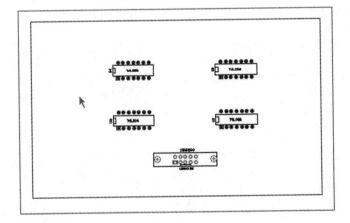

Figure D21. Full Adder Board Layout before Routing

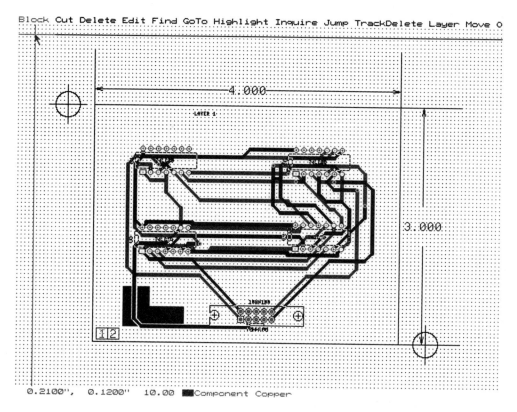

Figure D22: Full Adder Board Layout after Routing

Load Existing Board

Click on the **Edit Layout** icon located in PCB386_MENU. Select **Local Configuration** from the Transit Menu. Select the desired board file from the list box. If you are in the PCB386_MAIN menu, click on **QUIT→ Intialize Board File,** select the desired board file from the list box, and click on the OK icon.

E

ACTUAL SHAPE AND SIZE OF COMMON PCB MODULES

Appendix E / Actual Shape and Size of Common PCB Modules

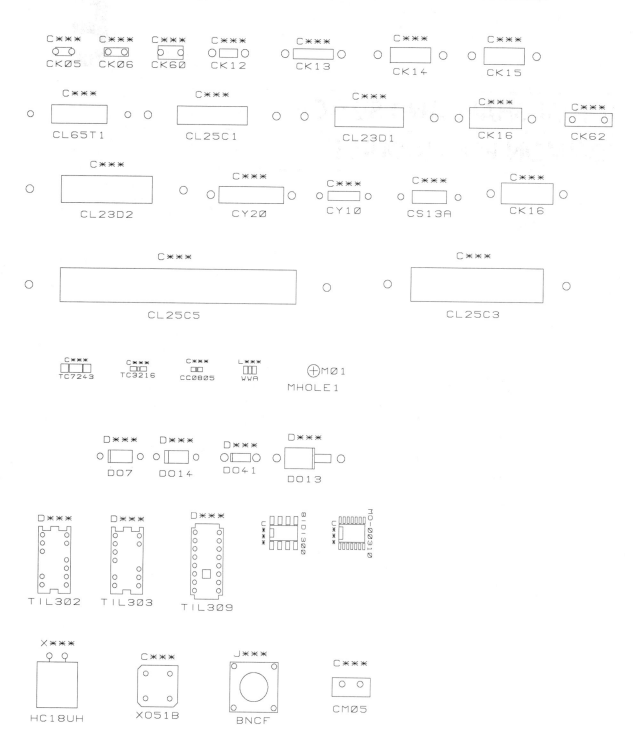

Appendix E / Actual Shape and Size of Common PCB Modules

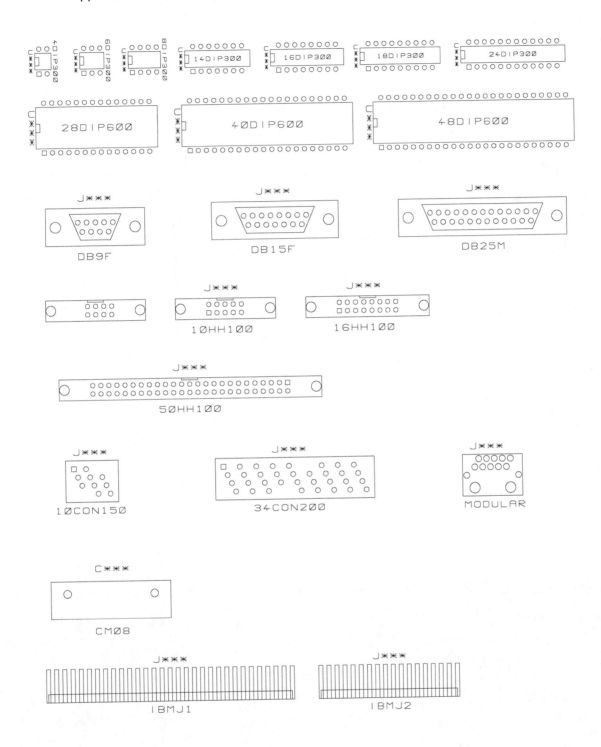

Appendix E / Actual Shape and Size of Common PCB Modules

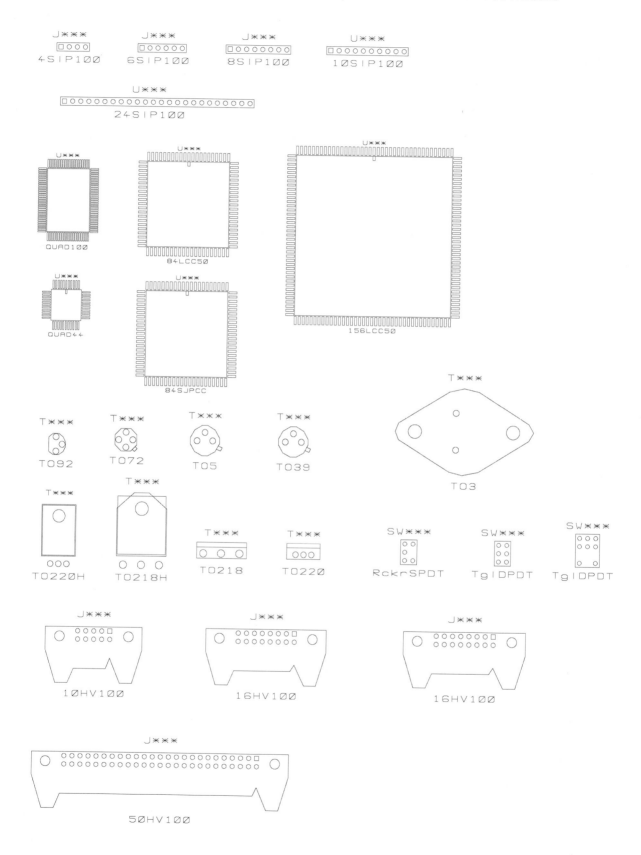

INDEX

A-H Tag, 170, 305
Abandon edit, 144, 169
Abandon program, 290, 291
Acid, 384, 388
AGAIN Command, 143, 144, 279
Alkali, 384, 385, 388
Ammonium persulphate, 397, 410
ANNOTATE Command, 125
    Annotate, 125
    Unannotate, 125, 197
Annotate Schematic, 72, 125, 193, 196, 226
    After, 127, 199, 229, 248
    Before, 126, 198, 227, 246
    Local Configuration, 196
Annotation, 103
ANVIL (AR) Tool, 422
Architecture, Board, 355
Archive Parts in Schematic, 88, 134, 209, 230
    COMPOSER, 209, 210, 232
    LIBARCH, 209, 210, 211, 232
Arrow buttons
    Single, 88
    Double, 88
Artwork, 82, 343
Artwork Master, 82, 343, 344
    Negative, 393
    Positive, 393
ASCII, 110
Auto panning, 144, 170, 302,
Auto Sel, 343, 344
AUTOEXEC.BAT, 76
Available display drivers, 89, 274
Available plotter drivers, 136, 274
Available printer drivers, 136, 274

Back Annotate, 128, 133, 200, 201
BACKANNO Command, 200, 201
Backup Design, 100, 101
Baud rate, 137, 274
Bed of Nails, 402

Bill of Materials, 8, 9, 135, 230, 232, 242, 260
    Local Configuration, 218
    Include file, 219
BIOS, 443
Blind Via, 14, 339
Buried Via, 339
BLOCK Command, 143, 280, 281, 282
    Drag, 145, 147
    Export, 148
    Fixup, 145
    Get, 148
    Import, 148
    Move, 143, 146
    Save, 148
Board
    Double-sided, 269
    Single-sided, 269
Board layout, 15
Board size, 30, 390
Breadboard, 33
Browse, library, 164
Buffer, Memory, 137, 274, 276, 278, 283
Burrs, Deburr, 392, 422
Bus Entry, 167, 185
Buses, 116, 117, 166, 167, 185,

Cancel Icon, 85
Capacitor, 441
Carbide drill bits, 17, 420
Capture Schematic, 180, 225, 226, 236, 240
Change directory, 444
Check Electrical Rules, 5, 7, 135, 139, 142, 191, 192, 194, 232, 456
    After, 195
    Before, 194
Chemical Disposal, 387
Chemical Requirement, 415
Chemical Safety, 385

Circolex, 394
CLEANUP Command
   Schematic, 132, 202
CMOS, 441
Color Table, 138, 141
Comment text, 119, 189, 190
Compile, 132, 134
Compiler, Netlist (INET), 213, 214
Compiling the connectivity database, 8, 213
Complex hierarchical, 102
Complex hierarchy, 122
Complex to Simple, 101, 103, 217, 456
Component Identification, 400, 402
Component mounting, 428
Component soldering, 428, 433, 434
Computer Numerical Control, 17, 383, 391, 400
Computer-aided design (CAD), 1, 79
Computer-aided engineering (CAE), 79
Computer-aided manufacturing (CAM), 79
CONDITION, 149, 283
Configuration
   Local, 84
   ESP Design Environment, 89, 90
   User button, 89
   Schematic Design Tools, 135, 175
Configured libraries, 87, 137, 140, 177, 178, 209
Configured Schematic Design Tools, 135, 175
Configured PC Board Layout Tools, 273, 310
Connecting Parts
   Busses, 116, 182, 185, 186
   Wires, 116, 183, 185
Connecting Power, 117, 118
Connectivity database, 213
Convert, 103, 159, 160
Convert Plot to IGES, 135
Copper electroplating, 392
Copper foil thickness, 30, 389
Copper plating electroless, 392
Copper-clad, 389, 390
Copy files, 108, 445

Create a part, 18, 20, 120
Create Bill of Materials, 8, 9, 135, 230, 232, 242, 260
Create Design, 97, 111
Create Hierarchical Netlist, 128
Creating a custom part, 18, 20, 120
Cross Reference Parts, 134

Deburring, 392, 422
Defects in PCB, 404
De laminate 412, 428
Delete, PCB 283
   Object, 283
   Segment, 283
   Track, 283
   Block, 283
DELETE Command, SDT, 149, 150
DELETE Command, PCB 283
Delete Design, 98
Delete File, 447
Deluxe Logic Probe, 34
   Annotate Schematic, 226, 227
   Archive Schematic Parts, 230
   Arrange Module on Board, 356
   Bill of Materials, 220
   Board Edge, 354
   Download Module, 354, 355
   Drill Template, 358
   File Organization, 224
   Final Hand Drawn Schematic, 41
   Generation of Netlist, 233, 234
   Hand Drawn Schematic, 37
   Load Module from Library, 354, 355
   Master Artwork, 356
   Module Report, 352, 353, 359
   Packaging Scheme, 38
   Parts List, 39
   PCB Layout Tool Configuration, 351
   Schematic Capture, 225
   SDT Configuration, 177, 224
   Set Board Layout Environment, 354
   Silkscreen, 357
   Solder Mask, 357
   Stuff File, 206, 230
   System Functional Diagram, 35
   Update Field Content, 227

Design
  Create, 97
  Load, 97
  Delete, 98
  Copy, 99
  Backup, 100
  Restore, 100
Design Conditions, 275, 278
Design Environment, PCB 386+, 464, 465, 466, 467, 468
Design Environment PCB II, 309, 311
Design Management Tools, 2, 80, 95, 110
Design methodology, 79
Design steps, 82
Designing and drawing, 175
Designing and drafting, 175
Designing Electronic Packaging, 29
Desoldering bulb, 33, 419, 433. *See also* Solder Sucker.
Developmental Plan, 33, 38, 43, 50
Digital Simulation Tools, 71, 72, 91
Diode, 440
Dip Soldering, 434
Dipping (Flux), 434
Directory, Make, 445
Directory structure, 75
Disk Format, 443
Display Driver, 88, 274
Display Drivers, 88, 274
DM_MENU, 71, 74, 96
DM_MENU1, 96
DM_MENU2, 97
DOS, Environment Variables, 76
DOS, 443
Double-Sided PC board, 269
Draft Command, 72, 81, 143
Drafting, 180, 181
Drill bit size, 32, 420, 421
Drill File, 348
Drill press, High-speed, 32, 415, 420, 425
Drill Report, 423, 424
Drill Template, 17, 345, 358, 364, 370, 377
Drill Template legend, 422
Drilling, 17, 391, 420, 425
Driver Options, 88, 136, 273, 274
Driver Prefix, 88, 136, 274
Dry film photoresist, 398

Dry transfers, labeling, 437
Du Pont, 394
Dynachem Corp., 396, 394

Edge, Board, 313
  Delete, 313
  Place, 313
Edit, 333, 334, 335
  Netname, 333
  Track, 334
  Via, 334
  Width, 335
EDIT Command, 150
  Convert, 159
  Ground, 152
  Label, 150
  Module port, 150
  Part, 155
  Power, 152
  Sheet, 153
  Title Block, 159
EDIT Command, 150
Edit File, 106, 125
Edit Library, 132
EDIT_FL, 107
Editing a Part, 150, 152, 155, 157, 187, 188, 189
Editors, 124
  Draft, 122
  Route Board, 271
EECELLON format, 17
Electrical Rule Check Matrix, 5, 139, 142, 192, 194
Electroless Copper deposition, 392
Electronic Cricket, 40
  Annotate Schematic, 237, 238, 239
  Archive Schematic Parts, 237
  Arranged Board, 362
  Bill of Materials, 237
  Bill of Materials, 237
  Drill Template, 364
  Final Hand-Drawn Schematic, 47
  Generation of Netlist, 237, 242
  Master Artwork, 362, 363
  Module Report, 365, 366
  Packaging Scheme, 44
  Parts List, 45
  Schematic Capture, 236
  SDT Configuration, 177
  Silkscreen, 363

Solder Mask, 364
Stuff File, 240
System Functional Diagram, 42
Unarranged Board, 361
Update Field Content, 237, 246
Electronic Design Automation
    (EDA), 2, 70, 80
EMS Memory, 137
Environment variables, 76
Epoxy ink, 14, 398
ERC, 5, 7, 135, 139, 142, 191, 192,
    232, 456. *See also* Check
    Electrical Rules.
ESP_MENU, 70, 88
ESP386+_MENU, 455
Etchant, 385, 387, 410
Etcher, 16, 386
Etching, PC boards, 410
Excel Process, 407, 394
Experimenting, 33, 40, 49, 52, 59
Exposing PCBs, 409
Exposure, PC board, 409
Exposure, time for PC board, 409
Exposure device, 409
Eyelets, 421, 426
    ANVIL (AR) Tools, 422
    FORM (FR) Tools, 422

Fabrication Needs, 453
Fabrication Process of PC Board,
    3, 21, 22, 383, 407, 415
FEDIF Netlist format, 459
Fetch part, 180
File format
    Include, 218, 219
    Netlist, 216
    Stuff, 230
    Was/Is, 200
File Organization, 95, 224
File Structure, 120
    Flat, 120, 121, 123
    Hierarchical, 120, 122, 123, 124
FIND Command, 161
Finger plating, 401
Flammability, 388
Flash plating, 395
Flat design, 121
Flat file structure, 121
Floppy Disk, 65
Flux Application, 433, 434
Foil, copper, 389
Foil lamination, 389

Force, Ratsnest Vector, 12, 296, 325,
    326, 327
FORM (FR) Tools, 422
FORMAT Disk, 443
Formatting a Netlist, 213, 214, 215
Formatter a hierarchical Netlist, 216
FR4, PCB substrate material, 389

Gerber File, 67, 69, 292
GET Command, 161, 180, 181, 182
    Place, 161, 181
    Rotate, 161, 181
    Convert, 161, 181
    Normal, 161, 181
    Up, 162, 181
    Over, 162, 181
    Down, 162, 181
    Mirror, 162, 181
Gold-plated finger, 401
Graphic symbol, 115
Grid, 312
Grid Parameters, 170
Grid references, 312
GROUND, 117

Hard disk, 65
    Installation, OrCAD, 66
HARDCOPY, 163, 221
Hierarchical file structure
    Complex, 102, 124
    Simple, 102, 123
Hand tool, 386
Hand tools
Hierarchical netlist format,
    125, 128
Hierarchy buffer, 137
Hierarchy Options, 137
High-level functional description,
    26
Highlight Command, 294

IFORM, netlist formatter, 213,
    215, 233
IGES, Convert Plot to, 135
ILINK configuration, 214, 233
Imaging, PCB, 394
Incremental netlist, 214
INET
    Configuration, 213, 233
    Netlist, 216
    File Options, 214
INF file, 214, 215

Index

Infrared Object Counter, 46
    Annotate Schematic, 247, 250
    Archive Schematic, 247
    Arranged Board, 368
    Bill of Materials, 247, 251
    Drill Template, 370
    File Organization, 240
    Final Hand-Drawn Schematic, 53
    Master Artwork, 368, 369
    Module Report, 371, 372, 373
    Packaging Scheme, 50
    Parts List, 51
    Schematic Capture, 247, 248
    SDT Configuration, 178
    Silkscreen, 369
    Solder Mask, 370
    Stuff File, 249
    System Functional Diagram, 48
    Unarranged Board, 367
    Updated Field Content, 247
Initialize Command, 169, 290, 291, 292
Inquire, 163
Inquire Command, 329, 330
    Module, 329, 330
    Net, 333
    Pad, 329, 331
    Track, 329, 332
    Via, 329
Inspection, PCB 400, 405
Installation, Or/CAD/SDT, PCB 66
Insulation, 387, 390
Integrated Circuit, (IC) 441
Isolation, 310
Item to Plot, 291, 293, 343, 344, 346
Include file, 219
International Eyelets, 426

Joint, solder, 433
JUMP Command, 163, 287
    Reference, 163
    Tag, 163
    X-Location, 164
    Y-Location, 164
Jumper wire, 412
Junction box, 184

Kepro, 410, 415, 453
Key Fields, 139, 141, 205
    Annotate Schematic, 139
    Bill of Materials, 139
    Update Field Content, 139
Keyboard, 63, 65

Labels, 118
Labeling buses, 118, 119
Laminate, Delaminate, 412, 428
Layer, PCB, 228
Layout directives, 163, 168, 169
Layout, PCB, 270, 309, 311
Layout Tools, 270
Labels, 118, 119
Librarian, 18, 132
Libraries
    Configured, 87, 140, 264
    Available Libraries, 87, 140
Library
    Options, 87, 137, 140
    Insert, 87
    Remove, 87
LIBRARY Command, 164
    Directory, 164
    Browse, 164, 221
Library files
    Compile, 134
    Decompile, 134
    List, 134, 264
Library prefix, 137
Linker, netlist, 214,
Load Module Command
    Using Netlist, 314
    From Library, 315
Local Configuration, 84
    SDT Tools, 84
    Layout Tools, 84
    Draft, 179
Level of Effort Requirement, 26, 36, 43, 48
Load Design, 97, 174
List, PCB Module, 449, 450, 451
Layout Design PCB 386+, 464, 465, 469, 470, 471, 472, 473
Layout Design PCB II, 309
Load Board PCB 386+, 475

M2EDIT Editor, 72, 106, 107, 203
Machine, CNC, 17, 383, 391, 400
Machinery Requirement, 415
MACRO Command, 164
    Capture, 165
    Delete, 165

Initialize, 165
List, 165
Option, 165
Read, 165
Write, 165
Macro Options, 137, 140
Magnifying lamp, 400
Main Menu
SDT, 81
PCB, 82
Make Directory, 445
Managing Files, 106
Manual Routing, 337, 338
Master Artwork Printing, 343
Material Procurement, 30, 36, 43, 49, 56
Memory Allocation, 274, 276, 278
Memory EMS, 140
Metallic conducting path, 14
Microcomputer, 443
Mini Stereo Amplifier, 52
　Annotate Schematic, 249, 255, 256
　Archive Schematic, 257
　Arranged Board, 375
　Bill of Materials, 257
　Drill Template, 377
　File Organization, 247
　Final Hand-Drawn Schematic, 58
　Master Artwork, 375, 376
　Module Report, 378, 379, 380
　Packaging Scheme, 56
　Parts List, 57
　Schematic Capture, 249, 255
　SDT Configuration, 178
　Silkscreen, 376
　Solder Mask, 377
　Stuff File, 258
　System Functional Diagram, 54
　Unarranged Board, 374
　Update Field Content, 257, 259
Modify Module, 315, 316
Module, Print, 448, 449
Module Buffer, 274
Module Library, 315
Module Ports, 118
Module Report, 19, 350
Mounting holes
Mouse, 63, 65
Moving machinery, 386

MS-DOS, 72
Mylar, 21, 407
MYORCAD, 80

Negative photoresists, 393, 394
NEMA (National Electrical Manufacturing Association), 30
Nbr of layers, 294, 305
NC DRILL file, 271, 348, 349, 423, 424
NCD file, 422, 424
Net Condition, 276, 279, 340
Net Pattern, 294, 340
Netlist, 131
　Compile, 213
　Create, 212
　Format, PCB, SDT, 454
　Hierarchical Design, 216
　IFORM, 213, 215
　ILINK, 213, 214
　INET, 213, 214
Netlist File, 11, 12, 131, 216,
Netname, Editing, 333, 334, 335
Nibbler, 425
No Via Strategy, 300
No Via Zone, 289, 321
Normal Strategy, 300
Number of Layers, 294, 305

OK Icon, 85
Op-amp, 57
Operating Environment, Software, 70
Operational check, 439
Optimize, PCB, 326, 327
　Force Vector, 327, 328, 341
　Ratsnest Vector, 327, 328, 341
Opto-coupler, 53
OrCAD Design Automation, 79
OrCAD Directory Structure, 75
OrCAD/ESP, 70, 71
OrCAD/PCB Ver 2.21, 2, 70
OrCAD/PCB 386+, 3, 454
OrCAD/SDT Ver. 4.22, 2, 70
OrCAD/SDT 386+, 3, 454
Orthogonal wires and buses, 170
Other, 288
Ounces-Copper-Clad, 390, 391
Exposure, time, 409

Packaging Scheme, 29, 36, 43, 49, 55

PADs, 310
Panel, 390
Panning, Auto, 144, 170, 302
Part, 115
Part fields, 141
Part Name, 141
Part Reference, 141
Part Value, 141
Parts list, 31, 39, 45, 51, 57
Parts per package, Which
    Device, 159
Pattern plating, 395
PC Board Fabrication, 390
    Material Selection, 390
    Drilling, 391, 420
    Deburring, 392, 422
    Plating, 392
    Artwork Transfer, 392, 393,
        394, 395
    Pattern Platting, 395
    Resists Stripping, 396
    Solder Mask, 398
    Solder Coating, 398
    Contact Finger Plating,
        400, 401
    Component Identification, 400
    Scoring, 400
    Inspecting, 400
    Laboratory Process, 402
    Shaping, 403, 402
    Tools and Machinery
        Requirement, 415, 419
    Chemical Requirement, 415
PCB Testing, 437, 439, 440
PCB Defects, 404
PCB Repair, 406
Print Modules, 448, 449
PCB Module List, 449, 450, 451
PC Board Layout, 270, 309
PC Board Material, 389, 390
PCB II, OrCAD, 69, 73, 82, 270
PCB_MENU, 71
Pen, 138
Personal Computer, 63
Photo-resists, 393, 394, 396,
    408, 409
Photodeveloping, 385, 393
    Photochemical, 387, 393
Photographic process, artwork
    transfer, 393
Photoplotter, 14, 343
Photosensitized, 14, 408

PLACE Command, PCB, 288
    Module, 288, 318
    Text, 289, 320
    Zone, 289, 320, 321
PLACE Command, SDT, 165
    Wire, 165, 183
    Bus, 166, 185, 186
    Junction Box, 166, 187
    Entry Bus, 167, 185
    Label, 167, 185, 190
    Module Port, 167
    Power, 167
    Sheet, 168
    Text, 168, 189
PLACE Command, 165
Placing a bus, 166, 167, 185
Planning, 25
Plated-through-holes, 392
Plating, 395
Plot Schematic, 8, 135
    Local Configuration, 221, 222
Plotter, 65, 221
Plotter driver, 136
Plotting a file, 343
Polymerization, 394
Positive photo-resists, 394, 395
Power, isolating, 117, 118
Power and Ground, 117
Power Strategy, 276, 300
Power Supply, regulated, 4, 6, 11
Print Artwork to a file, 344, 345
Print PCB, 344
Print Schematic, 8, 135, 222
    Local Configuration, 223, 345
Printed Circuit board
    drilling, 425
Printed Circuit Board Layout
    Tools, 80, 269
Printer, 65
Printer/Plotter
    Output Options, 140
    Artwork, 343, 344
    Component Legend, 345
    Solder Mask, 345
    CNC Drill Files, 348, 349
    Module Report, 350
Printing, Artwork, 343
Processor, 72, 73, 125, 271
Programmable Logic Devices
    Tools, 71, 72, 91
Project Description, 26, 34, 40,
    46, 52

Projects, 34, 111, 177, 178, 351
Punching Enclosure, 425

Quality Assessment and Assurance of PCB, 411, 412
Quality control, 412
QUIT Command, SDT
  Enter Sheet, 169
  Leave Sheet, 169
  Update Sheet, 169
  Write to File, 169
  Initialize, 169
  Suspend to System, 169
  Abandon Edit, 169
  Run User Command, 169
QUIT Command, PCB
  Abandon Program, 290
  Initialize, 290, 291, 292
  Library, 292, 293
  Report, 293
  Update File, 293
  Write to File, 293

REFERENCE designators, 8, 39, 45, 51, 57, 125, 155
Regulated Power Supply, 4, 11
Remove Directory, 447
RENAME File, 108
RENM_FL, 108
Repair PCB Defects, 406, 412, 414
REPEAT, 169
Reporters, 134, 272
Resistors, 8, 39, 45, 51, 57, 440
Resists, photo, 393, 394, 396, 408, 409
Resists Stripping, 396
REST_DGN, 102
Root Schematic, 104, 456, 458
Rotate Module Part, 158, 289
  Module, 289
  Part, 158
Route a Board, PCB II, 322
  Compile Netlist, 323
  Set Strategy, 323, 324
  Optimize Placement, 325
  Optimize Routes, 326
Route a Board, PCB386+, 473, 474
  Autorouter, 474
Routing, 294, 295, 322
  Other, 294
  Inquire, 294

Show, 294
Highlight, 294
Width, 294
Layer Pair, 294
Netlist, 294
Cleanup, 294
Autoroute, 294, 298, 300, 302, 339, 340, 341, 342,

Safety, 383, 385, 386, 387
Safety Practices, 385, 386, 387
Safety test, 388
  Chemical, 385
  Equipment, 386, 387
  Electricity, 387
  Disposal & Storage, 387, 415
Save, 169, 190, 291
  Schematic, 190, 223
  Artwork, 293
Schematic Design Tools, 80, 115, 122
Screen printing, 400
SDT_MAIN, 81, 144
SDT_MENU, 71, 72
SDT386+_MENU, 455
Seed, zones, 289, 290
Select Field View, 132, 141, 200
  Local Configuration, 202
SET, Isolation, 275, 302
SET Board Layout
  Environment, 311
  Grid, 312
  Layer, 312
  Frame X-Y-L, 312
  Isolation, 312
SET Command, SDT, 169, 170
SET Command, PCB, 302, 303, 304
SET Design Environment, PCB, 174
SET Design Environment, SDT, 309
SET Environment, 76
Shaping PCB, 422
Shear Tools, 387
Sheet Size, 140
Sheet Symbols, 104, 105, 118, 456
Show Schematic File Structure, 135
Silkscreen, 400
Simple hierarchical, 122
Single-Sided Board, 269

Index

Size of Worksheet, 140
Solder, 428, 433, 434
Solder Coating, 398
Solder Mask, 346, 347, 398
Solder Sucker, 419, 433
Soldering iron, 32, 419
Soldering PC boards, 428, 433, 434
Solderless breadboard, 33
Speaker, 54
Strategy, 276, 300
Stuff File, 206, 230, 240, 249, 258
Style of Cursor, 304
Suspend to System, 73, 74, 98, 109
System memory, 276

TAG Command, 170, 305
Template Table, 138, 141
Terminal Post, 427
Testing PCBs, 437, 439, 440
Text, 119
Theory and Operation, 36, 49
Time and Resource Requirement,
    26, 35, 42, 46, 54
Tin plating, 410
Tin/Lead solder, 433, 434
Tining, 385, 410
Title Block, 119, 120
To Digital Simulation, 71, 72,
    91, 272
To layout, 72, 73, 91, 272
To Main, 72, 73, 91, 272
Tools and Machinery Requirement,
    30, 32, 36, 43, 50, 56
Toxic chemicals, 402
Toxic Fumes, 397, 402
Track
    Buffer, 274, 276,
    Width, 334, 335, 336, 338, 339
Transfer Tools, 272
Transformer, 441
Transistor, 440

Transit Menu, 84, 91
Troubleshooting, 440
Try Again, 302, 340, 341
TTL, 18, 125
Turret terminals, 427

Ultraviolet (UV) Exposure System,
    16, 394, 409
UV Light, 16, 386, 388, 394, 409
Unannotate, 125
Update
    Schematic, 190, 191
Update ESP Data, 101
Update Field Content, 8, 10, 128,
    129, 130, 204, 228, 241,
    259, 459
    Local Configuration, 207
    Before, 229
    After, 208, 231, 259
Users Buttons, 91, 92, 272

Vector
    Force, 327, 341
    Ratsnest, 327, 341
Vellum, 21
Via, 14, 310
    Embedded, 14, 339
    Blinded, 14, 339
Via Pins, 427
View, Field, 132

Wire, 116, 182, 183
Work Sheet Options, 137, 140
Wire Connection, 183
Write to a File, 191, 293
Wildcards, 445

Zenith DOS, 72
Zener diode, 441
Zone, PCB, 289, 473
Zoom, 170, 305